The Western European Loess Belt

The Western European Loess Belt

Agrarian History, 5300 BC - AD 1000

by

Corrie C. Bakels

Faculty of Archaeology, Leiden University, The Netherlands

Springer

Prof. Dr. Corrie C. Bakels
Leiden University
Fac. Archaeology
P.O.Box 9515 2300 RA Leiden
Netherlands
C.C.Bakels@arch.leidenuniv.nl

Every effort has been made to contact the copyright holders of the figures and tables which have been reproduced from other sources. Anyone who has not been properly credited is requested to contact the publishers, so that due acknowledgement may be made in subsequent editions.

ISBN 978-1-4020-9839-0 e-ISBN 978-1-4020-9840-6
DOI 10.1007/978-1-4020-9840-6
Springer Dordrecht Heidelberg London New York

Library of Congress Control Number: 2009926975

Cover illustration: The picture represents the reaping machine from Gaul; it is based on fragments of tombstones found in Buzenol and Arlon in Belgium, reunited as suggested by C. Massart.

Printed on acid-free paper

Springer is part of Springer Science+Business Media (www.springer.com)

Preface

Many books on the agrarian history of Western Europe begin with the Middle Ages, which is quite understandable, because they are mostly based on written sources. But everybody who is interested in agriculture knows that agriculture started much earlier and also that information on that millennia-long period is provided by archaeology. Admittedly, there are books which deal with almost the entire story, from prehistory to well into historical periods, but the problem with them is that they focus on the past of modern nations, disregarding the fact that their boundaries were not necessarily the boundaries of former times. I wanted to write using another kind of unit: a region with only one type of soil and climate, as these two are all-important factors where farming is concerned.

I chose the region covered with loess, west of the river Rhine. The scientific background is explained in Chapter 1 of this book. Another reason was that I devoted and still devote much of my own research to this region. And, I have to admit that my interest was also triggered by the fact that I was born there.

Writing a book on a single region has, however, a distinct disadvantage compared with writing a book on a modern nation. There are always periods about which archaeology or other sources have nothing to tell. Authors dealing with a nation can always switch to a part of their country where information is not lacking. For instance, if information about farmers on loess soils is absent, there may be information available on farmers in sandy or marshy areas during the required period. This is often the case, but the practice of switching over leads to gaps in our knowledge being smoothed over. It is my opinion that switching is not always the correct approach, as the history of one region is not a priori identical to the history of another region. In my approach voids cannot be ignored. It lays bare how much we know, and how much we do not know.

This book starts with the first farmers and ends when food production is no longer the chief source of livelihood for the entire population. The long period, 5300 BC–AD 1000, is divided into six stages. Each stage has its own chapter with subchapters devoted to crops, crop cultivation, livestock and livestock handling, the farm and its yard, and the farm in connection with other farms and the outside world. Because the book is intended for a general public interested in the subject, every chapter starts with a short outline of the cultural context. After that the known facts are presented. The crop plants and animals are mentioned together with their origin.

The subchapters on crop cultivation deal with the operational chain from preparing fields to storage. The introduction of tools, such as the plough, the wheel and wagon, and the scythe is discussed. Farm buildings, or at least their ground-plans, are described. The clustering of farms into hamlets or the absence of such aggregations is mentioned. Two short chapters deal with the impact of farming on the landscape.

The information is drawn from my own work, but also to a large extent from publications written in French, German or Dutch, which are not easily accessible for a wider public. It was my intention to bring all this information together. But if I were to mention all my sources in the text, the book would have become unreadable. Therefore I have refrained from mentioning references and provide a 'select bibliography' instead. The publications mentioned there provide more details and more specific references.

Of course there had to be illustrations. Most of them are derived from or based on the multitude of articles read up for this book. I thank all the original authors for their willingness to allow me to use their intellectual offspring. I cannot mention them here, but their names are to be found in the list of 'Sources of figures and tables'. All the figures have been redrawn by one single person: Joanne Porck, who I thank for her great enthusiasm and care. As mentioned before, I wanted to make the agricultural history of the loess region west of the river Rhine known to a larger public. Therefore I wrote the book in English and this English had to be corrected of course. Kelly Fennema I thank you for this part of the work and also for assisting me with the editing of the manuscript.

Three referees have searched for scientific mistakes: Rose-Marie Arbogast of the University of Basel (Switzerland), Michael Ilett of the University of Paris I Panthéon-Sorbonne (France) and Willem Willems of Leiden University (the Netherlands). A fourth critical referee, who was not familiar with the subject at all, has read the text in order to see whether it was palatable to the kind of public I had in mind: Garbrand van Dijken, a Dutch agricultural engineer. I have learnt much from the comments of all four.

Contents

Chapter 1
The Loess-Covered Region West of the River Rhine, 5300 BC–AD 1000

1.1 Introduction

At first sight it may seem strange to have a book on agricultural history devoted to a region defined by its type of soil. But soil is, next to climate and availability of water, an all-important factor where farming is concerned. Thus, writing a history of farming on a specific class of soil makes perhaps more sense than writing an agricultural history of a present-day state.

My choice fell on the European loess belt. It is there that traces of the earliest farmers of Central and North-western Europe were discovered. These farmers settled almost exclusively on loess. This belt running roughly east-west covers a wide region, certainly too wide to be covered by a single book. Therefore I have concentrated on its western part, which I define as the part west of the river which runs from south to north (Fig. 1.1). This large river forms a natural barrier, but is not impossible to cross. In the past and into modern times its course has served repeatedly as a political frontier. During the period covered by this book this was most obvious during the Roman era. In addition I have chosen a western limit, namely the Channel. This wide stretch of water represents another natural barrier. As a result, the agricultural history of neither the English part nor the Channel Islands part of the loess belt is featured in this book. I also left out the north coast of Brittany, because this outlier of the continental loess belt seems to have had a rather different cultural development.

The first farmers west of the Rhine displayed a remarkably uniform cultural identity, though some slight subregional differences can be pointed out. The cultural cohesion was maintained during the following centuries. And even at the end of the period covered by this book cohesion was still present, firstly because the entire region became part of the Roman Empire and secondly because the region was the core of the early medieval Merovingian and Carolingian kingdoms.

C.C. Bakels, *The Western European Loess Belt*, DOI 10.1007/978-1-4020-9840-6_1,

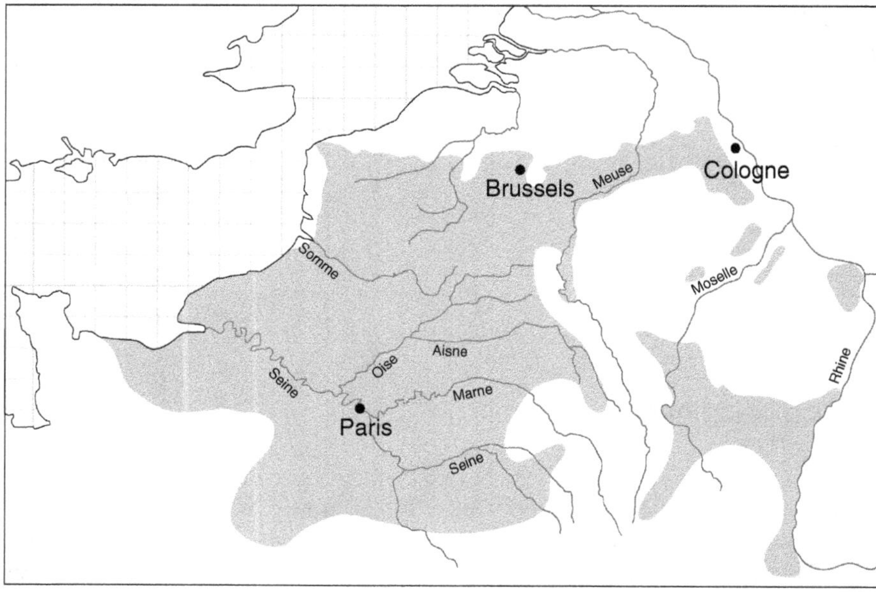

Fig. 1.1 The loess belt (*grey*) between the river Rhine and the Channel

1.2 Loess

Loess is an aeolian deposit, which means a sediment consisting of particles with sizes smaller than 60 μm transported by wind before settling down. Wind is able to pick up such material from deserts and carry it in suspension over considerable distances. The loess of the region in question was picked up in polar deserts bordering the ice-caps of the last two Ice Ages and was blown southward. The dust was either dropped by a decrease in wind velocity or washed down by rain. In the process the material was sorted out. Heavier particles of the size of sand settled closer to the deserts, and lighter particles were blown further away from the original source. Thus, a belt of aeolian sand lies north of the belt of loess. The east-west orientation of the belt is explained by the predominance of winds blowing from northerly points of the compass.

The deposit was prevented from being taken up again by wind because it was retained by vegetation, which was a steppe vegetation at the time. Loess is unstratified. Its particles are not firmly bound but loess is firm enough to maintain vertical exposure without immediate collapse. Drainage of surface water is good, but the capillary structure of the deposit is such that water retention is also good. Therefore loess is good for raising crops. It is also easy to till. Heavy rain in winter does not stay long enough on the surface to damage winter crops and the capacity for water retention is large enough to bridge dry spells in summer.

The loess deposits are oxidised to a pale brown colour. Originally the loess was calcareous, but during considerable stagnations in deposition, decalcification took place in connection with soil formation, resulting in several soil horizons separated

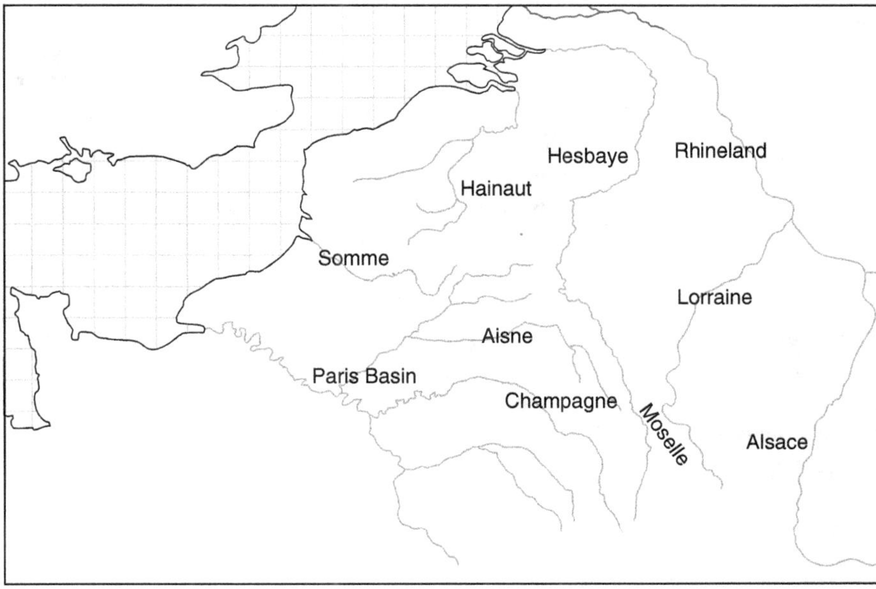

Fig. 1.2 The location of some important areas mentioned in the book

by unaltered loess. Such horizons are not important for agrarian history, except for the last one which developed after aeolian deposition had finally ceased.

Loess is very often associated with the soil type chernozem (black earth): a soil characterised by a thick, black, largely organic layer lying almost directly on top of the unaltered parent material. However, chernozem belongs to regions having a continental climate and limited rainfall. To what extent chernozem initially played a role in the region considered here is still a point of debate. Certain blackish deposits in prehistoric pits are interpreted by some researchers as relicts of a former black earth, but undeniable proof is still lacking. More probably, the amount of rainfall was too high to be compatible with the development of true chernozems, at least in the long run. The present-day soils show a displacement of soluble matter and the finest mineral particles from the top levels to a deeper level where they accumulated. The result is a 'brown earth' or 'grey-brown podzolic'. If the layer with enrichment of fine mineral particles (clay) is well developed this goes in certain areas under the name of 'brick layer', because it is more sticky than the parent loess and therefore suitable for making bricks (and pottery).

The exact process and speed of soil formation is still not very well known, but remnants of ancient soil trapped in deep pits or preserved by fire as lumps of burnt loam show that some soil formation had already taken place before the arrival of the first farmers. They must have found an already decalcified loess and something like a 'brick layer'.

As mentioned earlier, loess is very suitable for growing crops. Nevertheless, it is not optimal as far as nutrients are concerned. On the one hand loess is better than sandy soils, but on the other hand it is inferior to river loam. To maintain fertility it requires marling and manuring, but whether such techniques were already practised during the period covered by this book will be discussed in the appropriate chapters.

The thickness of the loess cover is very uneven. In some areas it may be as thick as twenty metres; in others the cover is very thin. The most important part with a thin cover is the north-western part of France, around the town of Amiens and further to the west. The rock immediately beneath this cover is chalk and this limits the water-retaining capacity of the overlying loamy deposits, making them less favourable to general crop growing. The same condition is found in the areas bordering on the French Champagne district (Fig. 1.2). In the Champagne district itself loess is only present in very small patches. It may be questioned whether these areas should be included in a book concerned with agriculture on loess, but I included them firstly because they may have lost some of their loess cover through erosion, and secondly because there were no indications during the writing of this book that they had a different agricultural history.

1.3 The Loess Region

The altitude of the region does not exceed 300 m above sea level. Essentially most of the region is a vast plain with a gently undulating surface, dissected by rivers. Stronger relief is mostly found at the edges, where loess accumulated only on the lee side of elevations in the original landscape. An example is the lower course of the river Moselle.

The rivers flow in valleys that were formed long before the deposition of loess. Some of the valleys are wide with gentle slopes, such as the valley of the river Rur (in Dutch Roer) in the German Rhineland, others have cut deep valleys with steep slopes, such as the lower course of the river Aisne in France. They divide the region into units which are known as 'plateaus'. It is the plateaus which are covered with loess. Gentle slopes have a loess cover as well, but steep slopes and valley floors do not. Loess is very prone to erosion. Slopes of 2% may already be too steep when not protected by a vegetation cover. Erosion may stop, if only temporarily, at the 'brick layer', which has more cohesion and only starts to give way at a slope of 8%. The eroded silt settles in depressions and shallow valleys as a sediment called colluvium. In valleys with active running streams the silt is taken up by the flow to be deposited elsewhere as river loam.

Erosion took place in the first periods after deposition and afterwards as a result of anthropogenic deforestation. In between the displacement of loess was kept in check by the vegetation cover. This vegetation also provided a good balance between rainfall and drainage. Small valleys did not have surface streams during this period. Investigations in the German Rhineland showed that in such valleys running water was present only before the return of the full climax forest after the Ice Age and after an advanced stage of deforestation by humans. During the period of full vegetation cover, stagnant water in depressions and on valley bottoms was a rare phenomenon as well. The result is that formation of peat was also rare. This fact restricts the possibility of a detailed reconstruction of the former vegetation (see Section 2.6).

In general the loess plateaus have a deep water table. Depths of over 10 m are common; and digging wells is therefore a strenuous task. Easier access to water is provided by springs occurring on slopes where impermeable rock underlying the loess is exposed, or, more commonly, by streams in the valleys. It is not surprising that most of the early occupation was confined to the edges of the plateaus and, in areas with steep slopes, also to the higher parts of valley bottoms.

During the period considered here the region had an oceanic version of a temperate climate, just as today. At present the mean annual temperature is c. 10°C. The mean rainfall lies in the range of 600–800 mm and precipitation is distributed rather evenly over the months. Of course, the temperate climate has not always been exactly the same, but reconstructions offer no indications of dramatic changes. The distribution of precipitation over the year may have altered too. Formerly there may have been slightly more rain in wintertime. Fluctuations occurred all the time. The first farmers arrived during what has become known as the climatic optimum of the Holocene, i.e. the period after the last Ice Age. Since then, summer temperature has decreased by c. 1°C. The worst dip in climatic conditions, a dip towards the cold and wet, is found to have occurred around 800 BC, but subsequently the climate improved again. Some changes may have had local and temporary impact, but they seem not to have affected rural life in the long run.

1.4 The Choice of the Period: 5300 BC–AD 1000

The beginning of the period was not hard to define. In many regions of Europe the integration of agriculture into the daily life of the inhabitants was a gradual process. The old way of procuring food by hunting, fishing and gathering was slowly replaced by the active production of food. Tilling the soil for growing crops was not adopted everywhere at the same time as the tending of livestock. But in the loess region, and certainly in the loess region west of the river Rhine, this slow process seems to have been practically absent. Agriculture arrived as one package to be applied in its entirety; whilst hunting, fishing and gathering lost much ground. The old way of life held out in certain places but was not mainstream anymore.

The beginning of agriculture is dated around 5300 BC. Like most dates in this book it is based on calibrated ^{14}C dates, which means that the values obtained by radiocarbon dating have been converted to their most probable calendar age.

The end of the period is set at the end of the 'period of direct agricultural consumption'. This choice was inspired by the book 'The Agrarian History of Western Europe A.D. 500–1850' published in 1963 by B.H. Slicher van Bath. He remarks that agricultural production before the nineteenth century was concentrated on satisfying the human need for food. Only from the nineteenth century onwards did raw materials other than food become important. He divided the agrarian history of food supply into three stages.

The first: Self-sufficiency, in which each household or small community produces all the food it consumes. This is called subsistence farming.

The second: Partial self-sufficiency, in which, while most people produce their own food, they also supply it as barter to the non-agricultural part of the population. This he calls direct agricultural consumption.

The third: Self-sufficiency, for a relatively small part of the population, while the whole non-agricultural population, and sometimes the agricultural population in part, satisfy their needs through a market where farm products are sold, mainly from districts with agricultural surpluses. This is indirect consumption. It is a society with an ever widening use of money.

The book by Slicher van Bath starts with his second stage, which he places between AD 500 and AD 1150. Its chief characteristic is that it concerns an agrarian society with an incomplete money economy. The consumption of agricultural goods is direct, i.e. there are no middlemen between producer and consumer. Everybody lives directly off the land.

I wanted to stop where this 'living directly off the land' ended. Following Slicher van Bath this should have been at c. 1150. But an important part of society that does not live directly off the land consists of the population of towns. As towns were already flourishing in the loess region before 1150 it was deemed reasonable to set the end of the period earlier. The tenth century is generally mentioned as the century of the rise of the medieval town. Therefore I decided to set the end at AD 1000.

The earliest farming communities belonged, as far as can be reconstructed, to the stage of self-sufficiency. In the course of time a small part of the population detached itself partly or entirely from food production. This has, more or less, been deduced from obvious differences in social status, such as can be detected by archaeological research. It must be kept in mind that the art of writing was not known during most of the period under review and that, therefore, written records simply do not exist. How the process proceeded through time is still a subject of investigation. But what is known is that social differentiation had its ups and downs. To set the lower limit of the second stage at c. AD 500, as Slicher van Bath does, is only half the truth. He chose this beginning because by then a new society was emerging after the collapse of the supply and demand system of Roman times. But in the long stretch of time since the introduction of agriculture, the Roman occupation represents only a short interlude. Elements of the second stage were already present before the arrival of the Romans.

1.5 The Framework of this Book

In the story presented here, the stage of self-sufficiency and the stage of direct agricultural consumption merge almost into one: everybody lived directly off the land. Nevertheless, the need was felt to divide the long period into sub-periods. I have opted for subdivisions with a cultural rather than an economic background. Introductory subchapters give brief outlines of this background, after which crops, crop production, livestock and livestock handling, farm buildings and yards, and the place of the farming communities in a wider context are described as far as known to me.

I have worked in the region for many years, compiled a lot of information and consulted many publications. Because of the profusion of both smaller and larger publications I have refrained from citing them in the text, in order to make the main text easier to follow.

The influence of farming on the natural environment is a subject that should not be neglected. The loess region has been farmed continuously from the beginning of farming onwards. This has changed the landscape completely and therefore two chapters have been devoted to this theme.

The book presents mostly facts. Our knowledge is still too patchy to allow more sophisticated economic or socio-economic analyses in the sense of a history of Longue Durée. An example of this patchiness is a complete millennium missing from the records.

Chapter 2
Sources

2.1 Information About a Distant Past

This book describes a long period during which writing was completely unknown or not commonly practised. Moreover, it deals with a time long ago and oral information is therefore also lacking. We do not know the people at all, except for roughly the last thousand years and then only fragmentarily. This implies that most of this book is based on information obtained through excavations. Information on crop plants, farm animals, tools, buildings, land, all rely on archaeological sources. This hampers our knowledge to a considerable extent as will be explained below. But even when written sources are available, these are of limited value.

2.2 Plants

The general fate of a plant is decay. Plants vanish by rotting away; but in certain environments plant material may survive and specific events may prevent decay. In an entirely dry environment plants are preserved by desiccation. Deserts, but also absolutely dry interiors of buildings or graves offer such conditions. In Western Europe these conditions have hardly ever been met with. On the contrary, the opposite, namely a waterlogged, oxygen-free environment is often encountered. Natural wet areas such as ponds and marshes, and unnatural ones such as wells, offer the right conditions. Preserved are sturdy parts of fruits, seeds, wood, and pollen. Fleshy fruits, tubers, bulbs and leaves become too slimy to recognise (Fig. 2.1).

Preservation by toxic substances is another way. It occurs in contact with, for instance, copper or bronze objects. This kind of preservation is rare though. More common is the replacement of plant tissue by mineral matter. Embedding in an environment rich in calcium and phosphorus salts provides such results. The organic substances become replaced by calcium phosphate. Pits dug in a calcareous soil and filled with dung, or pits filled with a large quantity of animal bone offer the right conditions (Fig. 2.2).

C.C. Bakels, *The Western European Loess Belt*, DOI 10.1007/978-1-4020-9840-6_2,

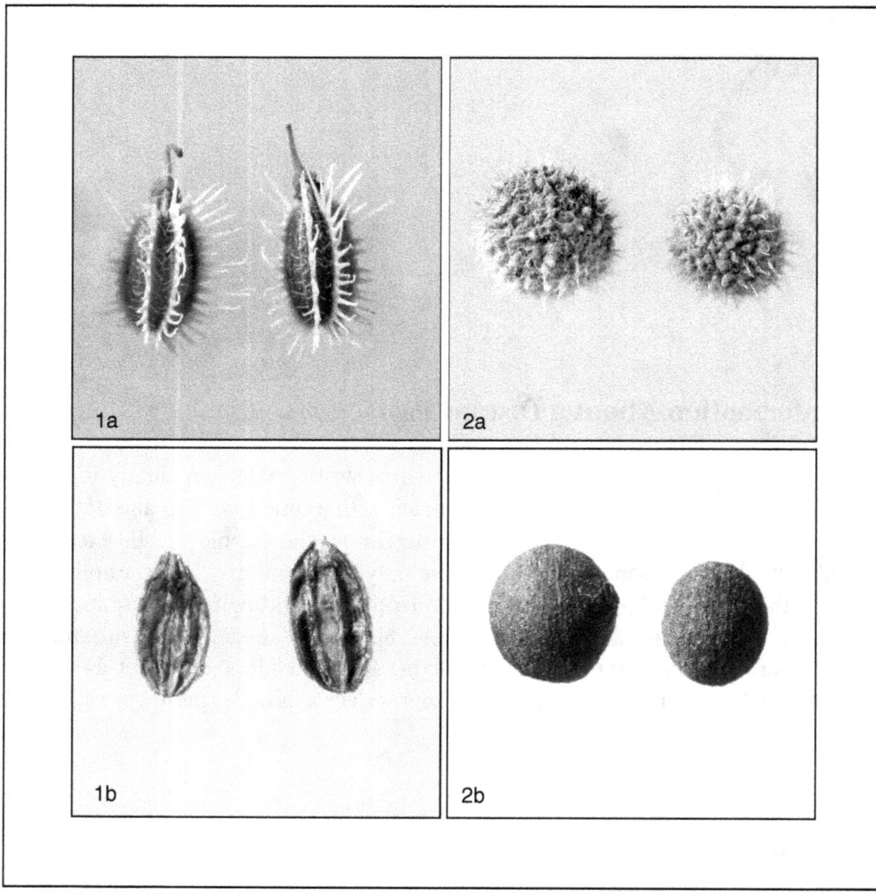

Fig. 2.1 Fruits of carrot (*Daucus carota*), recent (**1a**) and preserved by waterlogging whereby the spines are lost (**1b**); fruits of cleavers (*Galium aparine*), recent (**2a**) and preserved by carbonisation whereby the spines are lost (**2b**)

An event which prevents decay is heating to high temperatures in an environment poor in oxygen. It leads to charring. Especially dry, compact parts of plants are preserved in the resulting carbonised state. Examples are wood and dry seeds. Because most of the organic matter is converted into carbon, bacteria and fungi are unable to attack this kind of remains. The wood and seeds more or less keep their original form and can therefore still be identified (Fig. 2.3). Until quite recently it was thought that carbonised remains would keep forever. But it appears that environments with a high pH are detrimental to them. After burial for thousands of years in a calcareous sediment, such remains tend to fall apart.

Another event that preserves plants in a certain way is by impression. If parts of plants are, on purpose or by accident, mixed in with clay, for instance pottery clay, and if this clay is then fired, the plants burn away but leave an impression.

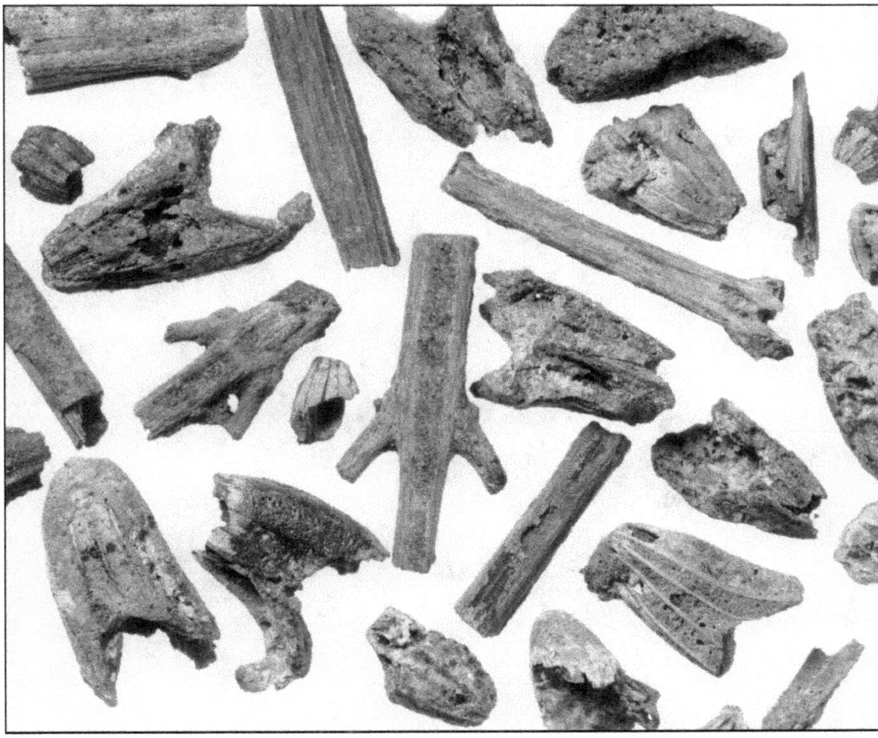

Fig. 2.2 Mineralised stems of sage (*Salvia officinalis*)

Impressions are studied by making casts which makes the identification of the original objects easier (Fig. 2.4).

In the case of the loess region it is the carbonised material that is commonly present. This kind of material contributes most to our knowledge of agricultural practice. Waterlogged material is less common. The water table in the region is rather low almost everywhere. Wells are scarce as will be seen in the following chapters. Moreover, people did not settle on the fringes of wetlands. Therefore, plant matter connected with rural activities had little opportunity to fall into the wet and become preserved by waterlogging. What is commonly found in the marshy parts in valleys was originally part of the natural vegetation. Such remains are used to reconstruct the environment. In this pollen plays a major part (see Section 2.6). Mineralised plants and impressions are not very important in the region.

Most plant remains are not readily noticed during excavations. Only fairly large objects or large quantities of seeds are generally observed and collected by hand. The reason is that plant remains are much smaller than the objects archaeologists are trained to look for. They have to be recovered by flotation or sieving. Carbonised matter tends to float in water whilst mineral matter tends to sink. If soil containing carbonised matter is poured into water and stirred, the plant parts will float to the

Fig. 2.3 Carbonised grains of barley

surface of the flotation basin and can be poured off. Unfortunately, some carbonised seeds are too heavy for this method and when the sediment sticks to the plants, as clay or loam will do, flotation may fail altogether. In such cases sieving of the wetted soil under running water is the only means (Fig. 2.5).

Waterlogged and mineralised plant remains are also retrieved by sieving, but not pollen. Pollen grains are recovered by a series of chemical treatments of the sediment encasing them. All matter other than pollen is dissolved or extracted by chemical means. This is possible because the outer wall of pollen grains is very resistant to all kinds of chemicals.

Identification of plant remains is achieved by comparing them with plates in atlases and by consulting reference collections of recent seeds, wood or pollen. Identification keys are of minor importance for seeds and fruits because they are often altered in the process of preservation. They tend to lose characteristic spines, for instance (Fig. 2.1). Most keys are based on recent seeds and fruits. The problem is smaller in the case of wood and pollen and keys are commonly used for the identification of these categories.

Summarising all that has been put forward so far, it is clear that not every plant can have survived down the ages. Sturdy, compact parts with a low water content

Fig. 2.4 Impression of a spikelet (two grains in their chaff) of emmer wheat in a pottery sherd (*left*) and a cast of the impression (*right*)

have had a better chance. The result is that such parts are the ones that are found during excavations. In the category crop plants cereals, at least their grain and chaff, and pulses are the best off, if deposited in a carbonised state. Oil-containing seeds tend to deform and even explode during charring, though they are quite often found carbonised, but they do better in a waterlogged environment. Waterlogging is also best for condiments and fruit stones. In contrast, waterlogging is detrimental to cereal grain and pulses. Only cereal chaff survives waterlogging.

The chances of survival considerably restrict our knowledge of the crops of the past. And, as remarked above, even carbonised cereals, pulses and oil seeds disappear from the records in the very long run if we are dealing with alkaline soils. This problem is felt in the southern part of the loess region, especially in the French Champagne, Paris Basin, and the areas near the Channel. It is striking that the settlements, occupied during the first millennia of farming in those areas, do not reveal much in the way of carbonised seeds and fruits. Although it has not yet been proven exactly, the calcareous subsoil may be the cause. As it is, most information on plants during this first stretch of time comes from excavations in the northern and eastern part of the loess region. From c. 1000 BC onwards the difference in preservation is no longer observed. From then on, all settlements contribute information concerning plants.

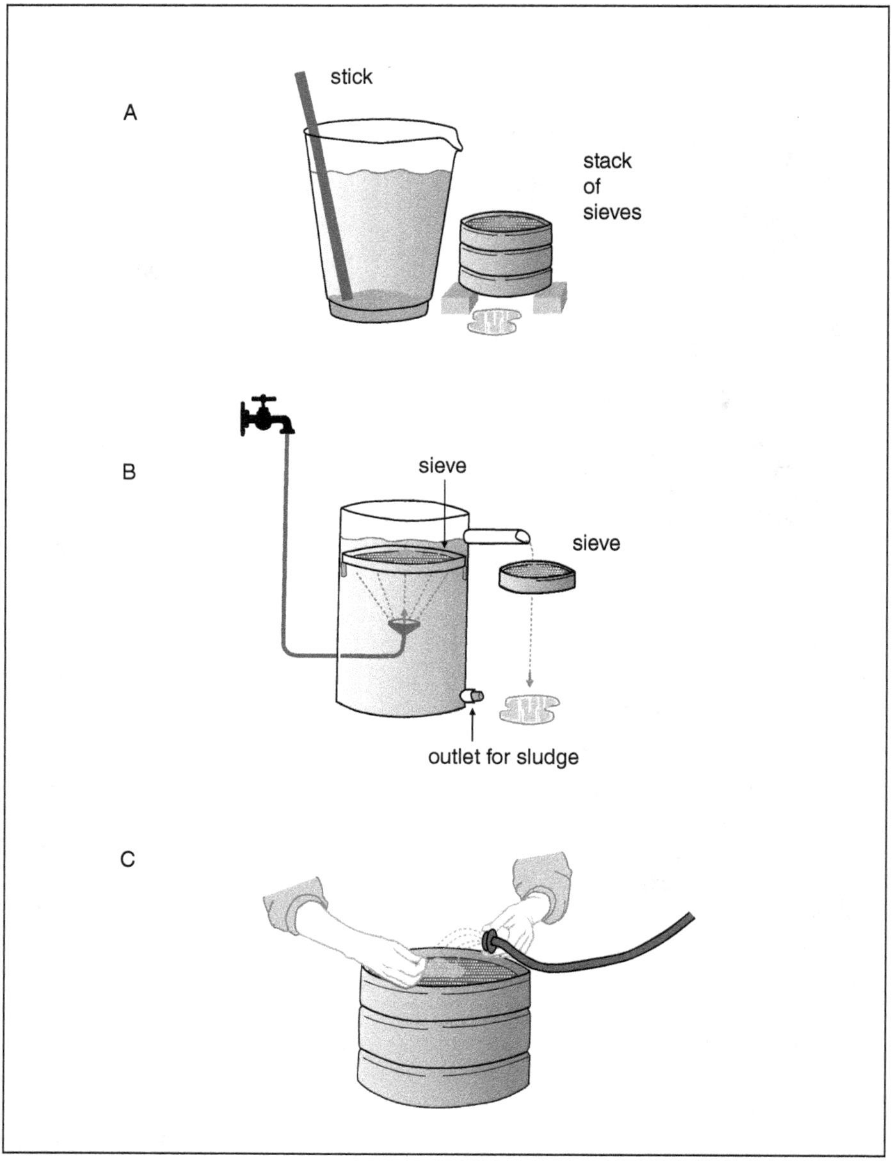

Fig. 2.5 Flotation and sieving. (**A**) A plain flotation method: stirring with a stick will bring the carbonised particles to the surface; the flot is poured from the vessel onto a sieve or a stack of sieves. (**B**) A more sophisticated flotation method: a layer of sediment is spread out on a sieve in the flotation basin and stirred by small jets of water coming from below; light particles swirl upwards and drift through the upper spout to fall on a sieve or stack of sieves. Sediment falling through settles on the bottom of the basin and can be discarded through the lower spout. (**C**) Sieving by hand

2.3 Animals

Just like plants, dead animals are subject to rapid decay. Only in year-round frozen or in desert-like environments can they be preserved as complete animals. Everywhere else their soft tissues rot away. This was the fate of the animals in the region under review. What remains depends on the characteristics of the soil in which they became embedded.

The bone skeleton of vertebrate animals will be preserved if buried in a calcareous environment. Bone is a calcified material. Laid down in it are salts in which phosphate and carbonate are combined with calcium. If a dead animal ends up in a soil poor in calcium its bones will dissolve and disappear from the archaeological record.

Dissolution is also the fate of bone deposited under water if this is not rich in calcium, but what may be preserved under wet conditions is the outer layer of bone which consists of another kind of material. When preserved, this outer tissue has a leathery appearance. Antler behaves in the same way. Not only the bones of vertebrates disappear by solution in a calcium-poor environment, molluscs disappear as well.

Burnt bone is more resistant to decay. It persists even in soils where normally bone dissolves. Unfortunately, this kind of remains is often rather fragmented, a state which affects the possibilities for identification. The enamel of teeth is resistant too, though it consists entirely of calcium salts. Its dense structure slows the dissolving process. In general, the compacter the structure, the slower the decay. Small, thin or hollow bones are the first to disappear. Large, massive bones are the last.

Skin may be preserved in bog environments, but has not been encountered in the region described in this book. The only exception is its derivative parchment, but that was kept under dry conditions. The same applies to hair, and therefore wool. Horn and feathers are not readily preserved either.

Remains of insects can be retrieved from waterlogged environments. Their chitin is not apt to rot. In some cases insects are also found charred, for instance when weevils in stored grain or pulses have become carbonised together with the crop (Fig. 2.6 and 2.7).

All kinds of remains are identified with the help of plates in atlases, but more so by comparison with collections of recent bones etc. Some animals which are of interest here, however, pose problems. It is more often than not impossible to distinguish between sheep and goats. Therefore, the following chapters refer often to sheep/goat. It is also difficult to distinguish mules from horses. And it is hardly possible to separate wild animals from their domestic relatives if the latter have not yet been altered by domestication or if they interbreed regularly with their wild kin. In our region this problem arises during the first stages of farming history where cattle and pigs are concerned. Wild sheep and goats are not native to the region. In later times, when keeping poultry becomes part of life on a farm, the distinction between wild geese and ducks and their domestic counterparts can be troublesome.

Bones provide more information than just names of animals. The morphology of bones and teeth reveal the age of the animals at death. Not exactly, but by age

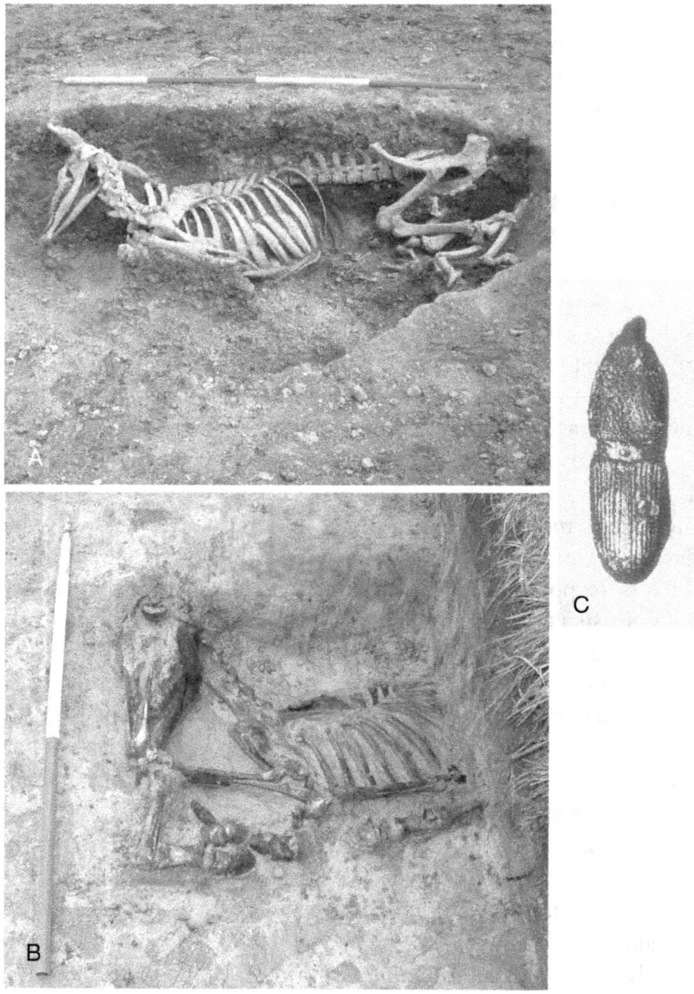

Fig. 2.6 Skeletons of a cow (**A**) and a horse (**B**) preserved under ideal conditions; **C** is an example of a carbonised grain weevil (*Sitophilus granarius*)

class. If, for example, most pigs were slaughtered as piglet or young animal, this will show. This means that pigs were kept for their meat. The few adult animals represent individuals required for breeding. Another example: if many sheep, both rams and ewes, were allowed to reach adulthood and even old age, they may have been kept for their wool. These are just some examples of what can be learned from an analysis of age.

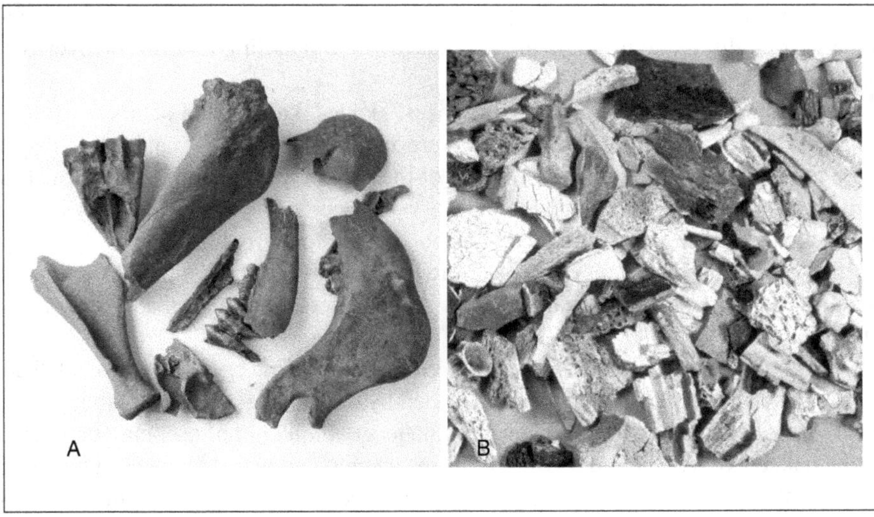

Fig. 2.7 Well-preserved bone (**A**) and burnt bone (**B**), both as found in common waste

Another aspect revealed by bones, is the size of the animals. Whenever possible the calculation of shoulder heights will be mentioned in this book. Castration is still another feature. The practice shows in the bones, and its recognition is of importance for the study of animal traction. Special traces of wear or deformation, caused by too often hauling too heavy loads is also sometimes noticed.

What animals exactly looked like is however almost impossible to reconstruct. If a characteristic is mentioned, such as the hairiness or woolliness of sheep, this is done by comparison with ancient races still living somewhere in Europe. In the case of sheep, a breed often referred to is the Soay breed from Scotland. Only from the Roman Period onwards is more known because of the existence of more or less reliable pictures. Depicted are, for instance, mules, recognisable by their long ears. Nevertheless, such important aspects as colour or spots remain largely unknown.

It is obvious from the above that bone reveals most of the information. But bone is poorly preserved in decalcified environments. Unfortunately such conditions prevail in the northern part of the loess region. Most of the information therefore comes from the southern part. This is just the part where information on plants is poor during the first millennia of agriculture.

2.4 Tools

Before mechanisation of farming, a large proportion of the farmers' tools were made of wood or other perishable material. What is said about plants in Section 2.2 also applies to tools. Turning up in the archaeological records are parts of implements with a non-wooden cutting edge, querns and large pots which have served as containers. In addition, some implements related to the handling and use of livestock have been preserved.

Tools intended for cutting are axes, adzes, sickles, knives, and scrapers. Before the introduction of metal, their blades were made of stone. Axe and adze blades were made of a tough crystalline rock or of the microcrystalline material known as flint or chert. Their handles were made of wood, as is shown by the axes found, with handle and all, in waterlogged sites outside the loess region. Sickles consisted of sharp pieces of flint wedged into a more or less curved wooden handle and fixed with a glue, commonly birch tar. Knives and scrapers were also made of flint set into a haft of perishable material.

Axe and adze blades are easily recognisable as such. The preserved parts of sickles, knives and scrapers are not only identified by their shape, but also by the kind of wear visible on their surface. Traces, specific to a certain kind of use, may be visible to the naked eye. An example is sickle-gloss, a shine caused by repeated rubbing of cereal stalks against flint (Fig. 2.8). But most traces are only visible with a microscope. They are called 'microwear'. Stone and flint as such are very resistant to decay. Nevertheless, long exposure to the air causes weathering of the surface. The surface of artefacts obtains a so-called patina. Weathered surfaces will have lost their traces of wear.

The introduction of metal led to a replacement of stone by metal for tools with a cutting edge. Bronze was the first metal to be used. This metal is moderately susceptible to corrosion. In contact with the air bronze develops a patina which protects the inner core. Nonetheless, this protection is not ever-lasting. In the end bronze objects disintegrate. After bronze came iron. Iron became more popular than bronze for tools. Before the arrival of iron many tools with a cutting edge were still made of stone, but after its introduction stone vanished almost completely as a raw material. Iron is unfortunately very susceptible to corrosion. Rust does not protect the inner core and finds of iron objects are rare. Whereas objects of bronze are fairly common in archaeological records, objects of iron are, if still present, found as clumps of rust. Laboratories, specialised in treating this kind of material, have to be resorted to reveal the inner core and the original shape of many such objects (Figs. 2.9 and 2.10).

Querns have always been made of stone. The kind of rock is of importance where their preservation is concerned. Some types of volcanic rock, much sought after for making querns because of the right kind of roughness, do not keep well after burial in the soil. They fall apart, and what is retrieved during excavation are just fragments. Other kinds of rock are very decay-resistant.

Pottery, if fired well, is decay-resistant too. Some farming communities used large earthenware vessels to store products. Many containers must, however, have been made of wood or basketry, and have not survived. Wooden casks have sometimes escaped decay, because they were used as the lining of wells. Below the water table their wood has been preserved (Fig. 2.11).

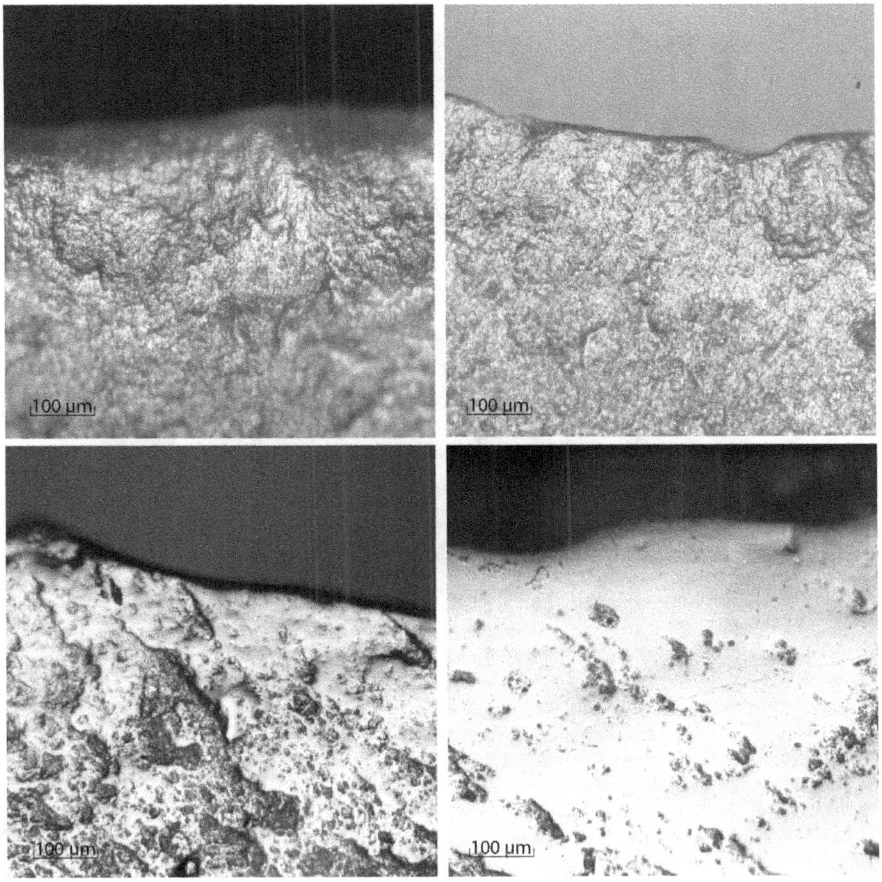

Fig. 2.8 Traces of microwear on flint artefacts; *upper left*: scraping of dry hide, *upper right*: cutting hide, *lower row*: two instances of cutting cereals. Provenance the *Linearbandkeramik* settlement at Geleen-Janskamperveld, the Netherlands

Implements related to the handling and use of animals were presumably also mostly of perishable materials. Yokes and harnesses, for instance, are hardly ever found. Sometimes their presence is revealed by their trimmings of bronze or iron. The same is true for wagons and carts. What is known from such things is mainly due to waterlogged sites outside the region and to special graves. Among the rare finds concerning animals, the horse bit is one of the commonest. The iron horse shoe is extremely rare, but as this invention was only introduced at the very end of the period considered in this book, this is hardly surprising. But other implements made of iron, such as shears, are also hardly ever found.

For the Roman Period and, to a lesser extent during the following Early Middle Ages, our knowledge is supplemented by images. Roman sculptures and medieval illuminations sometimes provide clear pictures of what tools looked like.

Fig. 2.9 A Roman axe, iron; from top to bottom: as found; after the first treatment by the company Restaura (Haelen, the Netherlands) and after a second treatment by the same company. Provenance Bocholtz, the Netherlands

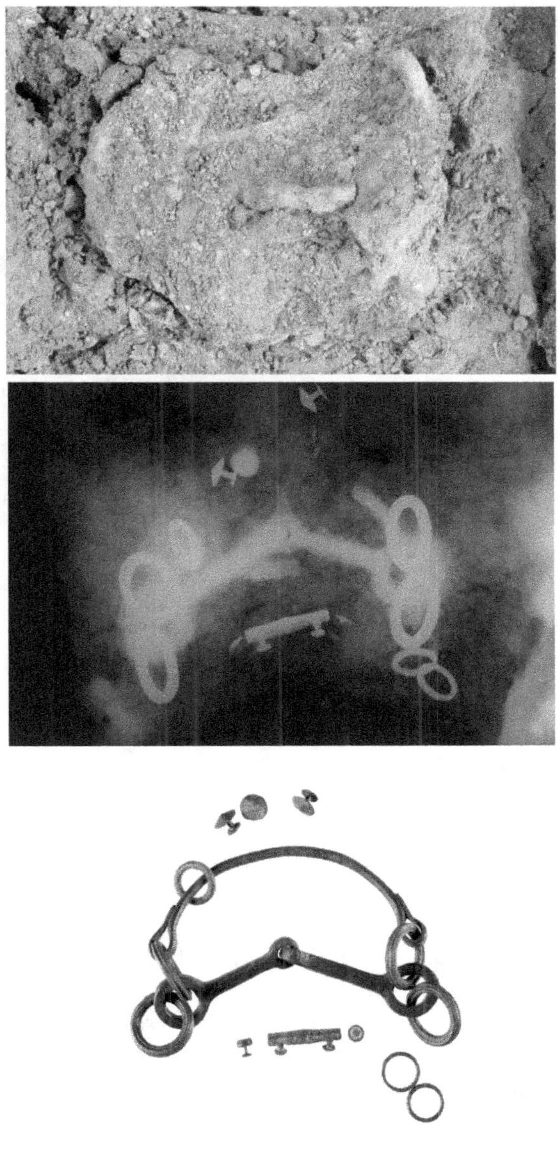

Fig. 2.10 A Roman bit, iron; from top to bottom: as found; seen in an X-ray and after restauration by the company Restaura. Provenance Bocholtz, the Netherlands

Fig. 2.11 Roman cask, re-used as lining of a well. The cords are modern. Provenance Oss, the Netherlands, a site on the river Meuse north of the loess belt

2.5 Buildings and Other Structures

During the period discussed in this book the main materials used in constructions were wood and loam. Buildings were not very durable. Upright posts were set directly into the ground. Rot was sometimes retarded by charring the lower ends, thus providing a kind of rot-resistant coating. In some periods wooden parts were placed on a stone foundation. But this was only done where stone was readily available in the environment, and even then not everywhere. Wattle-work was the common material for walls. The woven structure was plastered with loam. Such daub does not survive the decay of the building, unless the building was destroyed by fire and parts of the daub had a chance to get fired. Such orange to red coloured lumps are found during excavation. They sometimes bear impressions of the twigs of the wattle-work (Fig. 2.12).

Roofs were obviously also covered with a perishable material, but this is never found.

The result is that, of the buildings, only traces of the ground-plan have survived. These consist of the pits in which the upright posts were erected, the so-called postholes. The soil used to fill up the holes after the placing of posts had usually a different, slightly darker colour than the original subsoil. Therefore the original holes stand out against a lighter background. Sometimes, a still darker centre is seen in the postholes. This is the 'shadow' of the original post which owes its existence to

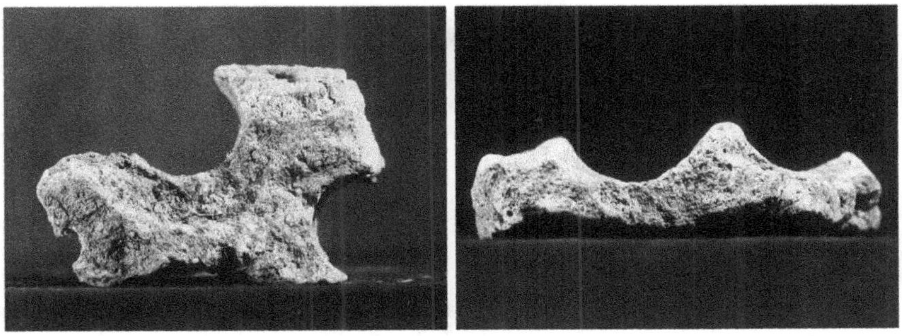

Fig. 2.12 Fragments of burned daub found in a *Linearbandkeramik* house destroyed by fire. The semicircular indentations are impressions left by the wattles

the fact that if, after the abandonment of a house, the posts were pulled out for secondary use, the resulting hole got filled up with the surrounding soil. This soil was usually rather polluted and therefore darker than the primary fill of the posthole. Or, if the post was not pulled out but left standing to decay, the organic substances of the wood led to a discoloration of the sediment which gradually replaced the post. The shadow often reflects the original size and shape of the post. Even split tree trunks may thus be recognised (Fig. 2.13, see also Fig. 3.9).

As the original surface on which the buildings were erected has almost everywhere been affected by erosion, the deeper the posts were dug in, the greater the chance to find them again. Postholes of shallow-founded posts have hardly withstood the ages. Roofbearing posts are found most often, because they were founded deepest. Uprights in walls, if not bearing the load of a roof, were set into shallower pits and are more often missing from the archaeological records. The same applies to inner divisions of houses and to fences on yards. Erosion is also the reason that the original floor is hardly ever present. This implies that the structures present on this floor, first and foremost hearths, are absent as well.

All in all, what is generally known is the approximate shape and size of buildings at ground level. The number of aisles can also be ascertained. The setting of the heavy uprights sometimes suggests the existence of an upper storey or a loft, especially if two of these uprights were inserted into the same posthole. One would then have supported the roof, the other the floor of the second storey. But the height of buildings, the nature of the roof and the presence or absence of windows remains elusive. Even the place of the door or doors is most of the time just conjecture.

Information is better where the Roman Period and, to a limited extent, the Early Medieval period are concerned. The houses of affluent people, such as the principal house on a Roman farm or the farmhouse of a medieval aristocrat, were partly

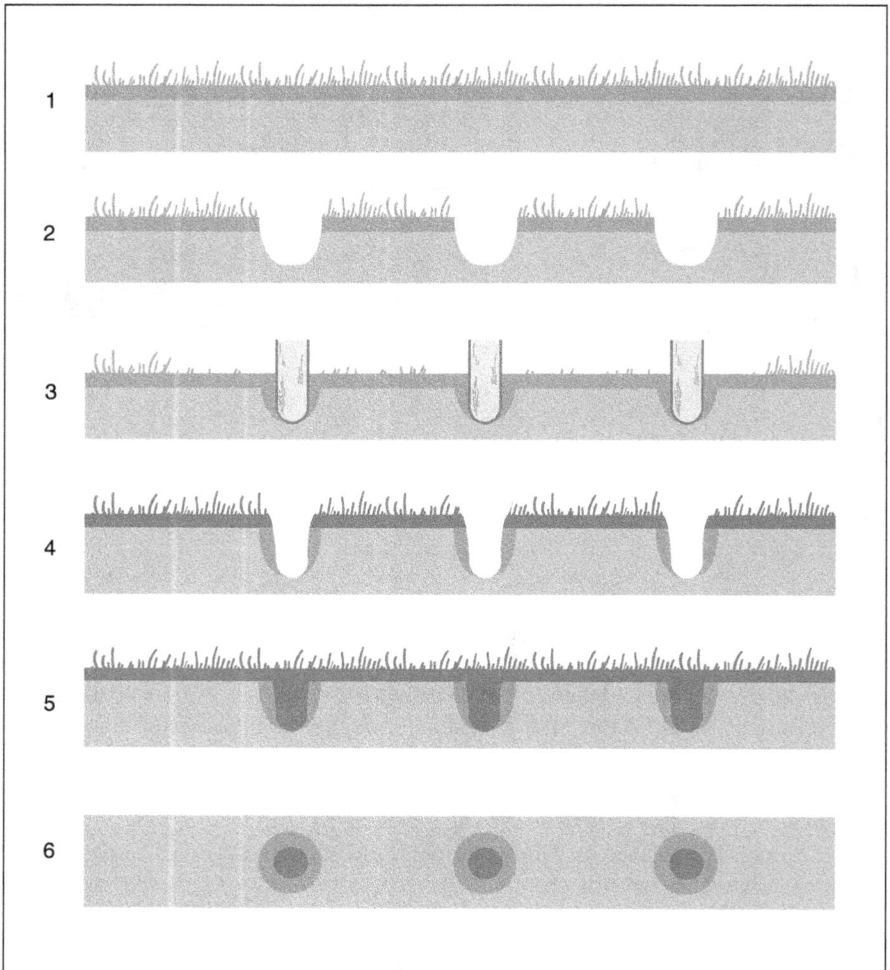

Fig. 2.13 The genesis of marks left by a row of posts. **1**: the original surface; **2**: holes dug for the erection of the posts; **3**: posts in position, the remaining open space is filled in with the topsoil which is darker than the subsoil; **4**: The post are drawn out and **5**: the remaining hole is filled with topsoil which is darker than the earlier topsoil due to pollution by human occupation; **6**: the same situation as in **5** but now seen from above during excavation when the topsoil has been removed. The ground-plans of houses shown in the following chapters depict this kind of marks

built in stone. The Romans even used brick. Stone basements and stone cellars are quite common. Floors were made of erosion-resistant materials and covered with mosaics. But even in such cases the upper storeys were built in wood and wattle and daub. Broken tiles show that the roofs of the rich houses were covered with ceramic material, though wooden shingles are known to have been used as well. Less important buildings were still erected using wood and wattle and daub and covered with thatch. What is known of the outside appearance of the houses is due to pictures, for instance pictures in floor mosaics.

Structures other than buildings are, for instance, underground storage pits, outside ovens and wells. Such structures are detected, because they are completely (storage pits and wells) or partly (ovens) dug into the ground. Their fill after abandonment stands out because of a different colour. If the water table has remained at the same or even higher level, the possible lining of wells may have been preserved too. Such linings often consisted of wood or wickerwork. After the invention of the cask, large casks were sometimes given a second life as well-lining (Fig. 2.11).

Ovens were built with a loam upper part covering a more or less elaborately shaped pit. The sides and roof of an oven were fired before its first use, or got fired during operation. Most traces of ovens contain lumps of red loam.

Considering everything, it is a structure with an underground part which is revealed during excavations. Other parts escape our notice. There must have been many more, but shallowly founded material will never be seen again.

2.6 Land and Countryside

A farm, at least a traditional farm, functions in close association with land, whether in the shape of gardens, fields, pastures or meadows. Land is a part of agriculture that is very difficult to reconstruct. Good agricultural land is used and re-used over and again. Millennia of farming have obliterated traces of earlier land use. Accumulations of sediments like wind-blown sands may preserve parts of soils or even ancient landscapes by burying them, but such instances are not known from the loess region. Only where plots of land were delimited by palisades set into deep foundation trenches or by ditches, may these boundaries have been preserved to give an insight into the kind of parcelling and the size of the parcels. Erosion of surfaces adds to the problem of reconstructing agricultural land. Some Roman parcelling is said to live on in the present parcels, but this is rather open to debate. The written sources of the Early Middle Ages provide some information on the fields and meadows of their period, but this is often not very precise.

The same problems are encountered when reconstructing the general countryside. The only aspect which can be reconstructed is the vegetation. In this reconstruction pollen plays a major role. Pollen grains are small, light and easily transported through the air. They are the only parts of plants that can be found far from their parent source. Pollen, released into the air, can travel considerable distances before settling down. When this 'pollen rain' happens to fall on a permanently wet surface, the pollen grains will be preserved. If this surface is growing upwards by the addition of new sediment, as for instance on the bottom of lakes, or by the addition of new plant matter as in living peats, a stack of individual pollen rains will be the result. Each level in this stack represents a time horizon or a restricted period. Pollen retrieved from these levels reflect the vanished vegetations (Fig. 2.14). Through the identification and counting of the pollen, a former vegetation can be reconstructed, at least to a certain degree. Unfortunately, wet environments where pollen might have been preserved are scarce in the loess region. Such environments owe their existence to the stagnation of water, either in depressions or in boggy vegetations drenched by a surplus of rainfall. The latter never occurred in the region. Most of the former came into existence only after deforestation, a process which disturbed the balance between rain and the retention of rainwater by plant growth, especially woods.

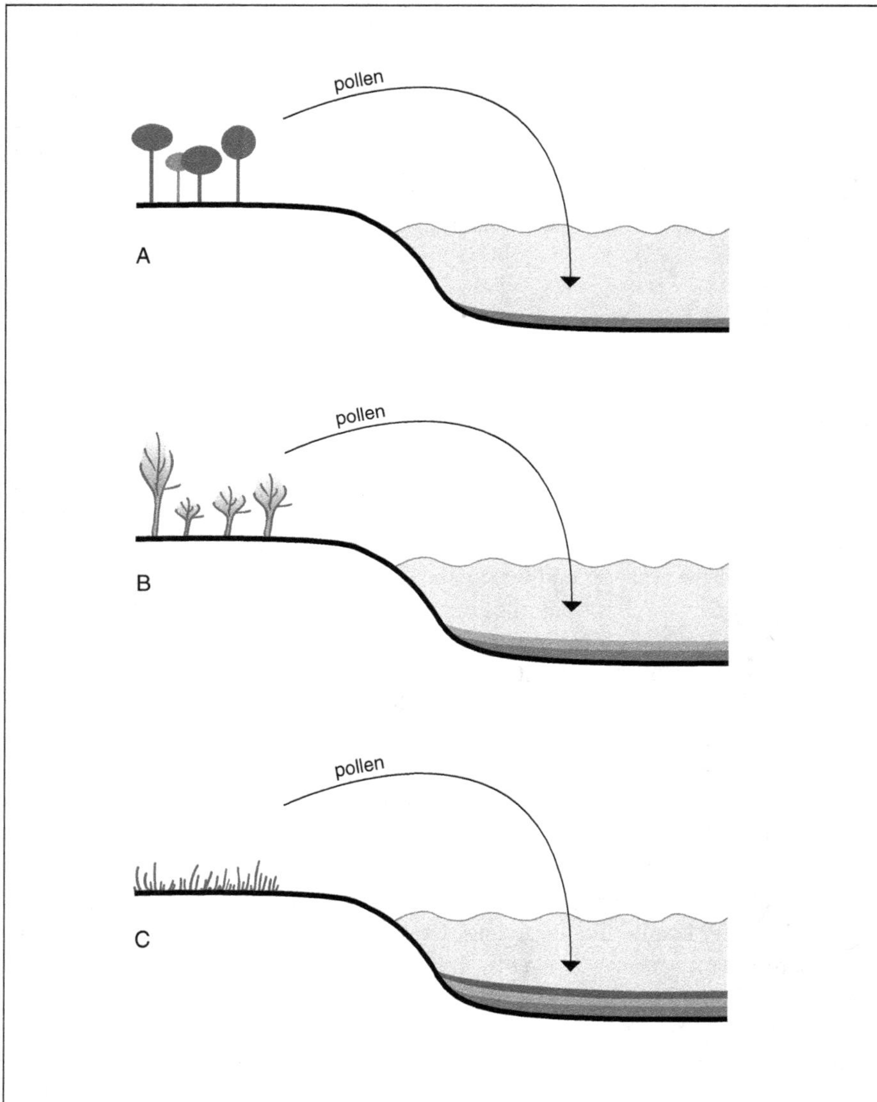

Fig. 2.14 The building-up of a pollen record in a lake. Pollen from vegetation (**A**) settles on the *bottom*, pollen from a later vegetation (**B**) settles on *top* of the sediment with composition A, etc

Other kinds of plant remains give additional information. For instance, if wood happens to be preserved, the appearance of its grain and tree-rings may give clues about the environment in which the tree grew up. Dense forests lead to tall trunks with a long, straight grain. Trees standing in open areas develop short trunks with much branching.

2.7 Written Sources

The introduction to this chapter mentioned the existence of written sources. It took a long time for the art of writing to reach the loess region. In the last centuries BC faint echoes of the barbarian North appear in the classical texts, written in the, in this respect more advanced, Mediterranean world. During the Roman occupation, which lasted from 50 BC to AD 407, the written word reached the region. But, unfortunately, nobody with an interest in farming left us any illuminating texts. Information starts with the Early Medieval period. Between the fifth and eighth centuries texts are, however, scarce. Only the last two centuries considered in this book, have more to offer.

Medieval sources can be divided into three categories: texts laying down rules, such as laws, texts dealing with administration, and texts with tales. All three give information on agriculture, each in a different way. The oldest texts which give an insight into rural matters are laws. In this book the Salian Law, the *Lex Salica*, is referred to. It is originally an oral law, written down at the beginning of the sixth century. Another important source within the rule category is the *Capitulare de Villis vel Curtis imperii*. This text gives instructions for the management of the royal estates. It was written between AD 792/793 and 800 on instigation of Charlemagne.

Texts dealing with administrative subjects describe the properties of large landowners. One of the most famous is the *Brevium Exempla ad describendas res ecclesiasticas et fiscales*. It provides an inventory of all things found during an inspection tour of five royal estates and was written in the same time as the *Capitulare de Villis*. Other important texts are the administrative texts called polyptychs which concern the large properties of abbeys. The oldest date from the beginning of the ninth century. A large proportion of these manuscripts is devoted to the enumeration of what tenants had to deliver and how much, both in kind and as corvée.

The tales give additional information. They comprise chronicles, saints lives and other stories. All texts are in Latin. The habit of putting things down, first on papyrus and later on parchment, was introduced by the Romans. With this heritage came the language too.

The subject of the administrative texts and the tales are the large properties and the upper stratum of society. Only the laws deal with other people. Most people still lived in a world where oral communication and transmission were the norm. Therefore, written sources are apt to provide a lopsided view on agriculture and rural life.

Chapter 3
The Beginning: 5300 BC–4900 BC

3.1 The First Farmers

The first farmers arrived around 5300 BC. They came from Germany and settled first in the eastern part of the loess region discussed here. During the next few centuries they extended their activities to areas further to the west and south-west. These people were named after their style of decorating pottery, which included a ribbon-like pattern, *Linearband* in German. Hence *Linearbandkeramische Kultur* or, shortened, *Linearbandkeramik* (LBK).

This culture originated in Hungary from where it spread all over the European loess zone, as far as Moldavia in the east and northern France in the west. There is a debate going on whether the culture with all its traits spread through adoption by indigenous hunter-gatherers or was introduced by migrating people. The latter hypothesis has the most adherents, because the archaeological record in the loess region does not reveal a gradual transition from the hunter-gatherer way of life to entirely sedentary farming communities. There are traces of people using pottery, a commodity unknown in hunter-gatherer societies, before the arrival of the LBK, but the status of this 'La Hoguette' group is not clear. It is not known if they were food-producers on any significant scale. At present it looks as though farming was not gradually adopted, but came as a kind of 'big bang'. LBK farmers were full-time farmers. In this the loess region differs from the neighbouring regions where the production of food was adopted gradually, often taking centuries to come to completion.

3.2 Crops

The LBK people west of the river Rhine cultivated seven crop plants. Three of them are cereals: emmer wheat, einkorn wheat and naked multirowed barley. Both wheats are hulled, which means that their glumes adhere to the grain, thus preventing the separation of grain and chaff by simple threshing methods. Emmer wheat is tetraploid, einkorn wheat is diploid. Very rarely has naked wheat been found in

C.C. Bakels, *The Western European Loess Belt*, DOI 10.1007/978-1-4020-9840-6_3,

the north-eastern part of the region but this is considered to be an 'accidental occurrence', not to be mistaken for a true crop plant. Next to the cereals two kinds of pulses are present: pea and lentil, and two oil seeds complete the list, linseed and poppy. The poppy is of the variety *Papaver somniferum* var. *setigerum.* Whether the flax, which provides the linseed, also provided fibres is open to doubt. From a site outside the region, in the Czech Republic, one thread made of flax has been reported, but implements for spinning fibres, such as spindle whorls, are remarkably scarce in our region. Such implements may have been made of wood, a perishable material, but as contemporaneous people elsewhere in Europe made them of pottery, this option is less likely. Therefore, the importance of flax fibre does seem to have been very minor (Fig. 3.1).

None of the crop plants were domesticated locally. Most species belong to the set of plants that was taken into cultivation in the Near East. The Near East is the centre of origin of many cultivated plants. Some of this set became lost in the process of the spread of agriculture to Central and Western Europe. Only six made it to the loess region west of the Rhine. The seventh, the opium poppy, surprisingly seems to have a different place of origin. Its predecessor, the wild plant, is considered to be native to the regions surrounding the Western Mediterranean Basin, ranging from western Italy to the coastal areas of Tunisia, Algeria, Morocco, eastern Spain, and southern France. The plant may have somehow reached the LBK farmers from there. Actually, poppy is only found in the western part of the LBK world; it has not yet been detected in the eastern part.

The potential availability of seven crop plants does not mean that every farming community grew them, or grew them on an equal scale. Emmer wheat, einkorn wheat, pea and linseed were obviously basics everywhere; but naked barley, lentil and presumably poppy were not. Naked barley was only cultivated in the two most western and rather isolated clusters of settlements in Belgium (those near Wange and in the Hainaut) and in the French Aisne valley. In the intensively studied areas of the German Rhineland between Cologne and Aachen and in the Dutch loess zone, barley is absent as a crop. Single grains turn up occasionally there but are seen as accidental, not as a true crop. Why naked barley was grown in the Belgian 'outback' is not clear. The local conditions are not that different from those in Germany and the Netherlands. The farmers in the Aisne valley tilled a calcareous sandy loam on top of a gravelly river terrace, which had good qualities though not quite equal to the loess of the other areas. In general, barley is hardier than wheat. Although the actual varieties of barley and wheat grown at the time are not known, it is quite possible that barley did better on these terraces than, for instance, emmer wheat. At least when conditions during the growing season were temporarily sub-optimal. This may explain the barley.

Lentil is, certainly during the first phases of LBK farming, found everywhere, but the sites where finds are abundant are located on the slopes of the valley of the river Moselle not far from Trier (Germany). The local climate there is considered to have promoted lentil growing. The plant is a heat lover and the steep, sunny slopes of the Moselle valley, at present renowned for their wine, may have suited lentil fairly well.

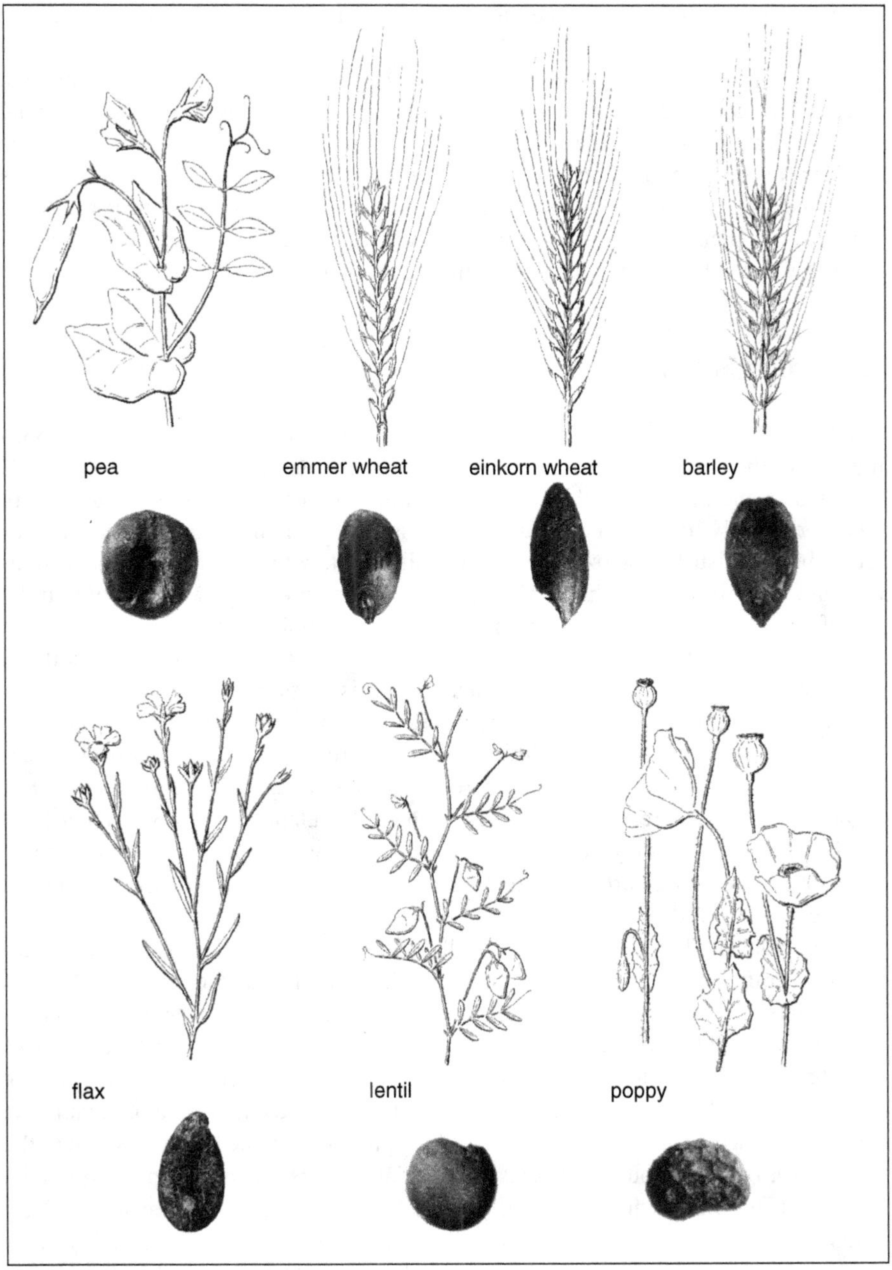

Fig. 3.1 The first seven crop plants. They are found as carbonised seeds photos of which are depicted under the plants which they represent

Opium poppy has not yet been found in northern France. Why poppy is absent there is not yet understood.

In addition to the seven crop plants, a near-crop plant should also be mentioned. This is rye-brome (*Bromus secalinus*), a wild grass species, which has been found in considerable amounts in contexts similar to those where cereals are attested. Its numbers exceed the numbers of other wild plants and it looks as if rye-brome was harvested for its own sake. It may have been a gathered product, but it may also have been a plant in a stage of incipient cultivation. If so, rye-brome is the first (semi-)cultivated cereal without an ultimate origin in the Near East.

3.3 Crop Cultivation

The LBK farmers were active during the Atlantic Period, which is the climatic optimum after the end of the last Ice Age so far. Summer temperatures were up to 2°C higher than the average of the twentieth century, whilst winter temperatures were equal or slightly higher. The annual rainfall is held to be more or less the same as today, but the distribution over the year was different, with less precipitation in summer and more in winter. The rainfall allowed for the growing of crops without the use of irrigation. Indeed, LBK traces of irrigation are unknown.

In the case of the Paris Basin, in the Aisne valley area for instance, the soil is a light river loam, while almost everywhere else it is loess. Soil formation processes had already altered the soils after their original deposition, but whether they had reached the present situation is difficult to reconstruct. Decalcification of the loess by dissolved CO_2 and organic acids produced by the vegetation and the decay of organic matter had already begun. Eluviation of clay elements from the upper level and illuviation of this clay in a lower level had also started. How far this process had advanced at the time of arrival of the first farmers is a question that has not yet been completely resolved.

The actual fields are unknown as well. No traces of them are left after millennia of farming the same plots. Some researchers think that the word field is not appropriate and that the word garden should be used instead. As the land was covered by primeval broadleaved climax forest at the time (see Section 6.1), clearances must have been made to allow the laying out of fields. The felling of trees was not achieved by wielding axes with a blade of flint, because these were unknown in the LBK. They used adzes with blades made of amphibolite (in the broadest sense of the definition), basalt or tough siliciclastic material such as a certain type of quartzite or lydite (Fig. 3.2). Similar implements can be found today in, for instance, New Guinea. Adzes were rather scarce in the Paris Basin settlements. Some roughly made axes of flint have been found in this area, but even those are rare. What people did use here to fell trees is unknown. A possible alternative is ringing, where a circular strip of bark is removed from the trunk to stop the transport of water and other fluids. It is supposed to kill the tree, but as experiments in Denmark have shown, ringing is not very effective in areas with sufficient rainfall. Burning is another option, but as the forests were composed of deciduous trees, which are hard to set fire to, this

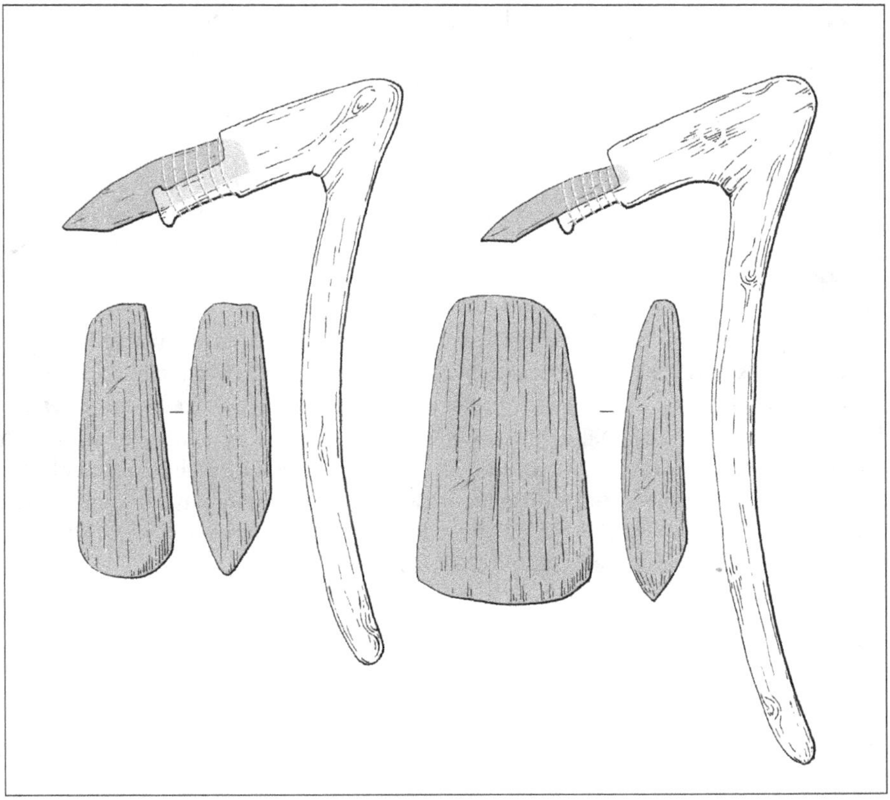

Fig. 3.2 Two adze blades and the way they may have been hafted. Scale of the blades 1:2, hafted implements not to scale

is also not as easy as it seems. Anyhow, there must have been ways to create clearances, unless suitable open patches were already available.

A hypothesis has been put forward that the climax vegetation was not as closed a canopy forest as is commonly assumed. Its supporters maintain that wild animals, especially large grazers, kept part of the landscape open, never allowing the climax forest to close. They see the land covered by a patchy kind of vegetation in which stands of trees alternated with open spaces. Such open places may have been chosen by LBK people to settle and farm on. This view is not shared universally and real proof is still lacking. If there was any natural open space, it would have been present especially in valleys (see Section 6.1). In the Paris Basin, settlement did occur on valley floors (see Section 3.6), and this is perhaps part of the background to the lack of great numbers of broken and subsequently discarded adzes in these sites.

Assuming that trees had to be felled, the resulting clearance must have been full of tree stumps. The removal of trunk bases and root systems requires a tremendous amount of work. The first European settlers in Ontario, Canada, left them standing until they rotted away. From the same European pioneers it has been reported that they did not need to plough during the first two to four years after clearing. Clearings in a primeval forest require a minimum of tillage, because the undergrowth is sparse. True ploughs came into use only long after the LBK period. It is even not yet certain that LBK farmers knew the ard, a kind of plough that cuts furrows without turning the soil (see Section 5.3). No remains of such an implement have been found and the ard may not have been very effective in fields full of tree stumps. The farmers did however have wooden hoes. Hoe-like implements have been found in a well at Erkelenz-Kückhoven in Germany (Fig. 3.3). The stone adzes mentioned earlier are sometimes interpreted as hoes, too. The use-wear pattern on their cutting edge however shows no traces of working the soil.

Hoeing may have been the only way to till the soil, but prior to that branches of felled trees may have been burned and their ashes spread over the surface. Remnants of ancient soils, connected with the original LBK surface, in the Aisne valley contain particles of charcoal. But the LBK farmers did not practise slash-and-burn cultivation. Their fields were more or less permanent. One of the arguments for permanency is that the average distance between contemporaneous settlements was limited to one or at most a few kilometres. Their economic territories were not large enough to allow agriculture based on slash-and-burn.

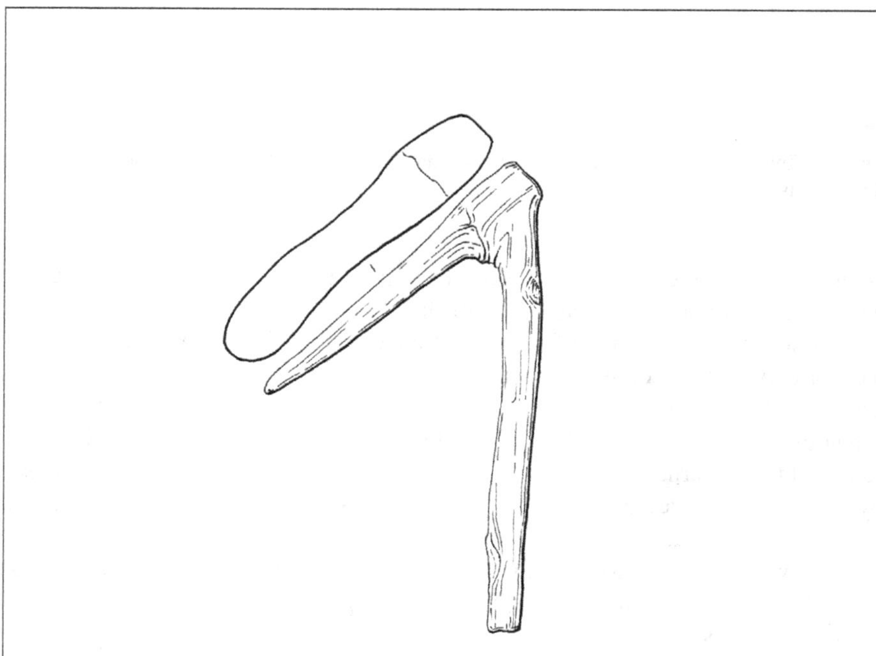

Fig. 3.3 A hoe-like implement from Erkelenz-Kückhoven, Germany. Originally the shaft was longer; length from point to butt 32 cm

The surface area of the territories varies between 60 to 250 ha. Calculations made for a small 'model settlement' with c. 50 inhabitants would have required around 11 ha of arable land, assuming that 65% of the daily food would have consisted of grain and the net grain yield amounted to 800 kg/ha. This guestimated yield is based on ethnographical, early historical and experimental data, which indicate that a gross yield of 1000 kg/ha of cereals is feasible. After the deduction of losses and sowing-seeds, a net yield of 800 kg remains. The postulated surface under cultivation falls within the labour capacities of farming families practising hand cultivation. According to current views a single family of 5 people can cultivate c. 2–4 ha and also harvest up to 4 ha. An average LBK settlement numbered 50–100 inhabitants (see Section 3.5). The model tells us that one to three years of crop cultivation followed by a full recuperation of the local forest, requiring decades of years if not more, is simply not possible in their relatively small territories. Another argument is provided by the study of LBK field weeds. These fit intensive gardening practices (see below).

With one exception, the fields have not proved traceable by analysis of pollen deposited in wet areas during LBK times. Waterlogged sediments, such as peats and lake bottom sediments, preserve pollen grains from past vegetations. In most cases they are the only remnants of this past. One of the problems is that lakes, accumulating sediment, and wetlands, furthering peat growth, were very scarce in the loess region during the Atlantic Period. The balance between precipitation and evaporation seems to have been in such an equilibrium that no surface water was left standing in mires or lakes. Another problem is that the fields may have been small and surrounded by high vegetation such as remnants of the original forest. In such cases pollen from low vegetation, crops or crop weeds for instance, do not rise sufficiently high into the air to be taken up by air currents and deposited elsewhere. The Wange pollen diagram provides an example. The core for this diagram was taken from a small peat deposit in a narrow valley at respectively 100 and 400 m from two LBK settlements in Belgium. This spot is one of the few places where peat formation took place during LBK occupation. Human activity is revealed by the pollen record but no traces of fields could be detected. An explanation could be that, although the settlements were situated at the edge of the loess-covered plateau bordering the narrow valley, and thus close to the peat, the fields were not. They were presumably laid out in the forest on the loess plateau behind the settlement and away from the edge (Fig. 3.4).

The one case in which pollen from fields have been detected is a peat near Kerkrade on the border of Germany and the Netherlands, where a field with cereals and opium poppy must have been almost bordering on the place of coring. By the way, pollen from crops and crop weeds has been found within settlements, but these do not count, as they attest to the handling of the plants rather than the place where they grew.

The remaining information on fields is provided by the seeds and fruits from the weed flora associated with crop remains. They are preserved in a carbonised state and turn up in waste deposited in pits dug in farm yards. The number of species is limited, and everywhere the same suite of species is found (Table 3.1). Some of

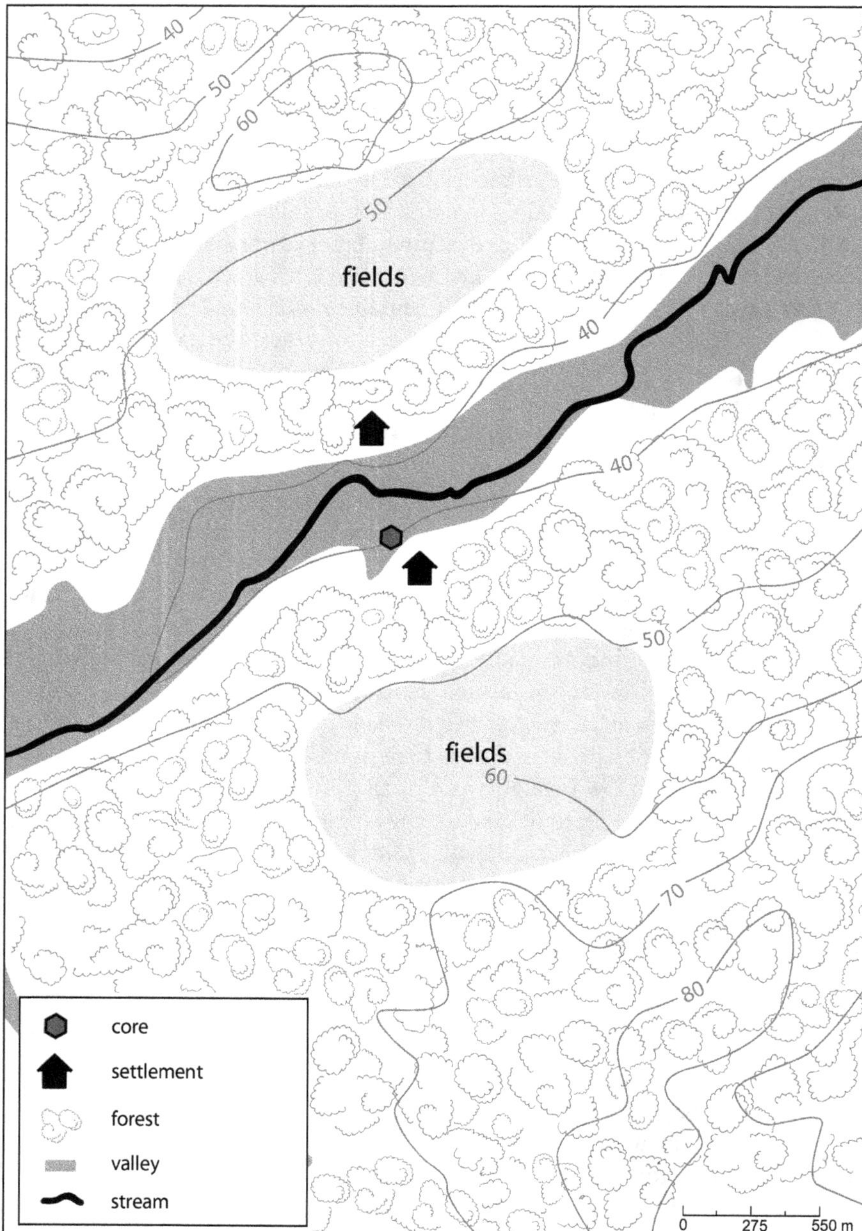

Fig. 3.4 The possible location of the fields belonging to two settlements in Belgium: Overhespen and Wange. The *shaded areas* are not meant to represent the actual surfaces tilled at one and the same time

Table 3.1 The weeds from LBK cereal fields combined with their characteristics. As no difference could be observed between the weed floras from the German Rhineland, the Netherlands, Belgium and the Moselle areas all data have been lumped together. R: Ellenberg value for soil acidity, ranging from 2 (mostly growing in acid soils) via 7 (mostly in neutral soils) to 9 (only in neutral to alkaline soils). N: Ellenberg value for mineral nitrogen ranging from 3 (mostly in poor soils), via 5 (mostly in intermediate soils) to 8 (nitrogen indicator). X in Ellenberg values means indifferent behaviour; values between brackets mean values not very well established: see the glossary for more information. S: summer annual, w: winter annual; height: height of the plant

LBK herbs				
Most common herbs	R	N	s or w	Height in cm
Bromus secalinus type	x	x	w	30–100
Bromus sterilis/tectorum	8	4	w	10–100
Chenopodium album	x	7	s	15–120
Echinochloa crus-galli	x	8	s	10–120
Fallopia convolvulus	x	(6)	s	> 100
Galium aparine	6	8	w	60–120
Lapsana communis	x	7	s	30–120
Persicaria lapathifolia	x	8	s	30–120
Persicaria maculosa	7	7	s	–100
Phleum pratense	x	6	–	20–100
Rumex sp. incl. R. sanguineus	7	(7)	–	60–120
Vicia hirsuta	x	4	w	30–60
Other herbs				
Atriplex patula/prostrata	x	7–9	s	30–90
Carduus sp.	7–9	4–8	–	–
Chenopodium ficifolium	x	7	s	30–90
Chenopodium polyspermum	x	8	s	20–75
Cruciata laevipes	5	7	w	15–45
Galeopsis angustifolia/ladanum/segetum	3–8	3–4	s	7–50
Galium spurium	8	5	–	10–40
Medicago sp./Trifolium sp.	–	–	–	–
Melilotus sp.	7–8	–	–	–
Plantago major	x	4–6	–	10–50
Poa annua	x	8	–	10–40
Poa pratensis/trivialis	x	7	–	15–100
Polygonum aviculare	x	6	s	10–100
Rumex acetosella	2	(2)	–	10–60
Setaria viridis/verticillata	x	7	s	3–100
Silene vulgaris/nutans	7	3	–	30–60
Silene sp.	–	–	–	–
Solanum nigrum	7	8	s	7–30
Stachys sylvatica	7	7	–	50–150
Trifolium sp.	–	–	–	–
Veronica hederifolia	7	7	–	7–30
Vicia cracca	x	x	w	30–120
Vicia sativa ssp. angustifolia	x	x	w	10–45
Vicia sepium	6	5	–	30–60
Vicia tetrasperma	5	5	w	15–50

them, *Lapsana communis*, *Rumex sanguineus* and *Galium aparine* for instance, are weeds that had their natural stands in forest edges. They thrive especially well on soils with a fast decomposition of humus. Newly cleared patches of forest would provide such soils. These plants are true representatives of one of the sources of the crop weed flora, the so-called autochthonous component. They like some shade, at least during part of the day. Shade may have been provided by tall crops such as emmer and einkorn wheat, but may also have been created by the forest at the edge of the small fields or by large trees left standing.

Other species arrived from other parts of Europe. As far as is known they were not growing in the region before the arrival of the first farmers. Examples of such allochthonous plants are *Bromus secalinus*, *Bromus sterilis/tectorum*, *Galium spurium* and *Vicia hirsuta*. Their natural stands are at present to be found in regions well to the east, and some even belong to sub-Mediterranean vegetation types. *Bromus sterilis/tectorum* is an example of the latter origin. It is of course possible that natural distributions were different in the past, but there is no evidence for this. It is more plausible that they were introduced by man together with sowing seed.

The limited number of weeds, occurring everywhere, has led K.H. Knörzer to the definition of a specific plant community, the *Bromo-Lapsanetum praehistoricum*. Perhaps this idea is too far-fetched, but it is true that such fixed combinations arise out of stable conditions. These fixed combinations and stable conditions concur with permanent fields. But it is also true that a fixed series of agricultural practices, repeated over and again in a slash-and-burn regime, may result in a well-defined combination of weeds as well.

Therefore a statistical test, comparing modern weed studies which encompass a range of growing conditions with the archaeobotanical data, has been used to assess the kind of crop husbandry. Four models were tested: shifting cultivation, extensive cultivation, floodplain cultivation on valley floors, and intensive garden cultivation. The intensive gardening model emerges as the most plausible.

Nevertheless, the weed species represent a range of requirements as to soil characteristics. Plants preferring neutral to calcareous soils prevail, but *Rumex acetosella* points towards the presence of decalcified soils in at least the German Rhineland and the Belgian Hainaut. Nutrient availability, as expressed in nitrogen values as given by Ellenberg, shows ranges between relatively poor (3) to high (8). Conditions were obviously not the same everywhere and differed between fields or patches within a field.

Many species are annuals. Some are known as winter annuals, which means that they germinate in autumn and hibernate above ground. Others are summer annuals which germinate in spring. In the past, winter annuals were considered as typical of autumn-sown crops and summer annuals of spring-sown crops. Today this strict division has less support than it used to have, and the modern socio-vegetational classification of field weeds does not consider it to be valid. Also, a field tilled intensively, garden-like, may show a mixed aspect. The presence of perennial plants like *Phleum pratense* is seemingly in contradiction with an intensive culture, if this implies a dense stand of the crop, weeded regularly. It can only be explained by some open space where this kind of plants was given a chance to survive. Perennials could, for instance, persist if the crops were sown in holes or in rows, not broad-cast. Such regimes would, in the case of autumn sowing, make it possible for perennials, winter annuals and summer annuals to appear together. Therefore, an autumn sowing of cereals is quite feasible. Pulses and opium poppy must have been spring sown, because they could not have hibernated above ground in the climate prevailing during the LBK period.

It is not known whether active manuring was already practised. The livestock may have been grazing on the stubble fields, leaving droppings behind and thus causing a kind of, uneven, enrichment of the soil. Nothing is known of any form of letting land lie fallow. There are no indications of crop rotation including cereals and pulses to enhance fertility.

The cereal fields were sown with a mixture of emmer and einkorn wheat. These two wheats are present as a mixture in burnt remnants of harvests (Fig. 3.5). Proportions may vary, but they are seldom found separately. Pulses, on the other hand, are not found as a mixture. They represent separate crops, as do the other crop plants. The most common weeds found associated with the cereals indicate harvesting at some distance above the ground, perhaps halfway the culm. Tall or climbing species prevail. Low growing weeds would have been present, but were obviously left behind on the fields, together with the high stubble. This implies that long straw was not a wished-for commodity.

Harvesting was done with flint sickles, made from pieces of flint presumably set in wooden handles. The flint blades, with their typical so-called sickle gloss, which

Fig. 3.5 A mixture of einkorn wheat and emmer wheat, provenance Geleen-Urmonderbaan, the Netherlands

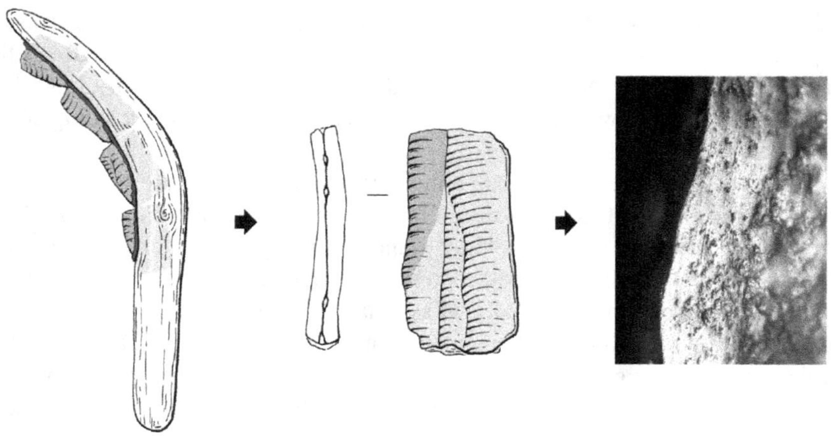

Fig. 3.6 Reconstruction of a sickle, a flint blade with sickle gloss (*shaded part*) and the gloss at considerable magnification

is caused by the cutting of culms rich in silica, turn up regularly in excavations (Fig. 3.6). No handles have been preserved.

The harvest was brought into the settlements to be processed there. Processing required several steps. The first series of steps was to prepare the harvest for storage. Emmer and einkorn wheat are hulled wheats, which implies that their husks adhere firmly to the grain. Threshing, winnowing and sieving result in still hulled grain together with weed seeds of the same size and the same weight. The wheats go as such into storage. Remains of this semi-clean grain have been found in LBK contexts. Storage took place in grain lofts and in underground silos (see Section 3.5).

The second series of processing is thought to have taken place on a more or less daily basis and includes dehusking, winnowing, sieving and cleaning by hand. Remnants of these processes, such as discarded chaff, have also been found. The procedure is known from ethnographic parallels. The barley, pulses and oil seeds were most probably processed during one single course of preparation.

Dehusking of hulled grain can be done with the aid of a quern. Querns of the saddle type belong to the frequently found LBK equipment, but whether they were used for dehusking has not yet been proven. They may have been used for grinding only (Fig. 3.7). Dehusking can also be achieved by pounding. Pestles and mortars made of stone have not been found, but they may have been made of wood. Other implements used in crop processing are not known either, as they too were probably made of wood and plant fibres.

The remains of stored wheat and discarded chaff reveal more than their stage in the chain of crop processing activities. The proportion of cereal to weed seeds, or chaff to weed seeds, falls into two classes. There seems to have been not-yet-dehusked wheat with few weed seeds, and chaff to match, that is, with also a few weed seeds mixed in. And there was not-yet-dehusked wheat with many weed seeds, and the same kind of chaff (Table 3.2). This implies that two kinds of crop arrived, or were stored, in the settlements: clean crops and crops with a lot of weeds. The conclusion may be that some crops were better weeded, or better cleaned after harvests, than others. A closer look, however, reveals that the clean product is found in

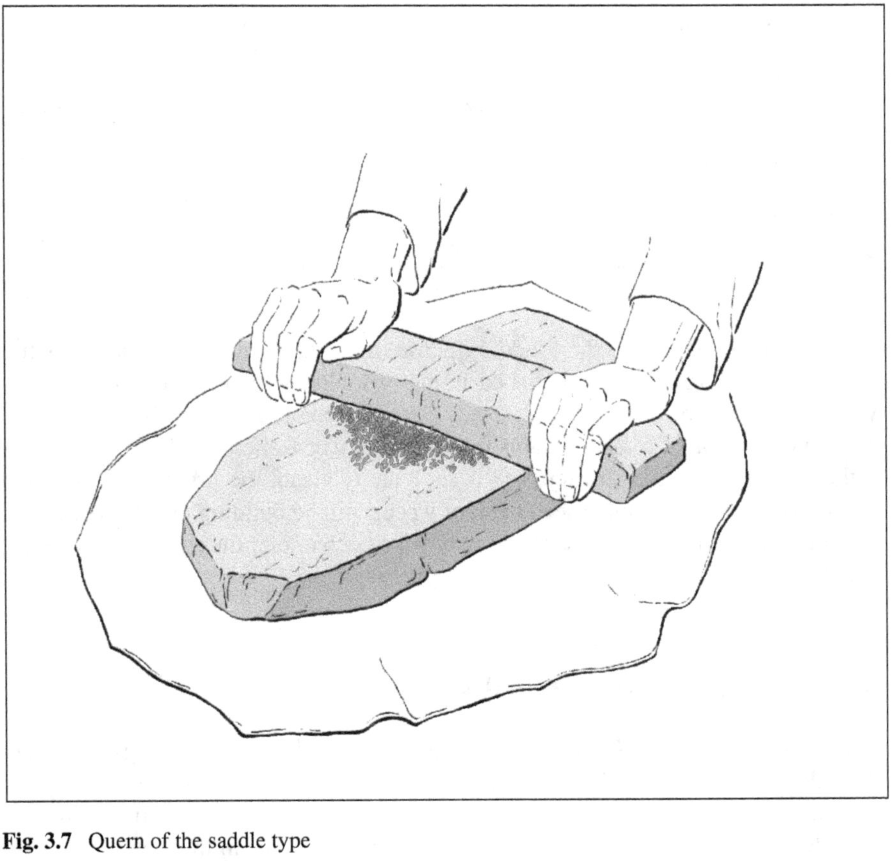

Fig. 3.7 Quern of the saddle type

Table 3.2 The four classes in cereal remains found in the German Rhineland and the Netherlands. Numbers are the actual numbers of specimens found in the lots. The grain is a mixture of emmer and einkorn wheat. The chaff consists of broken spikelet forks. In not-yet-dehusked grain the ratio of chaff: grain should lie between 0.5 and 2. Finds with few cereal remains but many weed seeds are considered to represent threshing waste and therefore 'chaff', even if the chaff is almost absent

Site and sample	Grain	Chaff	Weeds
Not yet dehusked, few weeds			
Crisnée	118	111	7
Geleen-Urmonderbaan 3	171	135	5
Geleen-Urmonderbaan 3a	367	239	7
Geleen-Urmonderbaan 9	150	34	2
Verlaine	76	99	2
Not yet dehusked, many weeds			
Beek 9-2	311	137	484
Langweiler-2 89	95	149	251
Langweiler-2 397	327	337	556
Langweiler-6 C6	137	73	723
Chaff, few weeds			
Beek 8-2	48	5000	85
Langweiler-9 146/108	39	1349	37
Langweiler-9 146/289	16	117	24
Langweiler-9 561B	4	128	4
Chaff, many weeds			
Bedburg-Garsdorf 28	442	2577	6161
Langweiler-2 306	29	412	141
Langweiler-3 Graben B	70	4422	331
Beek 5-2	21	30	2700
Geleen-Haesselderveld 1	2	3	682
Geleen-Haesselderveld 1a	8	7	990

contexts corresponding to the founding phase of a settlement, whilst the crops full of weeds came from long established farms. It looks as though the fields became weedier after years of use. This in fact does not come as a surprise. As argued above, newly cleared forest contains few potential weeds, neither above ground nor in the seed bank. If the sowing seed is reasonably clean, the first year will give a rather weed-free crop. But, a seed bank of weeds will be building up and weeds will become more and more a nuisance. It may have been one of the reasons why fields, and after that settlements, were in the end given up.

3.4 Livestock and Animal Husbandry

The livestock of the LBK was composed of cattle, sheep, goats and pigs. The only other domesticated animal was the dog, but the dog is not considered as part of livestock. Sheep and goat were not domesticated locally. This was impossible, as the wild species do not occur in the region. Their origin lies in the Near East. Wild cattle (aurochs) and wild boar, however, did form part of the local fauna. Most of the cattle and pig bones found in LBK settlements show that they are remains of animals which were in general smaller than their truly wild counterparts, suggesting that they were not close relatives. Intermediates do exist but they are rare. However, there are indications provided by DNA analysis that at least wild boar was also domesticated in Europe and that these European pigs rapidly replaced the domestic pigs of Near-Eastern origin throughout Europe. For cattle such European sources are not yet established and their provenance from Near-Eastern stocks cannot yet be excluded. However, notwithstanding the origin of the animals, it is clear that the first farmers arrived in the loess region with a complete livestock, cattle and pigs included.

The shoulder height of the cattle was 132 cm for bulls and 125 cm for cows. Oxen are absent. The size of the sheep is not known due to too few measurable finds, but both rams and ewes had horns. It is assumed that they were hair sheep. Some goats could be distinguished from sheep, and the few that could provide a shoulder height measured 59–60 cm. Pigs reached 70–93 cm with a mean of 81 cm.

The livestock varied as to composition. Cattle were the most important everywhere, but the relative importance of the other animals varies (Fig. 3.8). Two traditions are distinguished. In the main, 60% or more of the animal remains belong to cattle, followed by sheep/goat and slightly less pig. The second tradition is found in the lower Alsace, France, and has on average 40% cattle, whilst pig is slightly better represented than sheep/goat. Why the lower Alsace differs from the rest is not clear. The natural environment seems not to have played a distinguishing role. The other parts of the region, especially northern France because it is there that most bones have been preserved, show a relatively wide diversity of environments, ranging from the upper Alsace to the river valley floors in the Paris Basin. Nevertheless, they provide identical bone spectra. In the lower Alsace, a single group of farmers may well have had a cultural preference for pigs.

Domesticates were the main source of meat. Hunting and fishing provided additional food, but this food source contributed in most settlements less than 10% of the meat part of the diet. A striking exception is a settlement at Liège in Belgium. It is located on the Lower Terrace of the river Meuse and hunting and fishing seem to have been the main providers of meat. In this case, the environment may indeed have played a role. It has been remarked that in other parts of Europe, Bavaria in Germany for instance, hunting was more important in settlements along large rivers than in other settings. It is thought to have been due to the better availability of wild fauna in the riverine forests on the valley floor. Upland forests were darker and knew a lower density of game.

Cattle were slaughtered either before the age of three or as adults. Some animals reached an age of six years or more, and some were even over nine years old. The sex ratio in adult animals is three cows to one bull. Females could reach a higher age than males. This may imply the use of milk, but not necessarily so. Cows are also necessary for the continuation and increase of livestock. Implements with a straight connection with milking or cheese making have not been found in the region. The use of cattle for traction has not been established either. Nor have devices been

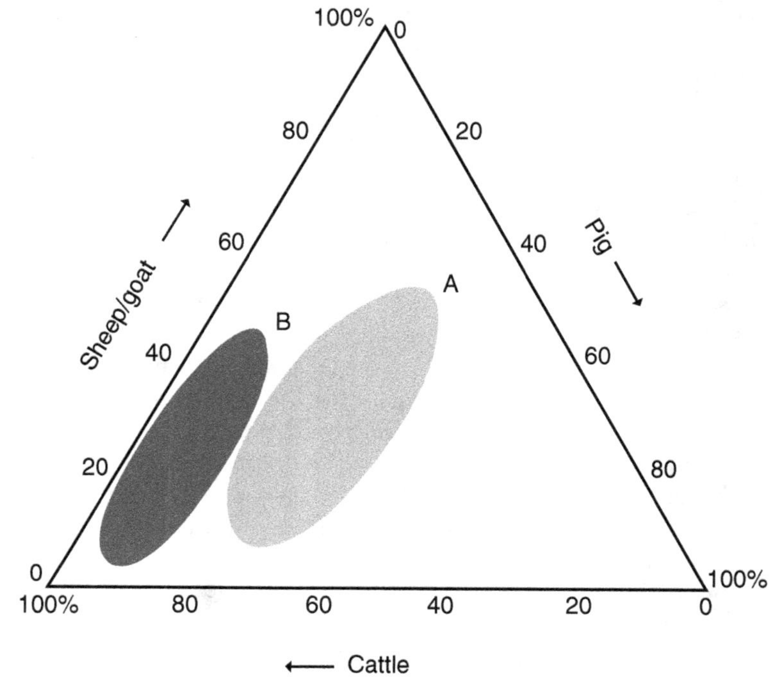

Fig. 3.8 Composition of the livestock. *A*: Lower Alsace, *B*: main tradition

found that had to be drawn or dragged, like the ard or the travois. The wheel was not yet known.

Sheep/goat remains show the same age distribution as cattle. The assumption that the sheep were still hair sheep is based on the near absence of spindle whorls (see Section 3.2). Of course, this is no actual proof of the non-availability of wool, but it is a hint. Forty to fifty percent of the pigs were slaughtered young, i.e. before the age of thirteen months.

The number of animals kept by a single household or a community is not known. Some calculations, based on estimations of the diet, suggest something between 50 and 100 for an average settlement. True stables are absent. In a certain type of farmhouse some animals may have been stabled under the same roof as the human occupants, but, as will be explained below, this 'stable-part' can also be explained differently. In any case, there was no place inside for all the animals. They must have been kept outside, but even open air pens have not been detected. The animals grazed and browsed on stubble fields, in natural clearings, and, presumably most often, in the natural forest. In the latter case, the livestock must have roamed over fairly large areas, because the original forest had a more or less closed canopy with sparse undergrowth due to lack of light. This was especially the case during the first phase of settling. The best places to find food were the slopes and floors of river valleys where the vegetation was more varied (see Sections 6.1 and 6.2). The livestock must have been herded and kept under control. This can at least be concluded from the fact that few cases of interbreeding with wild animals are known. This would have been quite possible for cattle and aurochs, and pigs and wild boar. The lack of pens near farm houses suggests that the livestock roamed freely all year round.

Food in winter must have posed a problem. Natural meadows providing hay were almost absent and remains of hay have not been found in the settlements. The farmers may have used leaf hay (small branches and leaves from trees) instead, as is the practice of many farmers in forested areas today. Or they may not have provided winter fodder at all. What is certain is that the livestock required large tracts of land to feed on.

3.5 Farm Buildings and Yards

The farm consisted of a single rectangular building surrounded by a yard dotted with pits and lacking a distinct fence. The house was laid out according to clearly defined architectonical rules. The width was more or less fixed at 5.5–6.5 m, but the length was highly variable, ranging from 6 to 35 m. In the north-eastern part of the region even longer houses did occur. The orientation of the long axis of the house was always southeast–northwest.

The skeleton was made of wood. The posts were erected in deep, narrow pits, dug for this purpose. The kind of wood used for heavy construction work is only known from the lining of the very rare LBK wells, where it is oak. Oak will also have been chosen for the construction of the house. Heavy trunks were split radially into halves or triangular lengths. This is evident from the clear marks left by such posts after decay (Fig. 3.9). The walls were made of wattle and daub, as is demonstrated by burned chunks of loam with impressions of wattle, which are interpreted as remnants of walls that had caught fire. Or walls were of boards of radially split wood set upright into a foundation trench as is proved by marks left in the soil. The roof was a gable roof. The kind of cover is unknown. Thatching with straw is a possibility, although long straw seems not to have been harvested. Reed is out of the question, as there was not enough reed growing in the natural environment to provide sufficient amounts to cover the large houses. Most likely the roof was made of branches or shingles, as is common practice in heavily forested areas to this day.

The main entrance was in the south-western gable wall. Another entrance, in the long wall, is sometimes postulated; the main argument for this entrance being a concentration of garbage outside the long wall, at least in some cases. This concentration outside the middle of the wall suggests an opening through which refuse could be thrown out. This may have been a door but also a window. But such centred distributions are not the rule, and the existence of a door or window has not been proven.

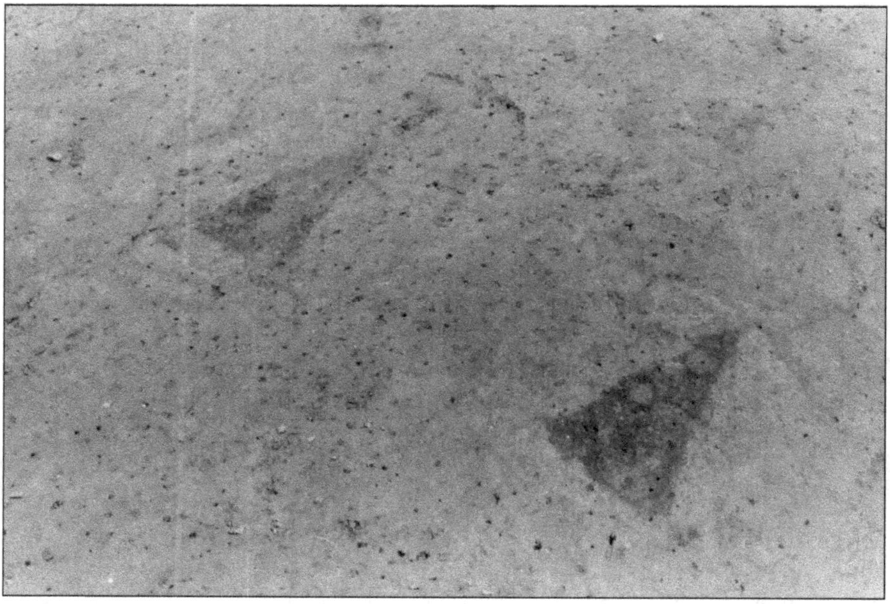

Fig. 3.9 Marks left by posts made by splitting a trunk. The long side of the triangle measures 50 cm. Provenance Stein, The Netherlands

Inside, the house was four-aisled. Transverse rows of three sturdy posts (so-called triplets), supporting the roof, divided the interior lengthwise into two main aisles. Slightly narrower aisles were situated between the wall and the roof supports. The arrangement of the triplets suggests a further division, crosswise, into two or three parts. The first part was entered through the door in the south-western wall. Additional sturdy posts, placed next to the triplets, indicate the presence of an upper floor, which is lacking in the remainder of the house. The upper floor is interpreted as a loft for storing grain. However, in a considerable number of farm buildings this part is reduced or even absent (Fig. 3.10).

From the first part, or, if this is absent, from the door, the main part of the house was reached. The hearth is supposed to have been situated here, although, due to the absence of the original floor level in all the excavated settlements, no traces of a hearth in situ have been discovered. Its presence is deduced from lumps of burnt loam in neighbouring postholes. This central part of the house is thought to have been the main living room, where also guests were admitted. A kind of cross-corridor blocked this part from the north-western part of the house. In a substantial number of cases, this part had a wooden wall made of boards set upright into a foundation trench. Some houses had an ample north-western part. In others its space was rather reduced or this part even absent. In the past this space has been described as the stable. The sturdy wall would have been constructed in view of moving and wall-butting animals. However, the space would not have accommodated many animals. Attempts have been made to discover the former presence of livestock and

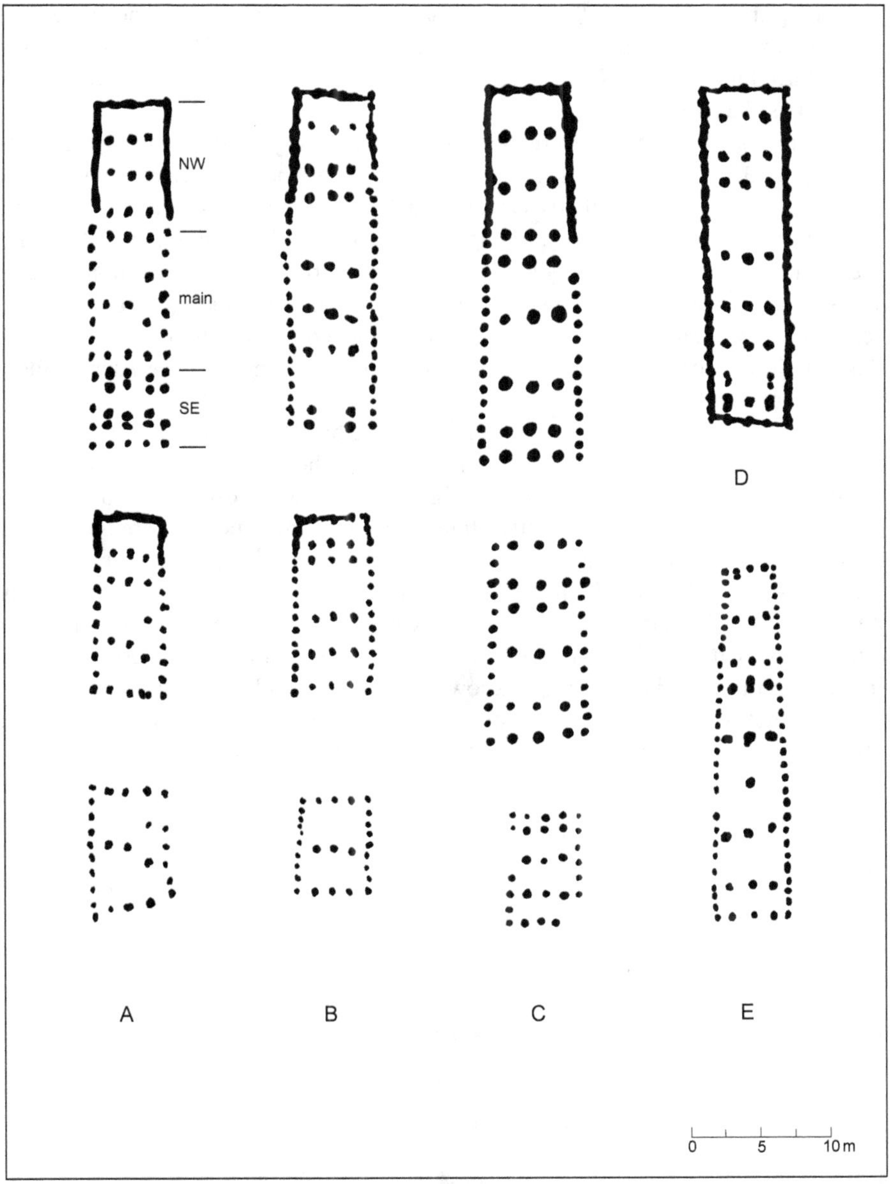

Fig. 3.10 Ground-plans of LBK houses. *Solid lines*: foundations trenches, *dots*: postholes. **A**: The Netherlands, early phase; from top to bottom: a house composed of a south-eastern (SE), main and north-western (NW) unit, a house without a SE unit and house consisting only of the main (central) unit. **B**: The same series, but from a later phase. **C**: The same series, but from Cuiry-lès-Chaudardes (Dept. Aisne, France). **D**: The Netherlands, a house with all three units, but with a wooden wall on four sides, made of boards set upright in a foundation trench. **E**: A plan belonging to a late phase of the LBK as found in the Dept. Aisne in France; the plan in question was excavated at Berry-ay-Bac and shows already elements of the house types discussed in Chapter 4

their dung through phosphate or nitrate analysis, but the results were negative. Not that this says much, because millennia of agriculture after the LBK occupation may have blurred the picture. Nowadays, however, the north-western parts of the houses are seen as sleeping-quarters.

As mentioned above, pits were the main feature of the yard surrounding the farmhouse (Fig. 3.11). Pits are scarce in the area bordering the south-eastern gable wall with the door. All pits ended up as refuse pits, but their initial purpose was different. Long, narrow pits accompany the long walls on both sides of the house. They are seen as the main source of the loam used in the construction of the wattle-and-daub walls. A confirmation of this viewpoint may be seen in the fact that such pits are scarce near wooden walls. Long pits are the most common kind of pits. The settlements in the southern part of the region, for instance those in the Aisne valley, almost exclusively have long pits.

The yards in the northern part of the loess region show, in addition to the long pits, scatters of pits at a greater distance from the house. One round type of pit stands out. It has a flat bottom and straight walls or walls diverging downwards, giving the pit a cylindrical or inverted funnel-shaped form. Such pits are interpreted as underground silos for the storage of grain or similar dry products (Fig. 3.12). Well-filled, closed by a lid, and sealed watertight with loam or dung, the grain is preserved in a dormant state, due to a lack of oxygen and the building up of an atmosphere rich in carbon dioxide. The originally present oxygen is consumed by initial respiration. It is a technique known from ethnographic sources, and verified by experiments. The pits only function properly if they stay closed, and are therefore only for long-term storage. Opening-up every day for retrieving grain would break

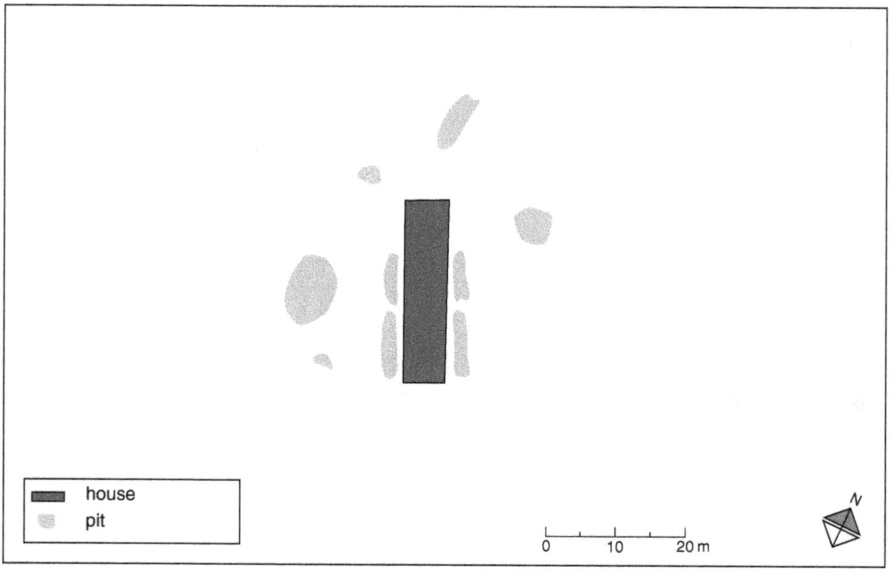

Fig. 3.11 The yard of an LBK farm

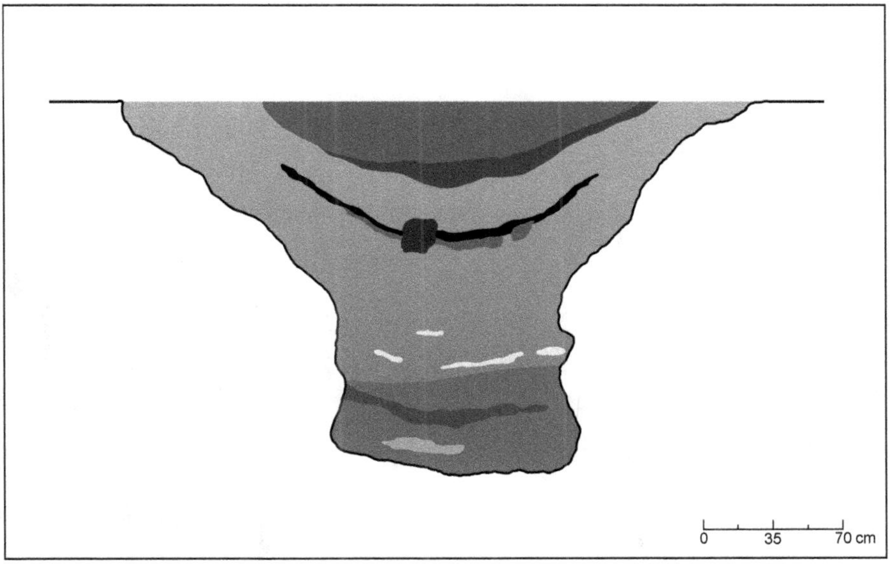

Fig. 3.12 Longitudinal section of a silo excavated at Elsloo, the Netherlands. After abandonment, erosion widened the *upper part*, whilst the *lower part* retained its original shape

down the conservational principle. Such silos could contain up to 2 m^3 of grain. It is supposed that the grain was not yet dehusked when put into storage, but a silo with grain in situ is a rare find. In such cases a layer of carbonised grain has been found at the bottom. How this grain could have become carbonised in a deep pit with a narrow opening is still open to discussion. Not every yard seems to have boasted a silo. There is no connection with the presence or absence of grain lofts in the houses.

Very uncommon is the presence of a well. In the loess region west of the Rhine some five instances are known. Best preserved is the well in Erkelenz-Kückhoven in Germany. It had a square lining made of split oak boards, worked with stone adzes and joined (Fig. 3.13). Narrow gaps between the logs were plugged with moss. The extreme scarceness of wells is explained by the use of streams as the common source of water. Settlements were never been founded on large plateaus devoid of running water.

Although the farmhouse is a long building, it was not a multi-family house. Calculations, following ethnographic models, and applied to the central part of the house, set the number of inhabitants at 6–15 persons.

3.6 The Farm in Its Setting

Farmhouses occur clustered in restricted, permanently settled areas, which can be described as hamlets (Fig. 3.14). These consisted of minimally 2 and maximally 11

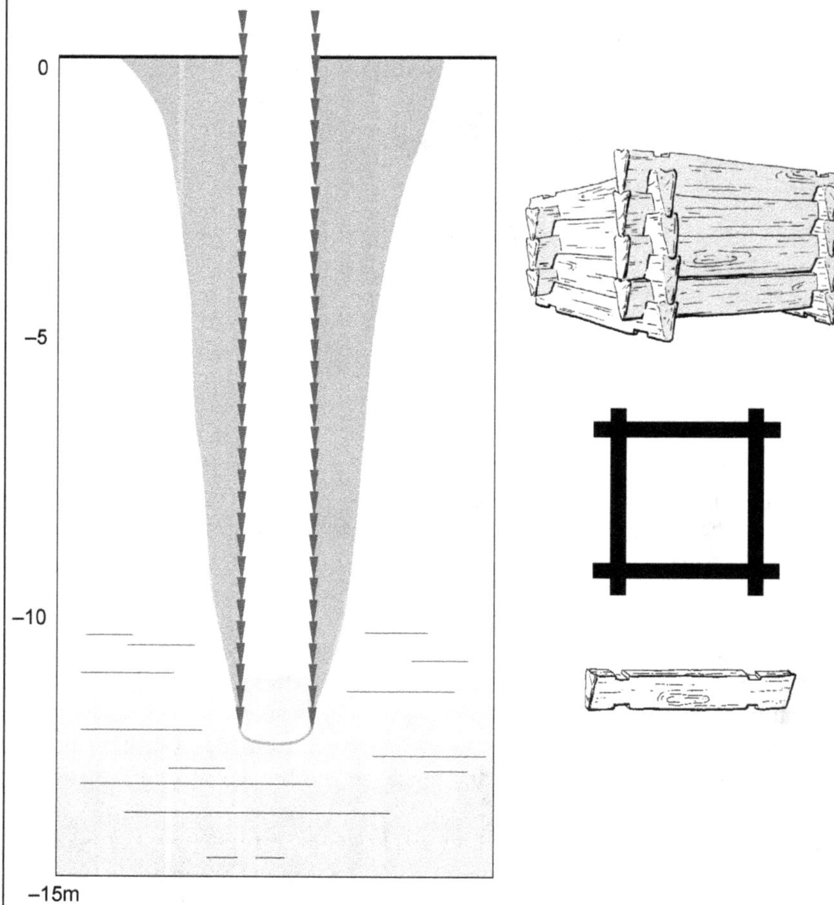

Fig. 3.13 A schematic section of the well. On the *right*: part of the lining, a schematic view from above and a single board

buildings, with an average of five farms. Distances between individual structures range from 50 to 150 m. As all LBK houses had the same south-eastern – north-western orientation, the buildings within a hamlet were parallel to each other. Why this strict orientation was followed remains unclear. One of the proposed explanations is based on the fact that the wall of many north-western parts is constructed using sturdy boards. This part should face the prevailing direction of strong winds.

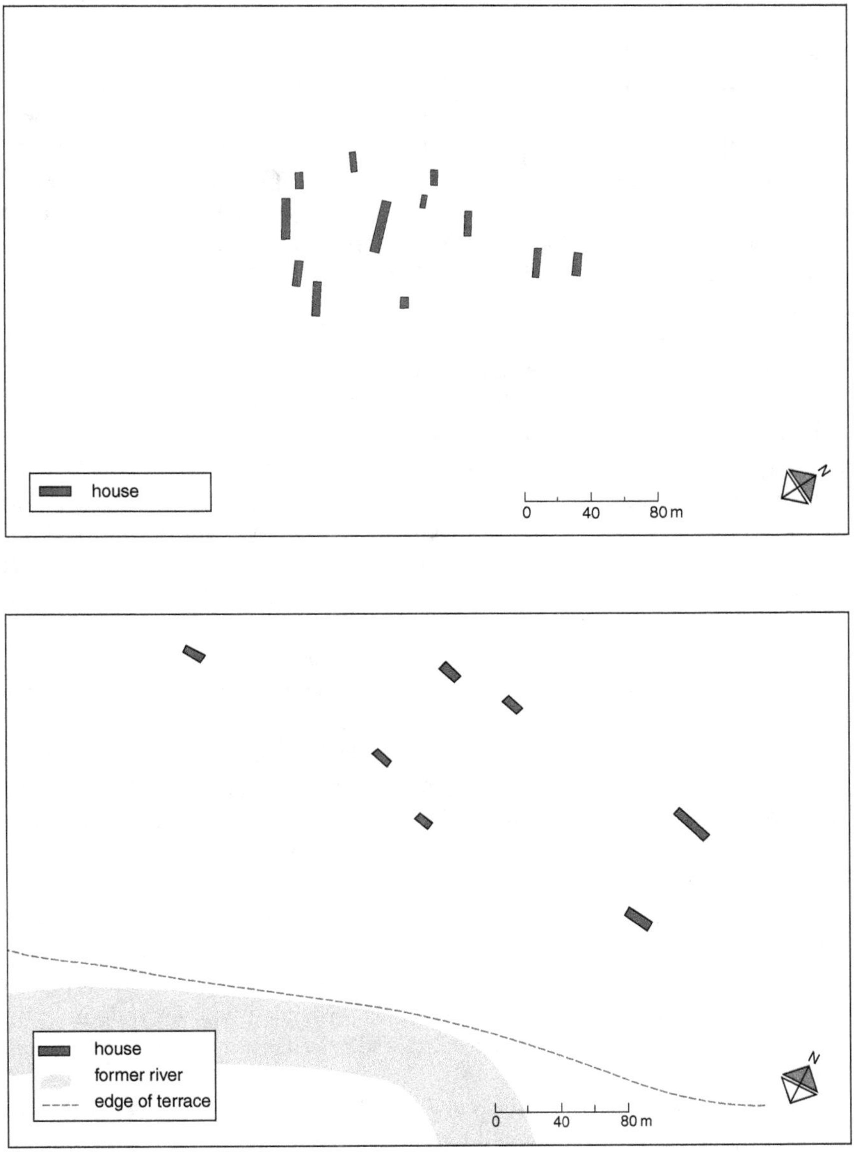

Fig. 3.14 Plans of settlements belonging to one single phase of occupation; *top*: Geleen-Janskamperveld (the Netherlands), *bottom*: Cuiry-lès-Chaudardes (Dept. Aisne, France)

But as the houses were erected in clearings in a forest, the effect of wind cannot have been as important as in the case of buildings erected in a vast open space. Others explain the direction as an orientation towards the nearest seacoast. But as this coast was far away, and perhaps never seen by the LBK builders, this explanation seems rather far-fetched. Nevertheless, there must have been a strong reason to have the gable-end with the door orientated towards the south-east. The practice may have its roots in a cultural factor rather than a physical one.

The locality chosen for the foundation of a settlement was usually the edge of a plateau, or a more or less horizontal part of a slope towards a watercourse. In this way, people had access to both the plateau and the valley with its more varied vegetation and supply of water (see also Chapter 6). Only in the Paris Basin, in northern France, did people choose to live on the higher parts of valley bottoms. A reason may be that the slopes are very steep here, making traffic between the plateau and the valley floor cumbersome (Fig. 3.15). During the last phases of LBK farming, settlement on valley bottoms also occurred elsewhere, perhaps as a result of population pressure.

Households belonging to the same hamlet shared customs in the use of space. This can be inferred from the habits in waste disposal. Burnt chaff, for instance, is found at fixed sides of the individual farmhouses. In some settlements chaff was always thrown away at the northern side, in others at the southern side. A related behaviour is seen in the discarding of bones. In the French site of Cuiry-lès-Chaudardes (Dept. Aisne), for instance, bones were thrown away preferably on a specific side. In some of the occupational phases of this site, disposal took place in the northern elongated pit next to the wall, in others in the southern pits. The choice seems to have been influenced by the distance to the nearest neighbour. The maximum distance seems to have been opted for. Flint waste follows the same pattern.

In principle, from a social as well as an economic aspect, households were equal. Every household is assumed to have provided for its own needs. Nevertheless, some inequality must have been present. Some houses lack grain lofts, for instance; others lack both grain lofts and sleeping quarters. In Cuiry-lès-Chaudardes (Dept. Aisne), it could be shown by the bones discarded by them that certain households, those occupying the smaller houses, had a higher proportion of hunted animals than others. However, not every difference in house-plan reflects a difference in subsistence economy. In the north-eastern part of the loess region, in the German Rhineland and adjacent Dutch area, some large houses had walls made entirely of upright boards. Their number was restricted to one per hamlet. In the other areas such houses are absent. Their internal subdivision and the finds associated with them do not differ from other constructions. Who lived there, and why these houses were different, remains unclear.

Based on the number of houses and the estimated number of people living in them, the population of a hamlet will have varied between 12 and 165 persons. An average settlement may have numbered 50–100 inhabitants. Except perhaps in the pioneering stage, such units did not live an isolated life. They were part of a cluster of several hamlets. Good examples are the Merzbachtal cluster in the German Rhineland, the Graetheide cluster in the Netherlands, and the Aisne valley cluster in France. These clusters represent the ultimate level of organisation of this early farming community. Land not settled on separated them. The distance between clusters varies from 20 km even to 100 km.

Single farms provided for their own needs, but the differences noted above suggest that single households within a hamlet were interdependent to some degree.

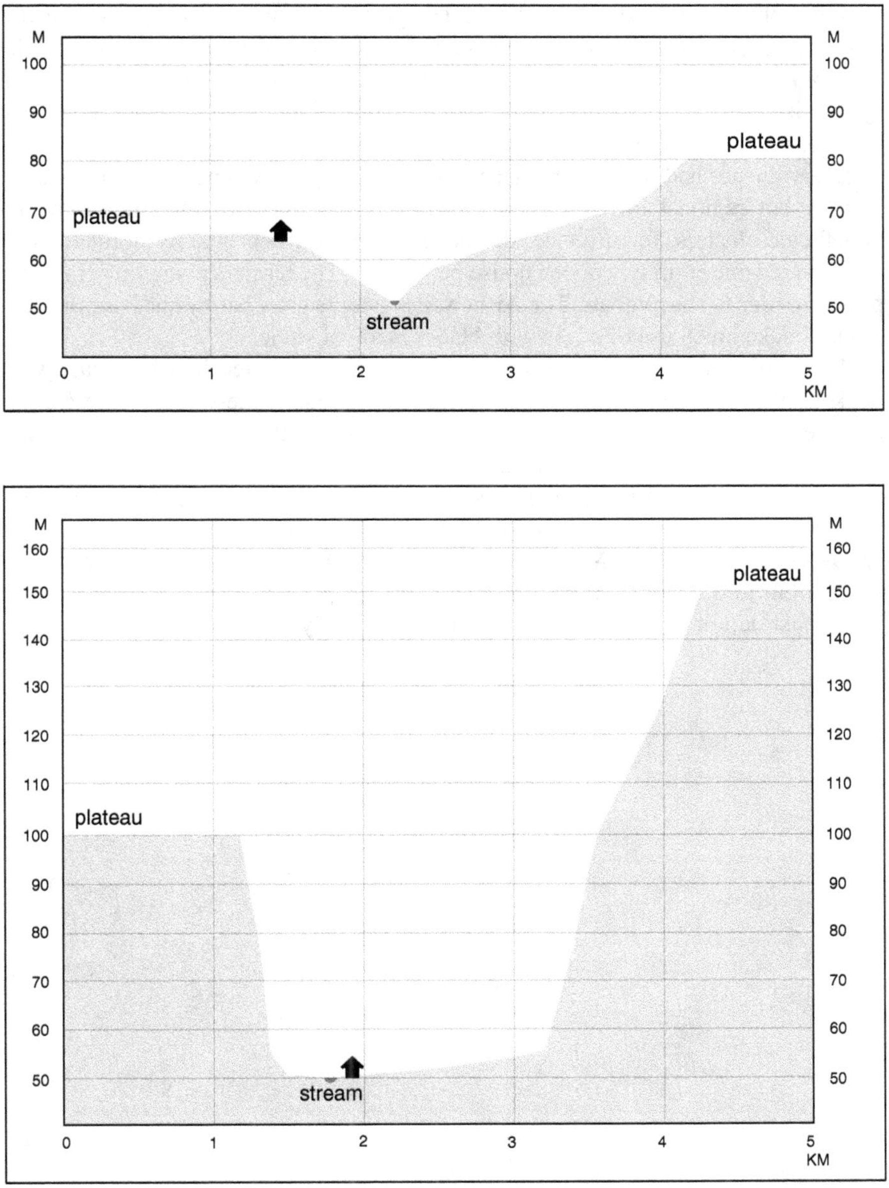

Fig. 3.15 *Top*: location of Geleen-Janskamperveld, the Netherlands; *bottom*: location of Cuiry-lès-Chaudardes (Dept. Aisne), France

The tilling of the land may have been a household affair. Fields were restricted in size. A larger stretch of land was required for livestock. Herding animals in the forest may have been a communal affair of the hamlet. Some basic raw materials that could be obtained without too much effort within one day's reach, may have been acquired on the same communal basis. The provision with more exotic raw materials, like good-quality stone for adze blades, seems to have been organised on the level of the cluster. Such goods could only have been obtained through an exchange system in which other clusters took part. Remains of this kind of material suggest that one hamlet in a cluster took the lead in this.

Every cluster had its own characteristics. Basically, they functioned in the same manner but regional differences are visible. The farmers of the Merzbach and Graetheide clusters, for instance, did not grow barley, kept mainly horned cattle, and counted one entirely wooden house per hamlet. The farmers in the Lower Alsace preferred pigs to sheep/goats. The Aisne valley people grew barley and, surprisingly, did not make much use of adzes with blades made of stone.

The arrival of the first farmers in the loess region resulted in a landscape with pockets of individual communities founded sometimes widely apart. These communities shared many cultural traits and functioned within an exchange network, but otherwise led a life of their own.

Towards the end of the LBK period, some hamlets are found to have been surrounded by a deep ditch and a palisade. Earlier hamlets seem not to have required strong enclosures. In the Merzbach cluster the enclosure does not show traces of houses, but in others the common house-plans are present. Why the need was felt for a construction of enclosures cannot be conclusively explained at present.

Chapter 4
Heirs to the First Farmers: 4900 BC–4300 BC

4.1 The Successors of the *Linearbandkeramik* Culture

The LBK farmers did not live in a static society; there were of course gradual changes in practices and habits. But around 4900 BC the rate of change saw a distinct acceleration. In the north-eastern part of the region the change even had the characteristics of a crisis. Farming communities disappeared temporarily from the Dutch Graetheide area and most of the Belgian Hesbaye. In the German Rhineland the population went into decline. In addition, people avoided the old settlement sites and chose to settle in areas on the fringe of the former areas of occupation. The Moselle area was also affected. Other regions do not show a similar crisis but are characterised by an expansion of farming communities into areas not occupied before. The result of the overall change was a splitting up of the farming society into different cultural groups. The earliest groups go by the names of Grossgartach culture, Groupe de Blicquy and Culture de Villeneuve-Saint-Germain. These developed later into the Rössen and Cerny cultures. Notwithstanding the cultural diversification, all of these cultures and groups are definitely heirs to the LBK farmers. The period lasted until about 4300 BC.

4.2 Crops

The farming system of the successors of the LBK had largely the same traits as the one practised before. Nevertheless, a change in the choice of crops can be observed. Both hulled wheats, emmer and einkorn, were still a main crop; but the wheat spectrum was enlarged by a free-threshing, or naked, wheat with compact kernels. In those cases where the remains of the central axis of the ear, the rachis internodes, have been found, this wheat could be identified as a hexaploid species, most probably bread wheat (Fig. 4.1). Only rachis remains allow the identification of naked wheats, because the kernels are too similar to be of use for allotting names. Admittedly, naked wheat has been observed in the north-eastern part of the region during

C.C. Bakels, *The Western European Loess Belt*, DOI 10.1007/978-1-4020-9840-6_4,

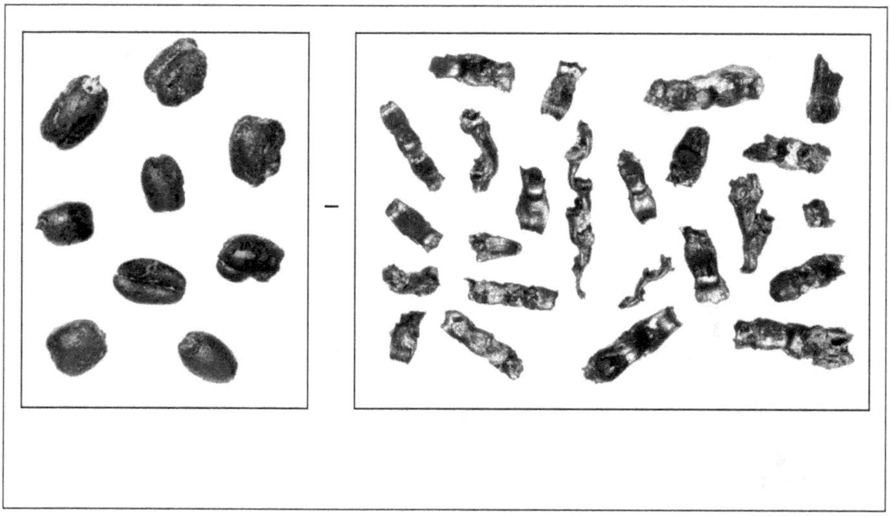

Fig. 4.1 Carbonised remains of naked hexaploid wheat; *left* the grain, *right* the chaff (rachis remains). Provenance Maastricht-Randwijck, the Netherlands

the LBK, but it was so scarce that it cannot have been a true crop plant, grown as a staple. After the LBK, naked wheat had obviously become one of the staple crops, not just in one area, but everywhere.

What was behind this development? One possibility is that the shift to new areas for settling and farming triggered the cultivation of the naked wheat, already known but not seriously considered before. But the new locations were not so very different from the former ones. Alternatively, the possibility that there was a climatic change may be considered, but such a change is already observed at an earlier date, around 5050 BC, as is shown by dendroclimatological research conducted in western Germany and based on the width of tree rings in oak. It is a change towards drier conditions, although some lowering of temperature cannot be excluded. Why this should have promoted the cultivation of a hexaploid naked wheat is not easily explained, because the climate had not changed so as to make growing emmer and einkorn unfavourable. This leaves a cultural explanation. Naked wheats were main crops in early farming communities around the Mediterranean as early as, or even prior to, the LBK farmers on the loess of Central Europe. It is possible that cultural contacts established ultimately between the Mediterranean world and the loess belt brought about a widening of dietary habits. During the post-LBK period, the successors of the early Mediterranean farmers had spread out to more northern regions in, for instance, France. It is known that both worlds did meet there somewhere. The occurrence of poppy in the LBK points to earlier contacts with an exchange of plants, but the naked wheat seems not to have been 'taken up' at that time. The contact is also indicated by other, more solid, evidence. At the end of the LBK, white limestone bracelets turn up in the Paris Basin, which, according to petrographic analysis, come from a zone in the lower Rhone area occupied by farmers of the Mediterranean tradition. This shows that objects were transferred from one world to another, so why not new kinds of food? The actual kind of naked wheat was presumably not the same, because, as far as is known, farmers in the Mediterranean world grew mainly tetraploid naked wheat.

The list of cereals is completed by multirowed barley. Its naked variety was no longer restricted to specific parts of the loess region, but was grown as a main crop everywhere. However, multirowed barley occurs also in a hulled variety, i.e. a variety in which the glumes adhere more strongly to the grain. This hulled variety turns up now and then in the south-western part of the region in Villeneuve-Saint-Germain and Cerny contexts. Where this hulled barley springs from is not yet quite understood. A find in an outlier of the LBK culture in Normandy (France), outside the region under review here, has revealed hulled barley as well. This find would suggest that this crop plant came in from areas located west or south-west of the loess region. But the history of agriculture there is not yet very well known.

Only one type of pulse seems to have been commonly grown: pea. The lentil disappeared already during the younger phases of the LBK and seems not to have retained a place in the spectrum of crop plants, although single finds show that its cultivation lingered on in some places. The reason for its decline may be the climatic one mentioned above. In the LBK, lentil was most abundant on the sunny slopes of the Moselle valley and it is possible that lentil had reached its northern limit in the region under consideration. Its culture was probably a relict of farming practices common to the Central European origins of the LBK and had to be given up because the yield was too uncertain. Lentil has not been found in any excavation dating to the next millennia.

Linseed has not yet been detected in any site of the successors of the LBK, but in this case the absence may be mere chance. Linseed has the disadvantage over cereals and pulses that it does not preserve well through carbonisation, which is the main form of preservation of seeds in loess regions. In the following periods linseed is present again, and it is unlikely that the species was absent in the intervening period. The other oil seed, poppy, is present in finds of the period.

4.3 Crop Cultivation

The fields on which the crops were grown are unknown for the same reasons as mentioned in Chapter 3. In several areas the settlements were not established on exactly the same terrains as their LBK predecessors. The farmers did not till land that had been tilled before. This is especially true for the German Rhineland. In other areas, first and foremost the valleys in northern France, with the Aisne valley as an outstanding example, the occupation of the land was more or less continuous. In the former (mainly German) case farmers needed to cut down forests to establish their garden-like fields. In the latter case farmers needed, presumably, only to clear secondary brush. In the north-east and east, the tool for felling trees and shrubs was still the adze with a blade of stone. In the south, people used a rather crude tranchet axe made of flint (Fig. 4.2).

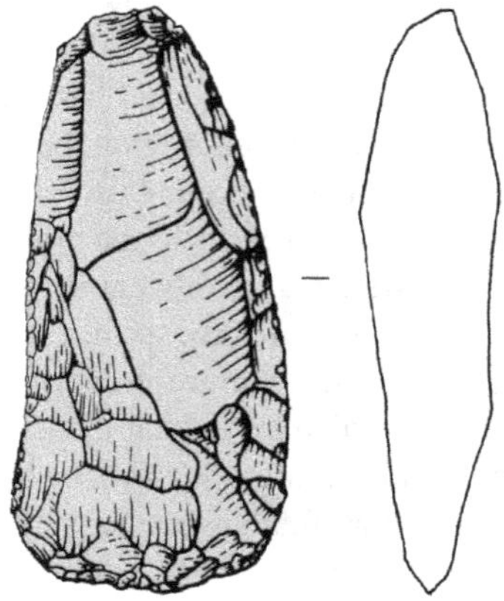

Fig. 4.2 Blade of a flint axe. Original length 8.5 cm. Provenance Blicquy, Belgium

Information about other implements used in tilling the soil is absent. It is assumed that the cultivation methods were similar to those of the LBK. This is, at least, suggested by the weeds accompanying discarded lots of cereals and chaff (Table 4.1).

The difference is negligible. The methods of crop cultivation and crop handling must have been comparable to create the same results. Not much is known about the subsequent treatment of the harvest, apart from the fact that the saddle quern was still the tool used to grind the grain.

4.4 Livestock and Animal Husbandry

With cattle, pigs and sheep/goats the list of animals raised is the same as before. Cattle remains are dominant in the livestock. The shoulder height of cattle has not yet been established, but as the animals of the period after this one are smaller than LBK cattle, it may tentatively be assumed that they became gradually smaller in the intervening period. A continuous shrinking in cattle size is characteristic of the millennia to come.

Pigs gained in importance and occurred in equal numbers, or even slightly more than the numbers of sheep/goats. As in cattle, the shoulder height is not yet known, but sizes did not decrease in the following millennia, and the pigs were probably not smaller than LBK pigs. The size of the sheep and goats is not known.

Why pig became more common is not understood, but the change is only a minor one. All in all, the livestock and livestock system did not differ much from the

Table 4.1 A comparison of LBK and Rössen weeds. See the caption of Table 3.1 for R, N, s, w and height. Frequency is the percentage of sites in which the plant was found

Subregion					LBK Netherlands	Rössen Netherlands	LBK Rhineland	Rössen Rhineland
Number of sites					8	1	10	4
	R	N	s or w	Height in cm	Frequency	Presence	Frequency	Frequency
Bromus secalinus type	X	x	w	30–100	100	+	100	100
Bromus sterilis/tectorum	8	4	w	10–100	62.5	+	100	100
Chenopodium album	X	7	s	15–120	100	+	90	100
Echinochloa crus-galli	X	8	s	10–120	37.5	.	60	100
Fallopia convolvulus	X	(6)	s	>100	100	+	100	100
Galium aparine	6	8	w	60–120	50	.	10	0
Galium spurium	8	5	–	10–40	12.5	.	70	75
Lapsana communis	X	7	s	30–120	50	+	80	100
Persicaria lapathifolia	X	8	s	10–120	25	+	0	25
Persicaria maculosa	7	7	s	20–100	50	.	60	25
Phleum pratense	X	6	–	20–100	75	+	90	0
Poa pratensis/trivialis	X	7	–	15–100	25	+	20	100
Rumex sp.	7	(7)	–	60–120	37.5	.	40	100
Vicia hirsuta	X	4	w	30–60	56	+	70	50

situation during the LBK. Domestic animals remained the main source of meat. Hunting was not very important.

4.5 Farmbuildings and Yards

The rectangular plan of the LBK houses has transformed into a trapezoidal plan tapering towards the north-west (Fig. 4.3). Ethnographic parallels show that trapezoidal plans go together with sloping roofs. The south-western gable with the main, if not only, entrance was higher than the one at the rear. The farmhouse had a true facade. Why the buildings changed is not quite understood. Some argue that a climatic change, with stronger winds blowing from the north-west, may have induced the construction of aerodynamic roofs. However, there is no proof of such a change in climatic conditions. Also, there is no actual experimental proof that gable roofs, sloping upwards in the main direction of the wind, are more resistant to damage. Another explanation which has been put forward is that the new design was the expression of a cultural change that asked for impressive facades. There may be a link with the crisis mentioned in the introduction to this chapter.

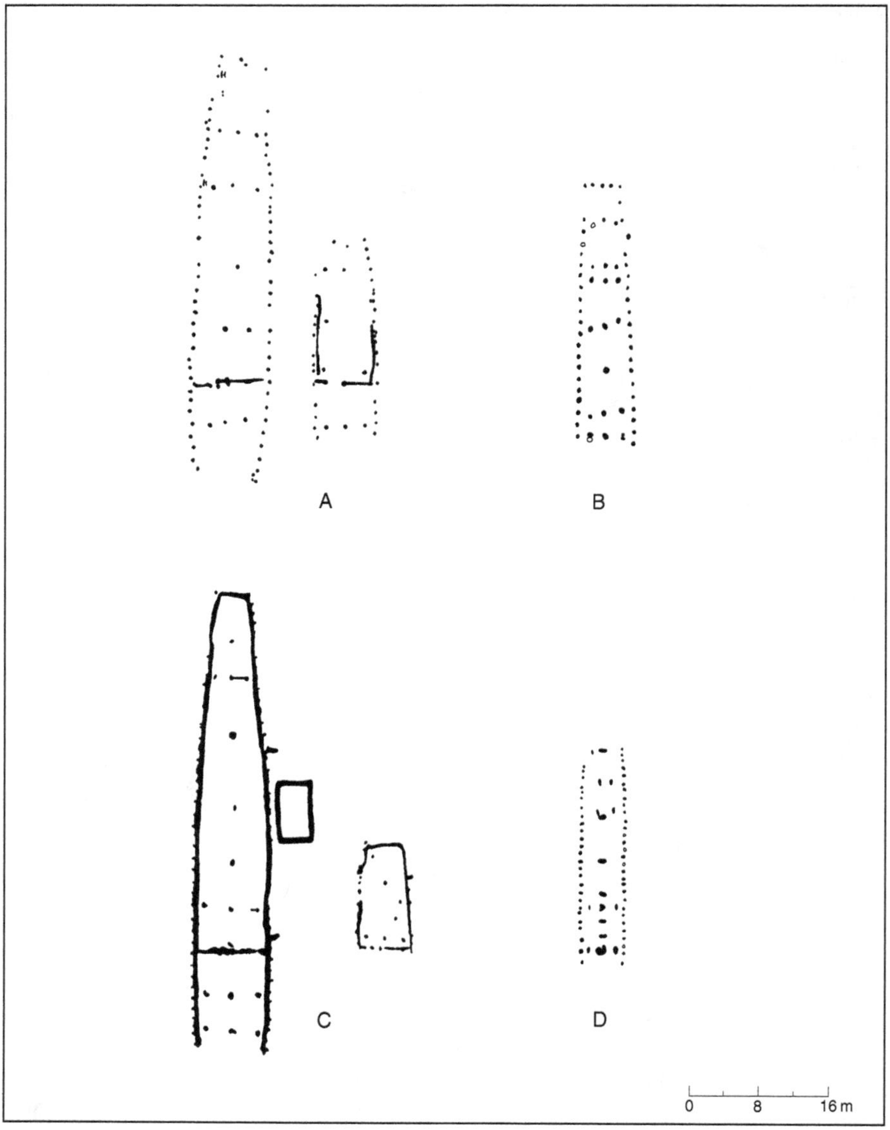

Fig. 4.3 Ground-plans of houses. **A**: Grossgartach, plans from Hambach 260, Germany. **B**: Villeneuve-Saint-Germain, plan from Poses, Dept. Eure, France. **C**: Rössen, plans from Inden 1, Germany, the longest house is accompanied by an outhouse. **D**: Plan from Berry-au-Bac, Dept. Aisne, France, attributed to Cerny but could be later Neolithic

The interior of the buildings echoes the triplets of roof-bearing posts of the LBK, organising the inner space into two main aisles and two narrower aisles running along the walls. The distance between the triplets is in many cases larger than in an LBK house, making buildings roomier inside. The two outer aisles tend to disappear, leaving a two-aisled interior, especially in the central part of the house. The weight of the roof is partly transferred to a more heavily constructed wall.

The first part of the house lost indications of an upper floor and with it its interpretation as a grain loft. The construction is much lighter, and this part is now seen as a kind of antechamber. A wall separated the first part from the central part of the house, as is suggested by a foundation trench, which could be observed in several cases. The central part could be very long, much longer than in an LBK house. This is also the part where the triplet arrangement of roof-bearing posts tends to disappear. The development resulted in houses that could be much longer than those of the LBK, whilst the width remained more or less the same. The longest house known is a 52 m long Rössen building excavated at Inden in the German Rhineland. Its central part has a length of 31 m.

Not much is known of the interior, but long Rössen houses do show transverse walls, which divide the central parts into separate compartments. In one case the central part revealed remains of an oven made of clay and supported by an internal wooden frame. The horse-shoe shaped structure was found in a Villeneuve-Saint-Germain house at Chambly (Dept. Oise).

The rear part of the house was the narrowest part. As in the previous LBK, it is seen as the sleeping quarters.

Whether the houses were inhabited by the same kind of family as during the LBK is open to debate. The large floor space of the central part of the longest houses suggests a larger number of inhabitants. Some researchers see them as multi-family houses, and the division into several compartments would agree with this view. But in that case the interpretation of the rear part as the sleeping quarters must be wrong, as that part would then really be too small. However, it is also possible that the roomiest houses were built by people who wanted to display their wealth in house length. The two interpretations would have consequences for the number of inhabitants. According to the first interpretation, the very long house at Inden mentioned above should have counted 141 inhabitants, based on a formula developed by A. Coudart. If just an extended family was living there, the number of inhabitants may have been only 15.

Although the general trends are the same, the gradual changes in house-plan traditions followed diverging paths during the post-LBK period, resulting in at least two 'provinces'. One developed in the north-eastern part of the loess region and concerned the Grossgartach and later the Rössen culture. This line ended with distinctly trapezoidal houses with wooden walls, built with boards with a narrow triangular section, split radially from large tree trunks, and set upright into foundation trenches. However, in the latest phase of the Rössen culture, called Bischheim, the construction of the houses reverted to the wattle-and-daub tradition. The second province is found in the southern part of the loess belt and concerned Blicquy, Villeneuve-Saint-Germain and later the Cerny culture. There, houses had a less pronounced trapezoidal plan and had wattle-and-daub walls, though truly reliably dated house-plans of the Cerny culture are still unknown.

The provinces differed also in the presence of outhouses. In the north-eastern province where such buildings have been found, they are situated next to the largest

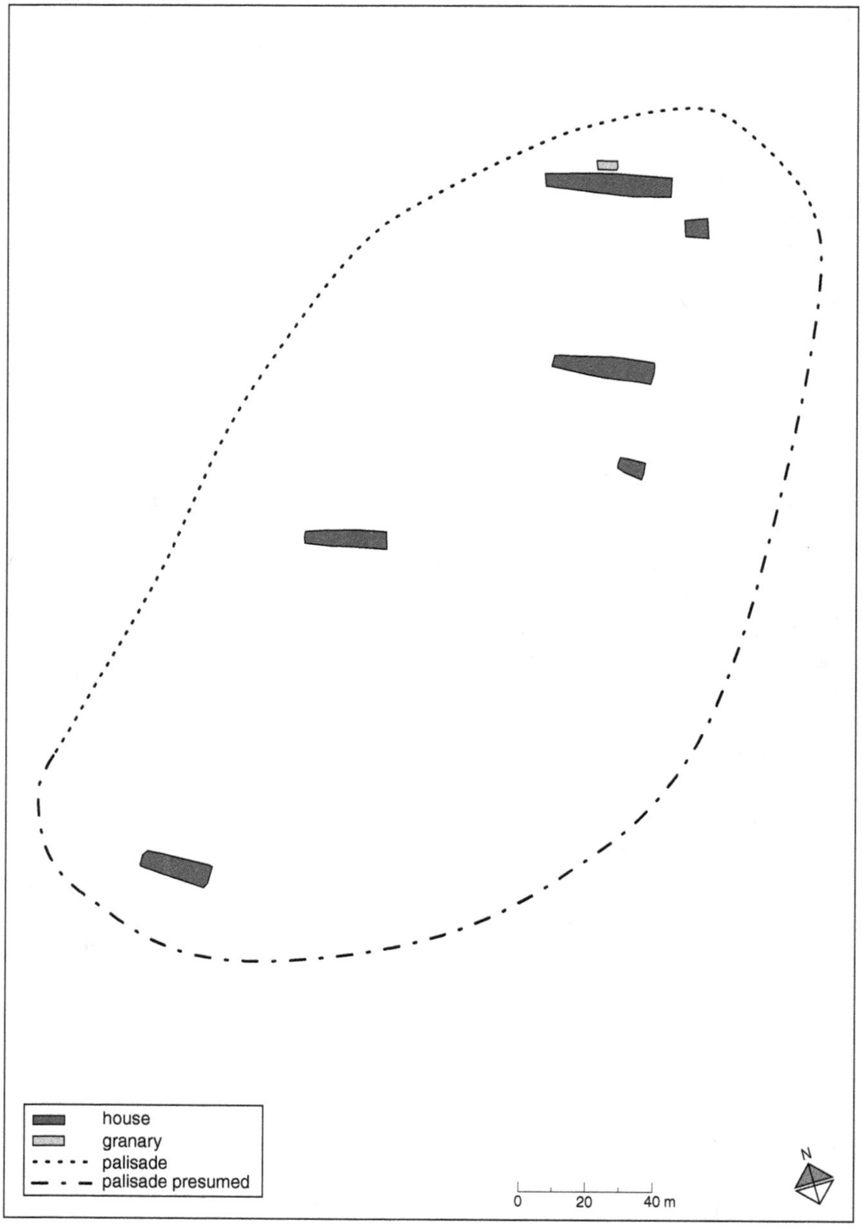

Fig. 4.4 The Rössen settlement at Inden 1, Germany, plan of a single phase of occupation

farmhouses. They are interpreted as granaries, which replaced the grain lofts in the south-eastern part of the main building. As in the LBK, not every house had a granary. Such structures are as yet unknown in the southern province. Where these farmers stored their grain above ground is not yet clear.

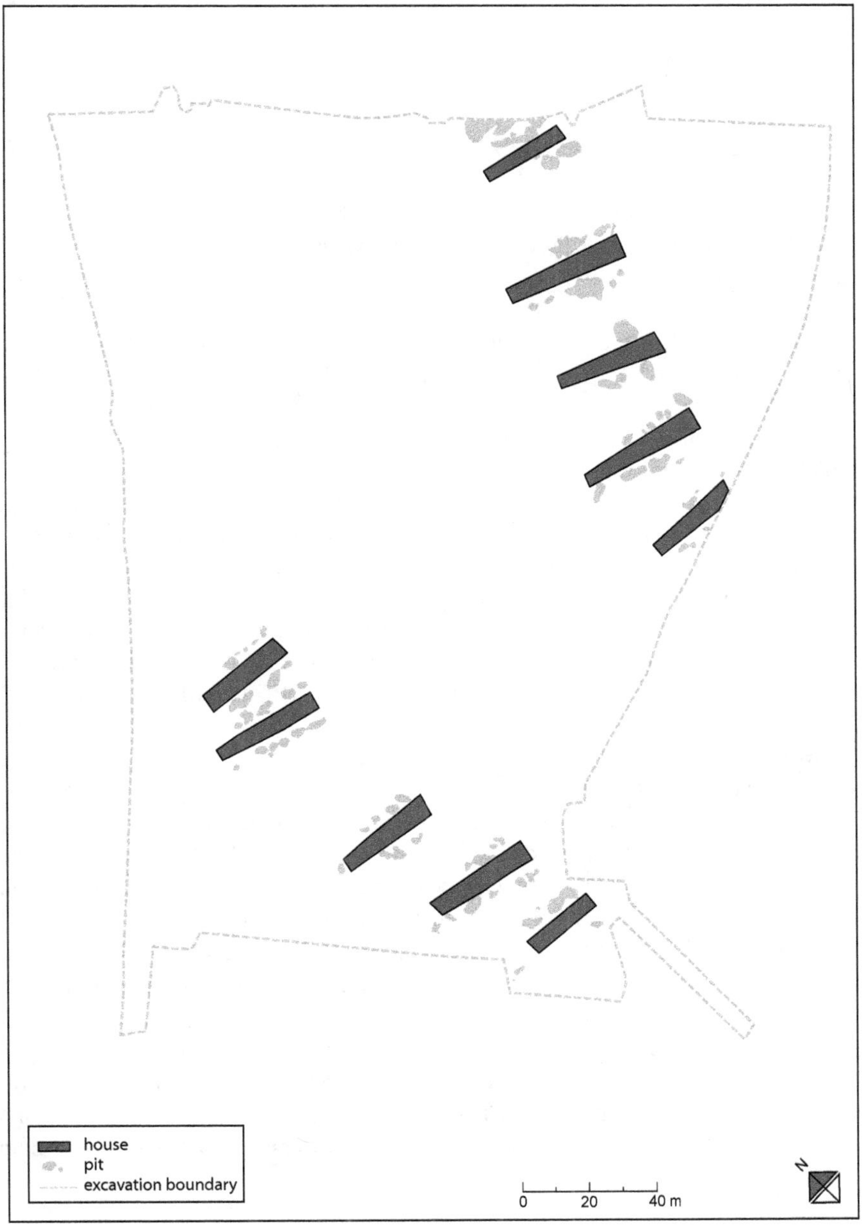

Fig. 4.5 The Villeneuve-Saint-Germain settlement at Poses, Dept. Eure, France. In this case it is not certain whether all houses are contemporaneous

The yard of the farms was, as in the previous period, dotted with pits. Pits along the walls were only present where buildings were made of wattle and daub, but other pits were common everywhere. Some of them have the characteristic form of subterranean silos.

Fences, defining yards, have not been observed. Wells were known but, as before, they are very uncommon.

4.6 The Farm in Its Setting

The individual farms occur in aggregations resembling those of the previous period. Some authors maintain that they are more compacted, but this impression is due to the fact that hamlets of only two or three farms seem to be very scarce. The distance between the main buildings within a hamlet has not significantly decreased. In the German Rhineland, for instance, the mean distance between two contemporaneous farmhouses in LBK Langweiler-9 amounts to 112 m. In the larger Langweiler-8 settlement it stands at 66 m. For the Rössen site Inden-1, in the same area, 74 m is given. The houses are still orientated in the same way, more or less parallel to each other. In the southern province, they even form short rows, as if they were built along a road. In the north-eastern province one of the most famous Rössen hamlets, Inden-1, was surrounded by a palisade, but enclosures seem not to have been the rule (Figs. 4.4 and 4.5). Inequality between the different households is most obvious in the Rössen culture, where the length of houses varies most. Every hamlet had at least one very long house and it is this kind of farmbuilding that is accompanied by a separate secondary building, interpreted as a granary. In spite of this incipient inequality, the societies of the successors of the LBK are still considered to be primarily egalitarian.

The number of inhabitants of an individual hamlet depends much on the interpretation of the dwellings, but in general populations of 75–200 people are mentioned. Clustering of hamlets is less obvious than in the LBK. Hamlets are also less numerous. In the north-eastern province, they lie at the fringe of the areas settled before. The setting is still the same: settlements were situated on the edge of plateaus, not far from streams. In the southern province, the lower terraces of the larger rivers were still occupied but settlement expanded to near secondary streams on plateaus. This process started with the Villeneuve-Saint-Germain culture and continued during the Cerny culture.

The pattern of regional organisation within the two provinces and different groups or cultures is not yet well described, as the period has been studied less well than the LBK. Part of the problem is that remains of these societies are less common than those of the LBK. That is to say that there are fewer excavations showing them.

Chapter 5
Innovation and Expansion: 4300 BC–2650 BC

5.1 A New Age

The last cultural heirs to the LBK people displayed new cultural elements, i.e. traits without roots in LBK traditions. In the north-eastern part of the region, for instance, a late Rössen group, called Bischheim, used for the first time true axes and, moreover, axes with blades made of flint. People before them used adzes with blades made of stone. In the south, the farmers of the Cerny culture had a different way of providing for their dead. Instead of burial in a simple pit, they erected elaborate grave monuments. Both are examples of new technical and cultural elements that heralded a new age. Suggestions have been made to connect the changes to a 'second neolithic crisis' – the first being the crisis at the end of the LBK – but facts indicating elements of crisis are mostly lacking. A better explanation may be that the rate of cultural change again underwent an acceleration after the kind of steady state which seems to have prevailed during the second part of the period described in the previous chapter.

The exact trigger of the cultural change is not very well understood, but one of the factors may have been the expansion of occupation beyond the land occupied previously. The expansion presumably brought contacts with new neighbours with different customs. These contacts need not have taken place within the loess region west of the Rhine. It is quite possible that the new elements arose at its borders, or even in more distant regions.

In any case, the cultural change resulted in several cultural groups, known as the Epirössen, Michelsberg, and northern Chasséen cultures. Notwithstanding their different names, the three are culturally quite related. After these cultures follows a new cultural horizon, with the little-known Stein group (possibly a variant of the German Wartberg group) and the Seine-Oise-Marne culture (SOM).

5.2 Crops

Of all the cultures mentioned above, the Michelsberg culture is the best known as far as plants are concerned. At first sight the list of crops grown by the Michelsberg

C.C. Bakels, *The Western European Loess Belt*, DOI 10.1007/978-1-4020-9840-6_5,

farmers looks identical to the list of the post-LBK cultures. This impression is however false, at least where cereals are concerned.

Wheat and barley are indeed still the main cereals grown, but the choice of species underwent a change. The main wheat was, in so far as can be deduced from the rachis remains, a naked tetraploid wheat. Its kernels were of the compact type. The preceding Rössen farmers grew a naked hexaploid wheat, also with compact kernels. In the late Rössen 'Bischheim' group, however, the tetraploid wheat already occurs next to the hexaploid wheat. In more than one respect this group is an intermediate between classic Rössen and Michelsberg.

The naked tetraploid wheat is held to be the same one as the wheat found in considerable quantities in contemporaneous lake-shore settlements in the Alps. In those sites, plant material is far better preserved and carbonised cereals have even been found as whole ears. They are much better for species identification than the fragments found in the loess region. An intensive study of the ears excavated at, for instance, Hornstaad-Hörnle IA at the shores of Lake Constance has shown that the wheat is not identical to any recent variety of the naked tetraploids macaroni wheat (*Triticum durum* Desf.) or rivet wheat (*Triticum turgidum* L.), which are the species best known today. The site produced at least six different types of ears, considered to represent six varieties of a single species. They may be races gone out of fashion and belonging to either macaroni wheat or rivet wheat, and are published under the name *Triticum durum/turgidum*. Macaroni wheat prefers a warm climate with rain in spring. It is susceptible to late frosts and requires good soils. Rivet wheat has, through its different landraces, a wider ecological amplitude, including climates with more rain.

It is impossible to tell which wheat was sown in the period under review and where it came from is unknown, too. The ultimate origin of the naked tetraploid wheats is thought to lie in the Near East. In the fourth millennium BC, the new wheat became one of the dominant cereals in Central Europe, as it was in the Michelsberg culture, and perhaps also in the Chasséen.

Next to naked wheat, the hulled wheats emmer and einkorn remained important as well. True, in some settlements einkorn is almost absent and seems to have been reduced to a very minor crop, but in others this wheat was still one of the main crops.

The culture of barley underwent less change. In most areas a naked, multirowed barley continued to be the one sown. However, Michelsberg farmers in the Paris Basin grew hulled, multirowed barley. As mentioned in Section 4.2, farmers in the south-western part of the region already grew this barley before Michelsberg times, though on which scale is unknown. It seems that the area where this crop was cultivated has expanded.

In addition to wheats and barleys, traces of a third kind of cereal have been found, i.e. millet (*Panicum miliaceum*). Its remains are so scarce that it is not yet clear whether millet was indeed grown as a crop on its own. Crop plants other than cereals are pea, linseed/flax and poppy. Spindle-whorls attest to spinning (Fig. 5.1). Yarn and textiles made of flax, found in the waterlogged lake-side settlements in the Alps, show that flax was indeed not only grown for its seeds but also for its fibres. There are no new categories of cultivated plants recorded.

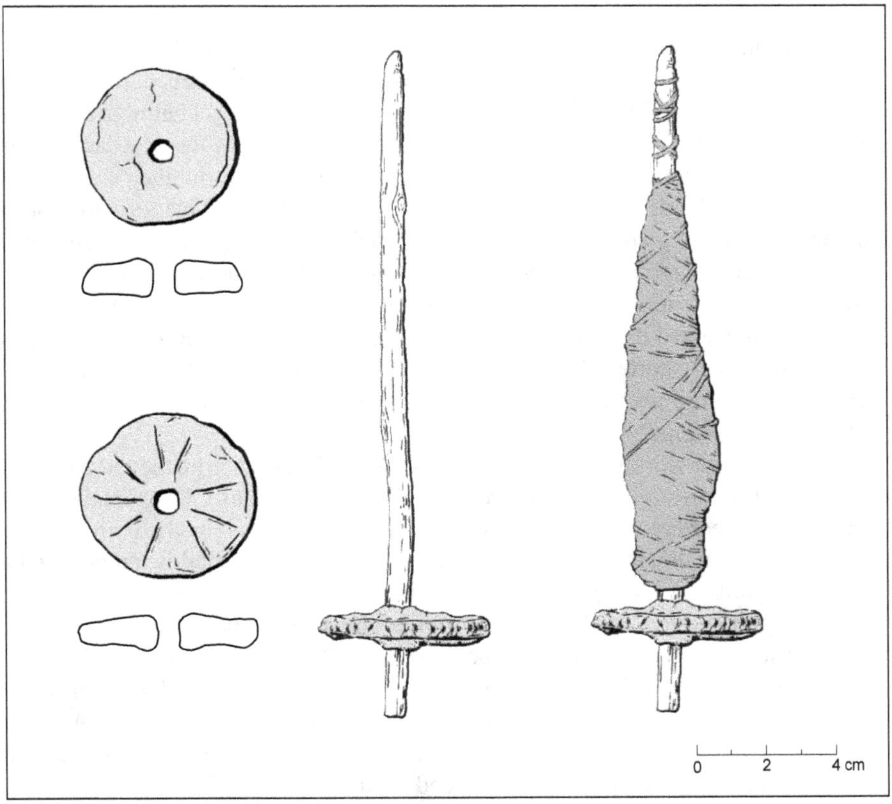

Fig. 5.1 Spindle for hand-spinning. As it is susceptible to decay, in general only the ceramic whorls, depicted on the left, are found

The crops of the cultures of the Epi-Rössen, which preceded the Michelsberg in the southern part of the loess region, were the same, as far as is known. The Stein group and SOM culture, which came after the Michelsberg and Chasséen, have hardly been investigated archaeobotanically, mostly due to the absence of suitable features. Some pits, filled with domestic waste, were available but produced meagre results. They showed only that hulled wheats were still grown. Nevertheless, the lake-side settlements in the Alps provide evidence that crop growing was an important activity in post-Michelsberg times. The lack of information in the loess region is merely due to a lack of suitable archaeological traces.

5.3 Crop Cultivation

The Michelsberg and Chasséen people spread farther over the countryside than their predecessors. Where they farmed areas that had previously been inhabited, presumably they had to clear secondary forest and shrubs at most. Where they took to farming new land outside the primary and secondary valleys and their fringes, such as on plateaus, they had to clear true forests. For both activities farmers certainly required an axe. The adze of the LBK and post-LBK was abandoned and replaced by true axes. As mentioned in Section 5.1, the late Rössen Bischheim group knew them already, but the people of the Michelsberg culture were the first large-scale users. Blades could be of stone, smooth and polished all over, and they could be of flint. Many flint blades are also polished all over, but there are blades with polish limited to the cutting edge, as well as unpolished blades (Fig. 5.2). The Chasséen people preferred unpolished blades. All these blades were inserted into a wooden haft. Towards the end of the period sleeves were placed between the blade and the haft, which prevented direct contact between the stone and the wood. These sleeves allowed the use of shorter blades and prevented damage by too strong or unluckily directed blows. The axe remained an important implement during the Stein group and SOM culture times.

Implements for working the soil are not known, at least not from the loess region. It can be safely presumed that the wooden hoe was still in use. Hoes made of the antlers of red deer are a possibility as well. Hoe-like pickaxes made of antler have been excavated in the underground flintmines for which the period is famous, and a related kind of tool may have been used for loosening the soil. Nevertheless, sometime during the period another kind of tool is presumed to have been introduced: the ard. The ard is a type of plough that cuts the soil without turning it. In its simplest form it is no more than a hoe dragged through the soil. The blade functions as share, and the handle as beam. This kind of ard is known as crook-ard. Another version is the spade-ard. It has a separate stilt with handle, a separate share, and a separate beam attached near the low end of the stilt (Fig. 5.3). All parts were made of wood. Ploughing with an ard results in a shallow, straight furrow. The distance between the furrows is about 30 cm and the depth of the furrow, and of the worked soil, is 15–20 cm. Fields were commonly ploughed cross-wise. Traces of this implement have not been detected in the loess region, and it will be difficult to detect them there. Such traces will only be preserved when a ploughed surface has become buried under wind-blown sand or silt, or man-made soil accumulations such as burial mounds. When old surfaces are preserved and excavated at the right plane, the criss-cross furrows become visible, and even then only if their filling and the original soil have contrasting colours. This is, for example, the case when a light-coloured sand has been ploughed and the filling of the furrows is a darker arable soil, darker because richer in humus and/or charcoal particles (Fig. 5.4). Such instances have not yet been found in the loess region. However, ard marks are regularly present in other regions in a period just after the one considered here and the ard is supposed to have been used even earlier. Therefore, it is not inconceivable that the implement was introduced in the loess region during late Michelsberg or SOM times. Moreover, the ard performs best on land free of tree stumps and tree roots. Such conditions were presumably met with in areas with a long history of agriculture, such as certain parts of the loess region.

Another innovation is manuring. The oldest fields known (not on loess soils) show scatters of potsherds, tiny fragments of burnt bone and charcoal. The charcoal

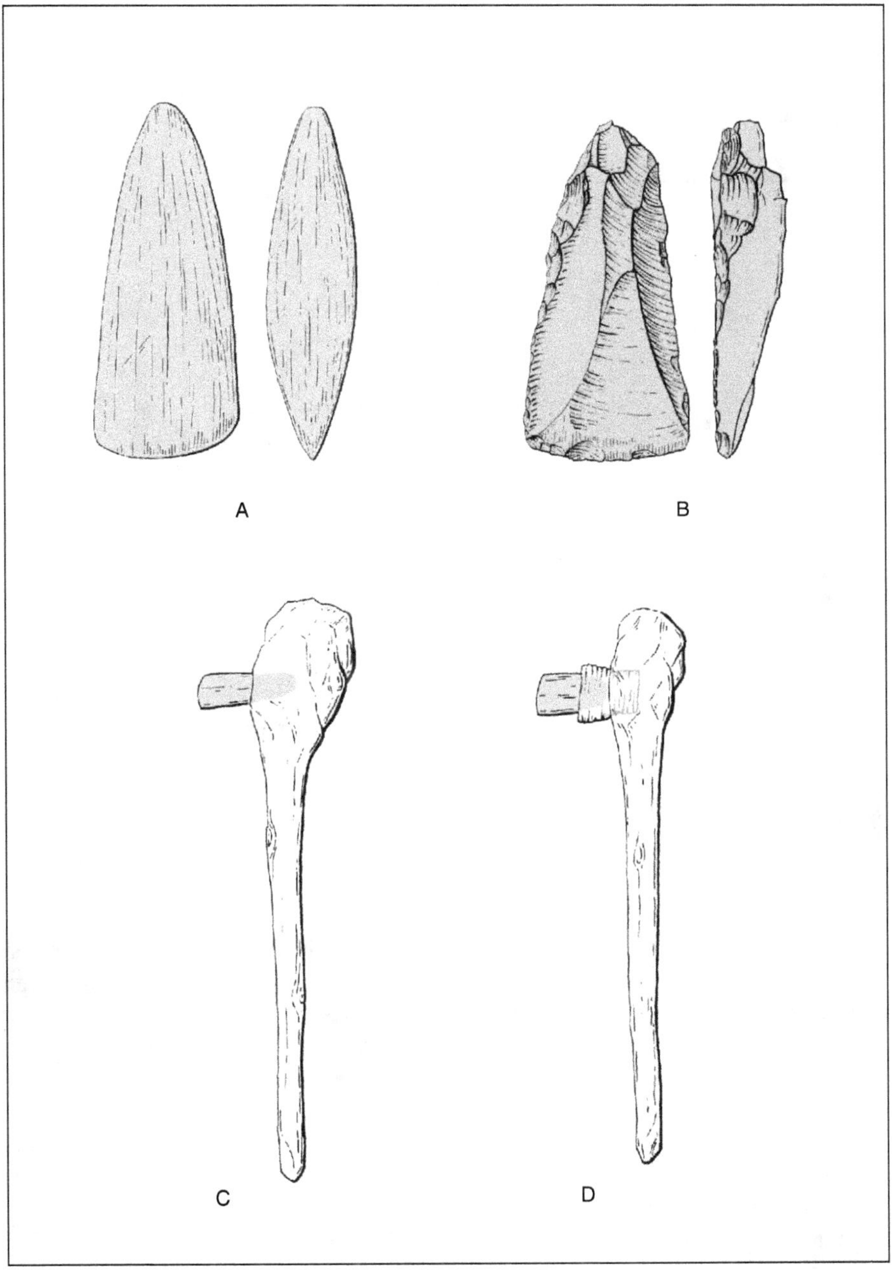

Fig. 5.2 Axes.**A**: blade of stone polished all over. **B**: blade of flint, unpolished. **C**: blade set directly into the haft, **D**: blade set in a sleeve which in its turn is set into the haft

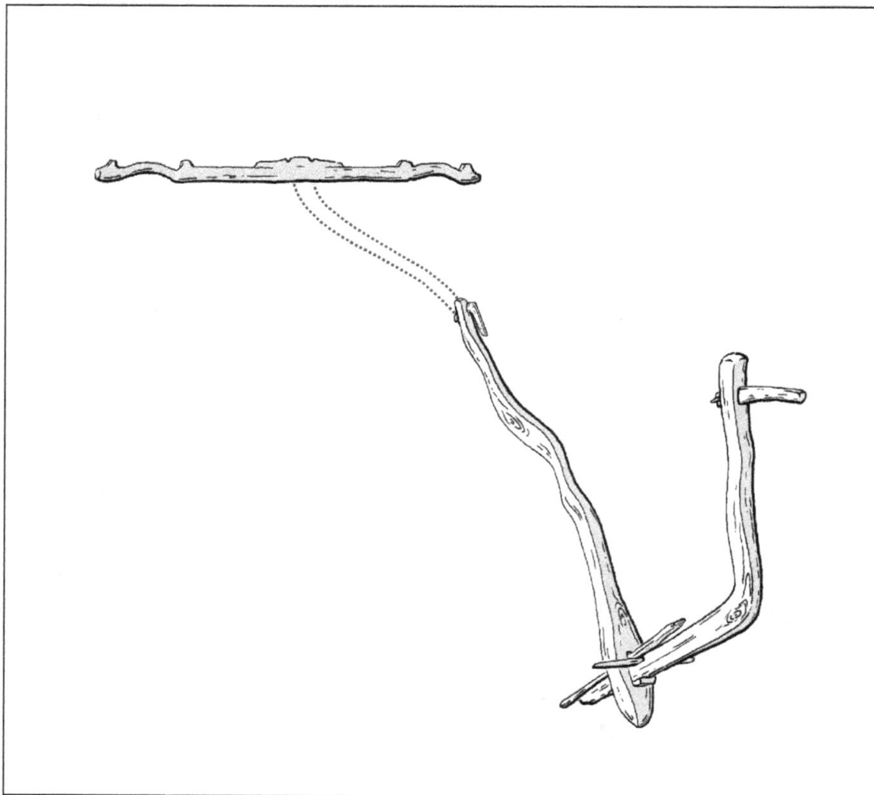

Fig. 5.3 Ard and yoke

may have come from the burning off of vegetation on the fields. Burning brushwood and branches provided ash which enriched the soil, a practice presumed to have been carried out already during the previous periods. But sherds and bone are a different matter. They might be the remnants of earlier settlement on land later converted into a field, but in most cases there are no other traces of earlier occupation. The sherds and bone must have been brought onto the fields and this would suggest that material from middens was added to the soil. The charcoal, or part of it, may also have come from middens. Domestic waste was rich in organic material, but the organic part has vanished of course. To which extent the manure held a component of animal dung is unknown. It does not have to have had any dung element. Indications of stabling livestock in places where dung could be collected are lacking in the loess region west of the Rhine. Not surprisingly really, as not much is known of the farm of that time (see Section 5.5). As a matter of fact, excavations of lake-side settlements in the Alps revealed parts of dwellings where cattle were stabled and had left dung, for instance in the Swiss site of Weier. Therefore, the use of animal dung cannot be ruled out completely.

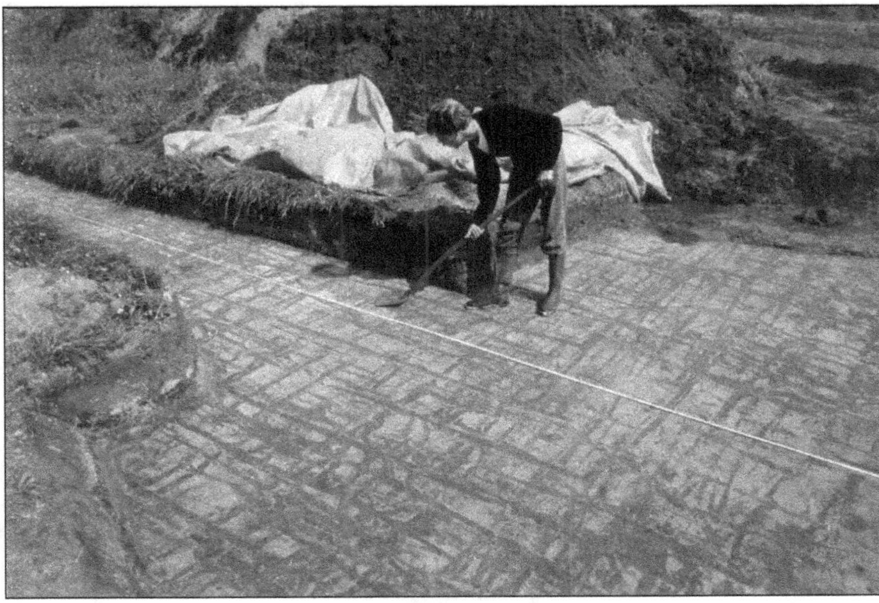

Fig. 5.4 Ardmarks. Provenance Haarlem, the Netherlands

The hypothesis that people did indeed manure their fields, or at least parts of them, at this early stage of agricultural history is difficult to test. Nevertheless, an answer may be provided by the composition of the weed flora. It is quite safe to assume that arable soils had become depleted in areas where the tilling of the land had taken place for over a millennium. A good example is the Aisne valley in northern France. Time and again the same parts of the landscape have been used, i.e. the higher parts of low-lying terraces in the river valley. This land may have regained some of its fertility during occasional floodings, but such floodings were not regular. Some soil exhaustion may be expected and the composition of the weed flora may reflect this. Unfortunately, the Aisne valley excavations have not yielded sufficient botanical material to allow verification of this assumption, but the scanty results obtained so far show that the Aisne list of Michelsberg species comprises as many weeds requiring nutrient-rich soils as the LBK list.

Archaeobotanical results are more numerous in the Netherlands. Of the 23 LBK 'species' (actually taxa) found in 11 late LBK settlements only 2 are not indicative of rich soils. The list of Michelsberg weeds, obtained from only four sites, numbers 16 'species', two of which also indicate poorer soils (Table 5.1). Therefore, the assumption that the soil lost much of its fertility during the ages cannot be maintained. Something must have added nutrients. As the LBK and post-LBK did not practise shifting cultivation, a long fallow period to 'rest the soil' seems to be out of the question. Because of the lack of data continuous farming of the same areas could not yet be established for the Michelsberg culture as far as the Dutch (or adjacent German) areas are concerned, but continuous farming is not completely out of the

Table 5.1 A comparison of Late LBK and Michelsberg weeds. See the caption of Table 3.1 for R, N and height. Frequency is the percentage of sites in which the plant was found

	Late LBK 11 sites Frequency	Michelsb. 4 sites Frequency	R	N	Height in cm
Tall and climbing weeds rich to intermediate soils					
Chenopodium album	100	75	x	7	15–100
Fallopia convolvulus	100	75	x	(6)	10–100
Bromus hordeaceus/secalinus	91	25	x	x	40–100
Lapsana communis	82	75	x	7	30–120
Phleum sp.	73	25	–	–	10–150
Bromus sterilis/tectorum	55	50	8	4	15–120
Echinochloa crus-galli	27	25	x	8	10–120
Persicaria maculosa	46	–	7	7	20–100
Galium aparine	36	50	6	8	15–100
Setaria verticillata/viridis	27	–	x	7	15–150
Silene sp.	27	–	–	–	30–100
Chenopodium polyspermum	18	–	x	8	10–80
Rumex cf sanguineus	18	–	7	7	60–120
Atriplex patula/prostrata	–	25	x	7–9	20–90
Solanum nigrum	–	25	7	8	5–60
Poa pratensis/trivialis	18	–	x	7	10–100
Avena fatua	9	–	7	x	6–120
Persicaria lapathifolia	9	–	x	8	10–120
Poor soils					
Vicia hirsuta/tetrasperma	64	75	x/5	4/5	15–70
Vicia hirsuta	18	–	x	4	15–60
Vicia tetrasperma	9	25	5	5	17–70
Low-growing weeds rich to intermediate soils					
Anagallis arvensis	18	–	x	6	5–50
Cruciata laevipes	18	–	6	7	15–45
Galium spurium	18	25	8	5	10–40
Stachys arvensis/sylvatica	18	–	3/7	6/7	7–30
Prunella vulgaris	9	–	7	x	7–45
Poa annua	–	25	x	8	15–50
Sherardia arvensis	–	25	7	5	15–25
Veronica hederifolia	–	25	7	7	5–30
Poor soils					
Scleranthus annuus	–	25	2	5	5–20

question. Pasturing of livestock on stubble fields may have restored some of the fertility, but it is questionable whether this practice would have sufficed. Additional manuring will have helped. It is almost superfluous to say that the size of the fields is unknown. Fields may still have been garden-like.

Weeds, found in combination with crops, also provide information on the height of harvesting. In the German-Dutch area the most frequent weeds are still tall ones or climbers. The list of plants more sporadically encountered comprises species with heights below 50 cm, and their number has not increased since LBK times. In the sites analysed for nutrient requirement, in both LBK and Michelsberg times only five species belong in this category (Table 5.1). In other areas there are unfortunately not enough data to prove that these results are valid elsewhere, but it is likely that long straw was a commodity not much sought after.

Continuity in the weed spectrum is not self-evident, as the harvesting technique had changed. This can be deduced from the absence of flint bladelets with 'sickle-gloss', common in LBK and post-LBK contexts. The curved sickle with a row of inserted flint teeth had been abandoned. It may have been replaced by a reaping knife, consisting of a long, straight blade of flint, inserted with its long side into a short wooden handle. Another version had the long blade set with its short side obliquely into the handle at an angle of c. 140 degrees between the working edge and the wooden stick. The cutting edge of this kind of knife had a length of between six and fifteen cm. Birch tar glue provided a good adhesion. Both types are known from lake-side settlements in the Alps and the Jura mountains (Fig. 5.5). Michelsberg and Chasséen settlements revealed similar blades, even with 'sickle-gloss', but they are far from numerous. From their scarcity it could be deduced that crop cultivation was not very important at this time, but other ways of harvesting should be considered, too. One possibility is uprooting plants. Against this possibility can be argued that uprooting causes a large quantity of low-growing weeds to be brought in together with the crop. But, as was put forward above, such weeds did not increase their numbers during this period. Another possibility is that harvesting was done by hand-picking, or with the use of perishable implements, such as mesorias. Mesorias consist of two sticks tied together with a piece of string at one end (Fig. 5.5). A small bunch of cereal culms, placed between the legs of the sticks, can be stripped off the ears by a rapid upward movement of the hand holding this implement. Both reaping methods work best with cereals that have a weak point between the ear and the stalk, allowing easy breaking off of the ear. Hulled wheats and barleys are such cereals. The two methods result in relatively clean harvests. In the few cases where concentrations of cereals have been found in excavations, the number of weed seeds was indeed low, but there are too few instances to serve as proof that such methods were practised. If they were, the straw must have been left on the field. For harvesting straw separately, a cutting implement would have been necessary, and it is precisely this that is lacking. The straw may not have been required in the farmers' households, except for limited use, such as the making of baskets.

The methods used in the processing of crops are not known either, due to the scarcity of data. It is, however, clear that underground storage in silos was important. The existence of granaries above ground is uncertain, but large earthenware vessels may have served to hold stocks of semi-clean or clean products. They seem to be especially abundant in the southern part of the loess region, where concentrations of sherds of such vessels were found. Some could hold 100 L or more of grain. In other areas, wooden containers or basketry may have served the same purpose.

As before, saddle querns were used to grind the grain. Fragments of querns are common finds.

It is assumed that much of the description and many of the suppositions offered above, apply to the Stein group and SOM culture as well.

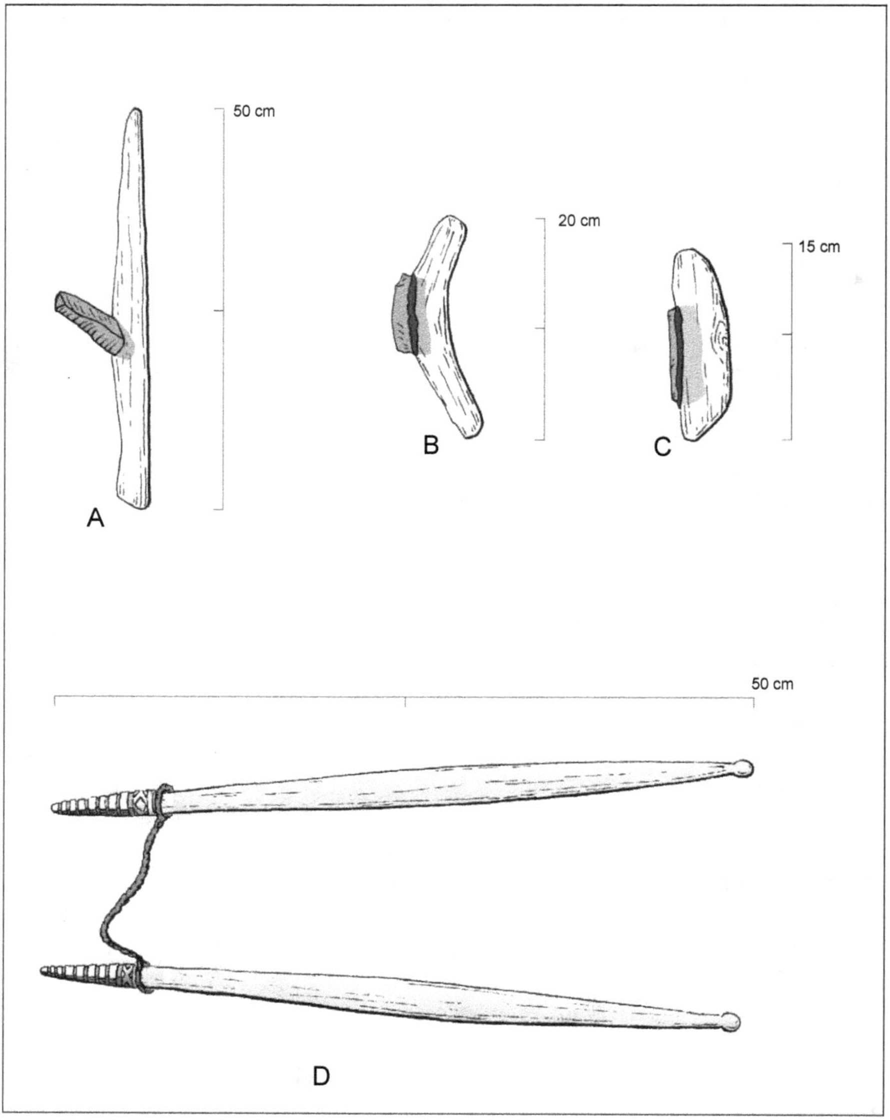

Fig. 5.5 **A**, **B** and **C**: reaping knives composed of a flint blade set into a wooden handle and fixed with birch tar (*dark grey* in **B** and **C**). **D**: recent wood-and-string mesorias

5.4 Livestock and Animal Husbandry

Cattle, pigs, sheep and goats were the animals kept. As in the previous chapters, information on livestock is restricted to those areas, where decalcification and subsequent disappearance of bone was not a serious problem. Domestic animals were the main providers of meat, though in the end there are differences between settlements.

On the plateaus livestock contributed c. 95% to the meat part of the diet, at least as far as can be deduced from sites around the river Oise. In settlements on river terraces in valleys such as the Aisne, the share of domestic animals in the bone material amounted to a mere 60% and though part of the hunted animals was not destined to be eaten, hunting seems to have had a greater role in providing meat. Nevertheless, this is not true for every valley site. The important site of Mairy (Dept. Ardennes) in the French Ardennes revealed almost no bones of wild animals, but this was perhaps a special site (see Section 5.5).

Cattle had the largest share in the livestock, but pigs were an important second. Sheep/goats came third (Fig. 5.6). Pigs have gained in importance since the LBK. This must have been a deliberate choice, based on changed human preferences, as there is no environmental trigger. The quota of pigs increased in both areas settled of old, i.e. the river valleys, and the newly settled areas on plateau edges and plateaus.

The information referred to so far concerns residential areas with common domestic waste. However, some sites are different, such as the Michelsberg site of Mairy and the Chasséen site of Boury-en-Vexin (Dept. Oise). In Mairy more than a

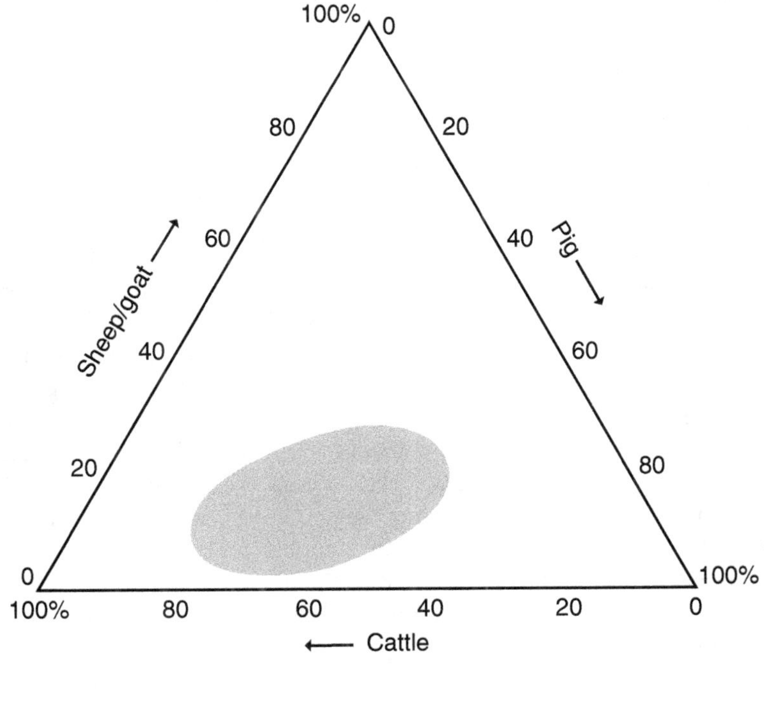

Fig. 5.6 Composition of the livestock

hundred abandoned silos were found, filled with bones, rarely broken and often still articulated. The bones mostly came from cattle, while pig bones were almost absent. Boury-en-Vexin revealed deposits of thousands of bones and even entire skeletons in the ditch bordering the site. Sheep/goat bones were predominant in these finds, with sheep far more numerous than goat bones. The bone concentrations at these two sites probably do not reflect the common composition of the livestock. Ritual choices must lie behind the deviation from the normal situation.

The shoulder height of cattle was some 1.20 m. At Boury-en-Vexin the shoulder height of bulls was 1.21 m and the height of cows 1.18 m, thus a sexual difference was present but not very pronounced. These animals were considerably smaller than LBK cattle. Pigs, on the contrary, had not become any smaller and with 80–91 cm to the shoulders they were almost bigger. The height of sheep generally varied between 52 and 61 cm. Ewes at Boury-en-Vexin had a shoulder height of 55–60 cm and rams measured 66–68 cm, which reveals a distinct sexual difference. Most ewes had horns, but some skulls lacked them. The few goats found had a height to the shoulders of 59–60 cm.

Because the number of domestic sites with suitable material is rather limited, the age at slaughter is not yet fully established. At the domestic Chasséen site of Catenoy (Dept. Oise), 94% of cattle were slaughtered rather young, by which is understood: before an age of 6.5 years. The peak lay around 2–4 years. At the presumably not truly domestic Michelsberg site of Mairy 87% of cattle were slaughtered young, with a distinct preference for calves up to 1 year old, both male and female. It looks as if adult cows, able to give milk, were not especially valued. Dairy products, at least based on cows, may not have been part of the farmers' output.

Most pigs in Catenoy were slaughtered before they reached their second year, that is before they reached an age at which their rapid growth stopped. Only a few adults were left, presumably for reproduction. In Mairy it was almost exclusively piglets, of both sexes, that were found. As to sheep, most were slaughtered between an age of 1.5 and 3 years, both at Catenoy and Mairy. Nevertheless, the proportion of rams to ewes, in the age class of 2 years, was 1:13, which implies that many rams were killed before they reached that age. In Mairy, the proportion in adult animals is 1:15. This means that many ewes were allowed to live to higher ages. They may have been kept for reproduction. Milking of the ewes is also a possibility, but as cows do not seem to have been kept for this purpose, this is hardly more than a theoretical option. Wool is another product for which sheep, and especially the more docile ewes, may have been spared an early slaughtering. Spinning and weaving were known, witness the textiles made of linen mentioned in Section 5.2. But as wool is a material not often found in excavations, due to its special requirements for preservation, it is not known if wool was already being processed in this period. The sheep of this time are considered still to be hair sheep.

Milk and wool are examples of products of the so-called 'secondary products revolution', a term coined to describe the introduction of the use of products for which animals need not be killed. The use of living animals to produce something turns up in a specific time horizon, which is generally placed after the period discussed here. Nevertheless, one aspect deserves some attention at this point, and that is the use of oxen. During the Michelsberg and Chasséen cultures, the first oxen are observed in the bone material. It could be that bulls were castrated because oxen provide much good quality meat. But another reason may be added: the secondary product 'animal power for traction'. If the ard was indeed introduced during this period, traction, and preferably animal traction, was certainly required.

Towards the end of the period under review another important innovation became known, namely the wheel, and with it the wagon. As the horse was not yet a domesticated animal in the region in this period (the very few remains of horse are considered to represent wild animals), cattle are the obvious providers of animal power. Of course, cows can pull ards and wagons, as they still do in many human societies today, but oxen are the stronger animals. Unfortunately, the remains of the Michelsberg and Chasséen oxen do not give evidence that they were used for traction, and this function cannot yet be considered as proven.

Early devices requiring traction in addition to the ard were the sledge and the travois. The travois consists of a harness attached to two poles with ends trailing the ground. The poles support a platform, on which loads can be carried (Fig. 5.7). That the travois was known in Europe is deduced from ruts, for instance those worn into the bedrock on Malta in the Mediterranean, and from finds in waterlogged settlements in the Alps. Sledges have been found in Northern Europe. Wheeled vehicles appeared in Europe already during the fourth millennium BC and are considered to

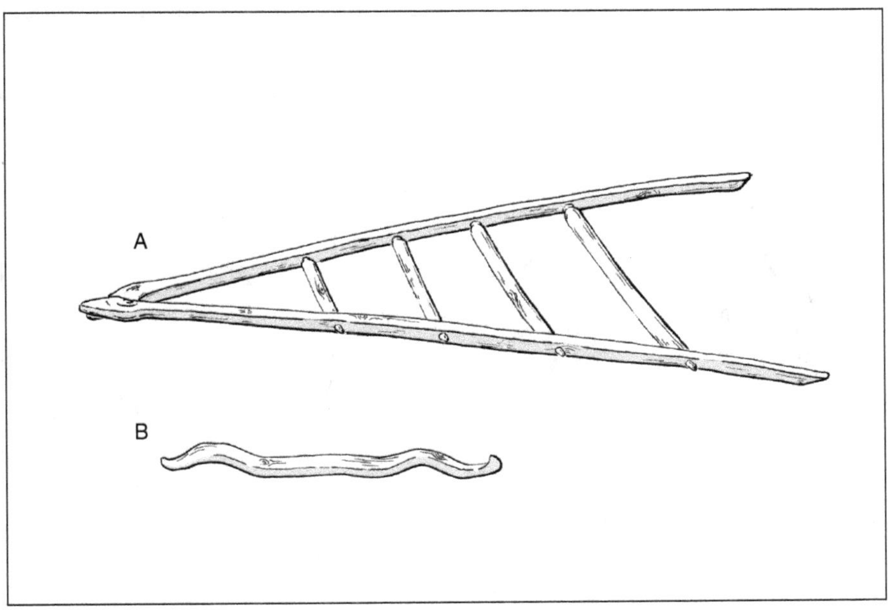

Fig. 5.7 **A**: an implement interpreted as a travois (after a find at Chalain Dept. Jura, France). **B**: a yoke, found at Vinelz, Switzerland

be a major socio-economic development. The oldest evidence comes from northern Germany and southern Poland.

Near Kiel, in Germany, wagon-ruts were discovered under a barrow. The traces consisted of 5–6 cm wide ruts, belonging to vehicles with a gauge of 1.10–1.20 m. These ruts were found in groups, which would indicate repeated passage. The wavy bottom of the ruts points towards true wheels and not to something like a sledge. The Polish evidence consists of pottery, decorated with wagon motifs, found in Bronocice near Krakow (Fig. 5.8). Actual finds of wheels come only slightly later. They are made of wood. The oldest specimens are single-piece discs, such as the wheel excavated in Zürich, Switzerland, which is made of a single piece of maple. Soon afterwards, such wheels were used simultaneously with, or were replaced by, composite disc wheels made of several boards joined together (Fig. 5.9). Wheels with spokes are a much later development. The Swiss wheel has a square hole in the middle, and was obviously attached by this mortise to a rotating axle. This wheelwright tradition seems to have been a South-central to Southern European affair. Elsewhere in Europe wheels had raised tubular naves turning around a fixed axle. The vehicles themselves are not known, but pottery models of the period show that they were four-wheeled (Fig. 5.8).

The ultimate origin of the wheeled vehicle is considered to lie in the Near East, where, for instance, wagon pictographs found on clay tablets from Uruk-Warka in Iraq testify to their existence. However, the Near-Eastern evidence is not older than the earliest European evidence. Wheeled vehicles may have been invented more or

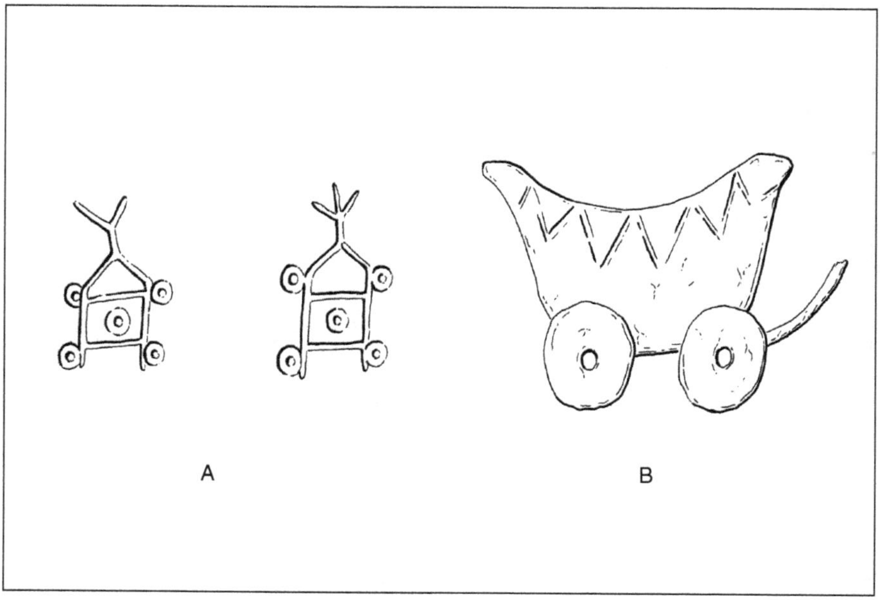

Fig. 5.8 **A**: depiction of wagons on a vessel found at Bronocice, Poland. **B**: pottery model of a wagon, provenance Budakalász, Hungary

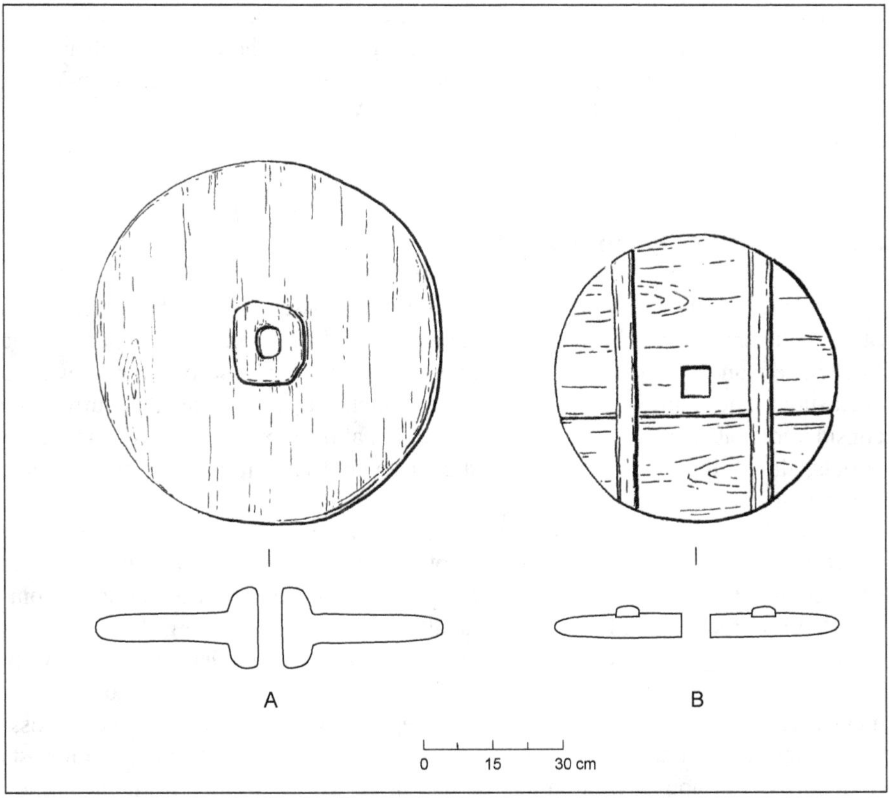

Fig. 5.9 Examples of disc wheels. **A**: a single disc wheel with a tubular nave fit for a fixed axle, provenance Emmen, the Netherlands. **B**: a composite disc wheel with a mortise fit for a rotating axle, after remains found in Southern Germany

less simultaneously in both areas, but the possibility of a very fast technological transmission from the Near East to Europe is the favoured view.

The early wheeled vehicles were drawn by two animals. They were attached to a single draught-pole by a yoke lashed to their horns, in the same way as animals were harnessed to an ard. In the loess region west of the Rhine, evidence of wheels and wagons is lacking but this should be attributed to the absence of suitable conditions for the preservation of such perishable objects. Wagons are very useful in agrarian societies, for instance for carting manure to fields and carting the harvest to the farms. Nevertheless, they are of limited use in heavily wooded areas and very hilly country. The latter was no problem in the loess region, but the first may have been. It is perhaps no surprise that wagons were adopted only when centuries of agriculture had removed obstacles like tree stumps, which hampered the movements of the rather clumsy vehicles.

How many animals the livestock of a single farm numbered is hard to reconstruct. One of the problems is that it is difficult to get a clear picture of a farm belonging to any of the cultures described here, because of the lack of suitable traces (see Section 5.5). Presumably the livestock was kept outdoors. The way in which animals were used on a grand scale in rituals, as detected in Mairy and Boury-en-Vexin, shows that livestock played an important part in Michelsberg and Chasséen society at least.

5.5 Farmbuildings and Yards

It is strange, but the farmhouse and yard of the Epirössen, Michelsberg, Chasséen, Stein group and SOM culture are almost unknown. The Epirössen and especially the Michelsberg and Chasséen cultures are better known for their so-called causewayed enclosures. These are systems of interrupted ditches and palisades that surround a considerable piece of land. The later cultures are almost exclusively known by their burials, although pits filled with domestic waste are found, too. Nevertheless, some house-plans have been recovered.

The best known example of a farming community that looks 'normal' is provided by Berry-au-Bac 'Le Croix Maigret' (Dept. Aisne). It is dated to the Epirössen and belongs therefore to the beginning of the period under review. The hamlet is composed of four rectangular buildings not at all resembling the buildings from previous periods. The largest house measures 20 × 10 m, and the smaller ones 10 × 10 m. They are two-aisled. The larger building has an inner division into three compartments, realised by cross-walls, which are deeply founded in trenches. These cross-walls may have acted as extra support for the roof, in addition to the central posts (Fig. 5.10). The houses lie in a curved row at 10–20 m distance from each other, and parallel to the surrounding palisade and ditch. Not all of the enclosed area could be excavated, but the total would have covered 2–3 ha (Fig. 5.11). The large house may have been the main farmhouse and the other three secondary constructions, but in general it is thought that they represent four individual dwellings. Some silos and other pits are the only other elements detected. A well is absent, but the settlement is close to the river Aisne, which would have provided the necessary water. Traces of a similar Epirössen settlement have been excavated near Osly-Courtil (Dept. Aisne), also in the Aisne valley.

True Michelsberg is the site of Mairy (Dept. Ardennes). It covered at least 11.5 ha and is therefore a much larger site. At least 23 main buildings have been excavated (Figs. 5.10 and 5.12). They resemble the Berry-au-Bac houses, but are much longer and look more like a stringing together of the largest Berry-au-Bac structure. The longest measures 60 × 13 m and is divided into four compartments, connected by central doors in the centre of the dividing walls. Two others measure 45 × 9.5 m and 42 × 9.5 m respectively. A fourth has an extension of two other rooms perpendicular to the main axis. The traces are here and there incomplete, but some structures were indeed smaller than these four and had perhaps only two compartments. It is not known whether all buildings were contemporaneous. The houses are not arranged in neat rows. The surrounding terrain is dotted with over one hundred underground silos with diameters of 0.70–1.00 m and depths of 1.35–1.65 m. Sets of four post-

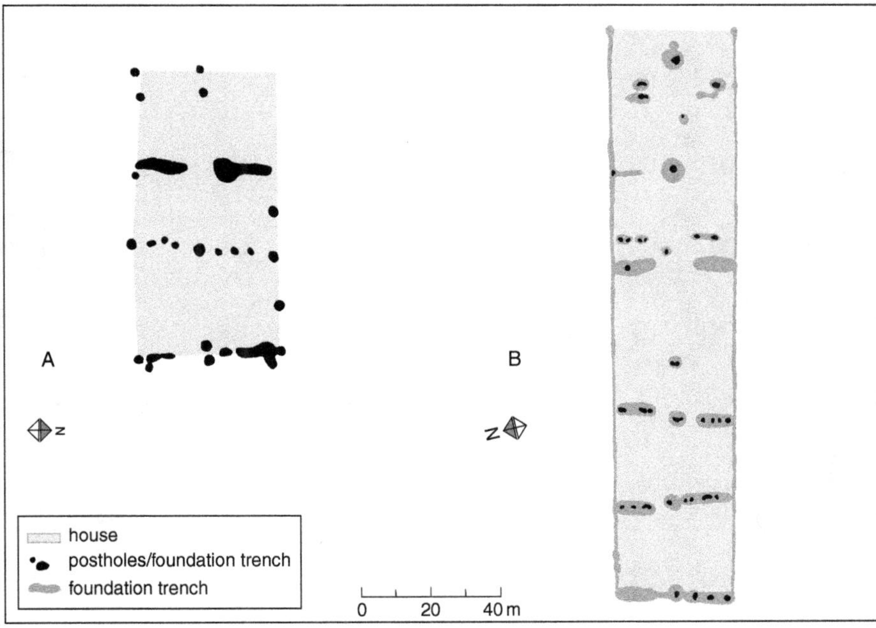

Fig. 5.10 **A**: ground-plan of a house at Berry-au-Bac (Dept. Aisne, France). **B**: ground-plan of a house at Mairy (Dept. Ardennes, France)

Fig. 5.11 Plan of the settlement at Berry-au-Bac (Dept. Aisne, France)

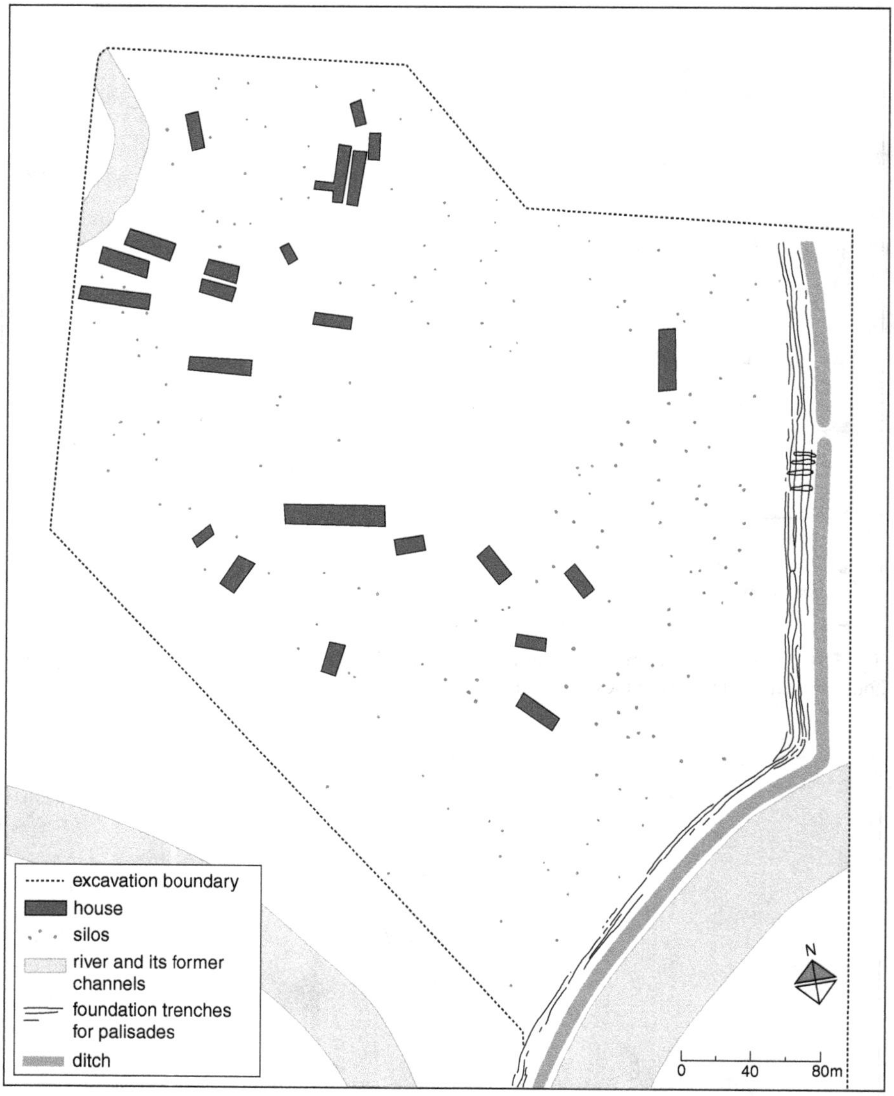

Fig. 5.12 Plan of the Michelsberg settlement at Mairy (Dept. Ardennes, France). It is not known whether all buildings are contemporaneous

holes set in a square suggest the presence of above-ground granaries, but these are not well dated and may belong to the much later Bronze Age or Iron Age occupation of the area. The area with the dwellings was enclosed, partly by an abandoned meander of the river Meuse, partly by a ditch and triple palisade. As indicated in Section 5.4, Mairy was possibly not a normal farmers' village but may have been a ceremonial site.

A more or less related type of building has been found in Lantremange, in the Belgian Hesbaye. It has been dated by thermoluminescence to the first half of the fourth millennium BC. Sherds and other artefacts, which would have helped in assigning the structure to a specific culture, are absent. The building is slightly trapezoidal, with a width of 7.7 m at one end, and 6.3 m at the other. The length is set at 21 m. It is two-aisled and comprises two rooms. The house looks like a kind of hybrid between a Villeneuve-Saint-Germain house and the Berry-au-Bac house. Traces of a second house were found nearby. Some pits were found around the structures. Although the place did not provide much litter to go by, it may have been the remnant of a farming community.

There are two more instances of houses reflecting earlier building traditions. One is situated in Dampierre-le-Château (Dept. Marne) and is built more or less in a Villeneuve-Saint-Germain tradition, though fully two-aisled. It is dated as late as the second half of the fourth millennium. The other is a 15 × 8 m, very slightly trapezoidal structure excavated in Jüchen-Belmen in the German Rhineland. It is dated to the Michelsberg culture, but reveals a Bischheim background.

The last three instances show that the style of building represents not always and everywhere a sudden break with former customs, as would be suggested by the Berry-au-Bac and Mairy structures, even though those are commonly depicted as exemplary for the period.

All houses mentioned above are of the gable-roof type. With them the list of known houses is more or less complete for the time being.

As mentioned earlier, the best-known feature of the period is the causewayed enclosure of the Michelsberg and Chasséen cultures (Fig. 5.13). Areas surrounded by ditches and palisades are known from the later phases of the LBK onwards, but in the Michelsberg and Chasséen they had their apogee. Cutting off promontories, surrounding parts of plateaus or closing off land surrounded at three sides by river bends, by palisades and/or ditches became common practice. The surface of these enclosures varied from very large, 20 ha, to small, 1 ha. Multiple interruptions of the palisades and ditches allowed access, hence their name of 'causewayed enclosures'. The ends of the ditches near the entrances are usually full of waste and ritual depositions of pots and complete skeletons. In many cases the interior is void of features and finds, but in others it reveals silos and other pits. There are also a few that show postholes, indicating that some structures had been erected inside the enclosure, but so far Mairy is the only case where true buildings could be discerned.

The largest enclosures are the ones with least finds in their interior. Their purpose is the subject of debate. Some attribute an economic use to them, as central market or a place to keep the livestock. Others see in them a ceremonial centre. Whatever they may have been, the large enclosures do not represent common farming units.

The farming population lived perhaps in the smaller enclosures or at those places where no traces of palisades or ditches have been found, only sets of pits. Why their dwellings are so elusive is not clear, but they may have had such shallow foundations that even a slight erosion of the surface has erased them.

The dwellings of the succeeding cultures are still more enigmatic, because completely unknown.

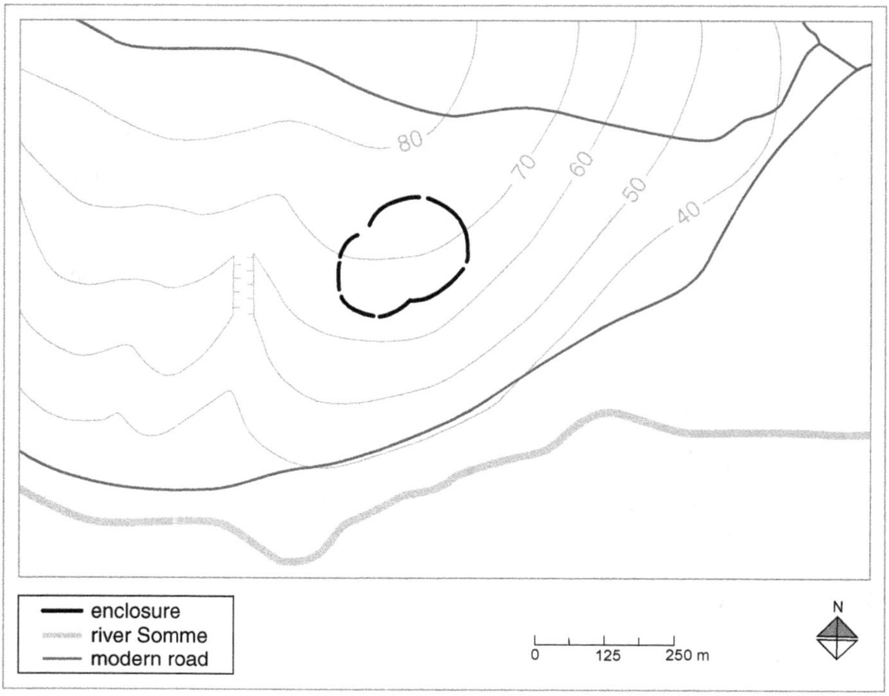

Fig. 5.13 The causewayed enclosure at L'Étoile (Dept. Somme, France)

5.6 The Farm in Its Setting

In the long period covered by this chapter, the farming communities displayed two traits, which distinguish this period from the previous ones. For the first time it is obvious that communities worked together to achieve communal goals. The second is that they swarmed all over the country, completing the 'agriculturalisation' of most of the loess region.

Working together in units going beyond the single core family or extended family or small cluster of families, is expressed in several ways. During the Michelsberg and Chasséen cultures, it was the large enclosure which, as can be concluded from their number and distribution over the landscape, played a part in the life of several hamlets. Although little is known about these hamlets, being only represented by a scattering of pits, they must have housed the main part of the population. A certain number of these hamlets shared one large enclosure, which must have been built and maintained by the joint forces of their occupants. It was obviously the focus of a territory, shared by several settlements (Fig. 5.14).

Another feature of this period is the flint mine. Although the Cerny culture already exploited flint mines, it is during the Michelsberg and Chasséen cultures that they had their heyday. Whereas in previous times flint was extracted from surface sources, flint was now obtained from its origin, namely bands of flint nodules

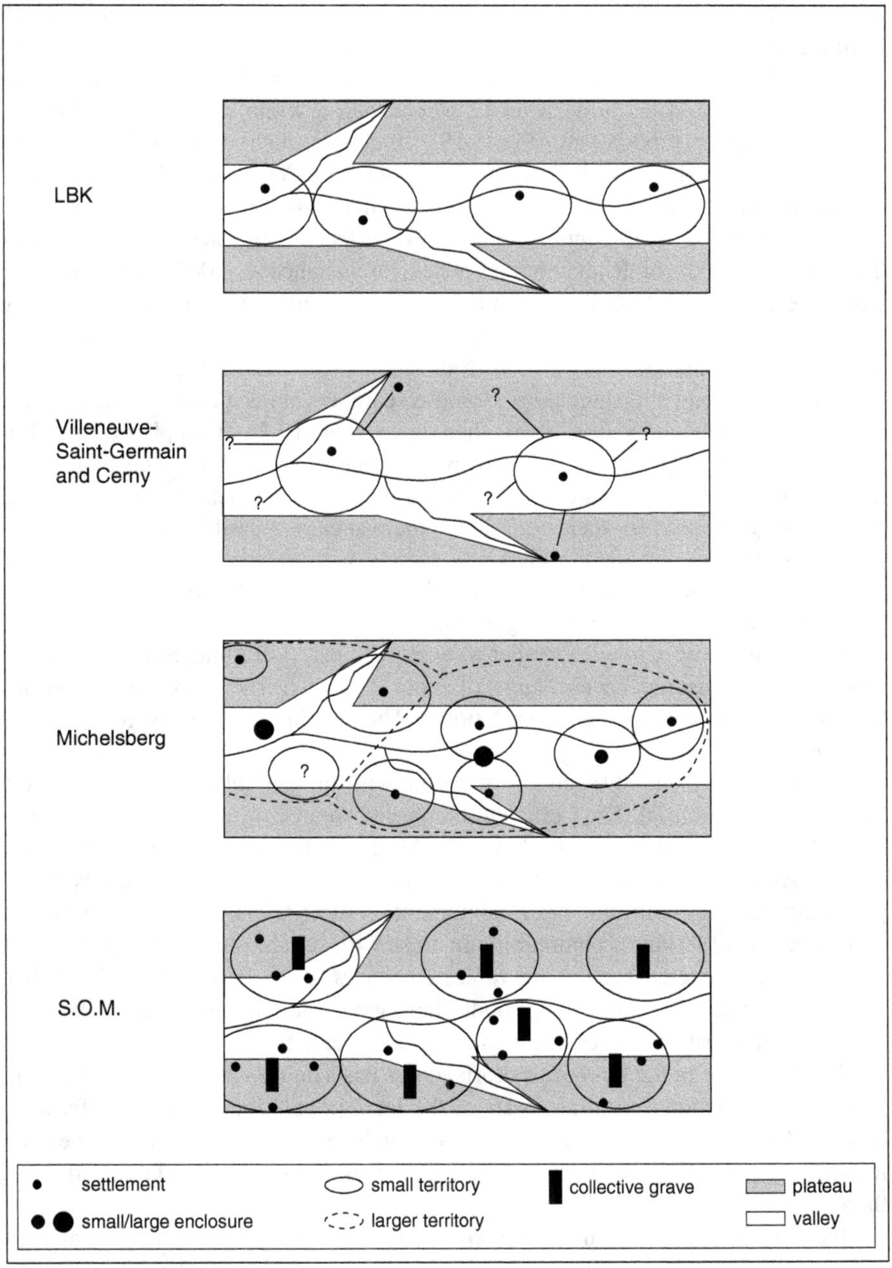

Fig. 5.14 The evolution of the settlement system in the Aisne valley

in limestone. To reach these bands, vertical shafts were hewn into the bedrock until the good-quality flint was reached. From there, horizontal galleries were hewn from which the flint nodules were extracted and hauled to the surface through the shafts. Shafts could be as deep as 12 m and had a width of 1–1.5 m. The galleries had heights of 60–80 cm (Fig. 5.15). Of course, such mines could only exist where suitable bedrock with flint was available. They occur in a limited number of mining centres, such as Rijckholt in the Netherlands, Spiennes in Belgium and Jablines in France. Such centres occupy vast surfaces with traces of many shafts. The mine complex of Rijckholt comprises for instance c. 2000 shafts. But the extensive areas with traces of mining offer a false impression of the actual size of the mines. Rijckholt was used for at least 500 years and perhaps even for 1300 years. This implies that only two to four shafts with attendant galleries were in use during one and the same year. Some researchers think that such mines were worked by people who visited them when they required a fresh supply of flint. Others maintain that they were exploited by full-time miners who traded their products for food and other things. The territorial life of the Michelsberg and Chasséen farmers, as expressed by their communal enclosures, is possibly evidence against the first view. And the second implies a society with members specialised in one single craft, which in this region is thought to have come about only much later. More in line with the level of organisation of this time is the view that a local farming community worked together to exploit 'their' flint mine, as an extra commodity which ensured an exchange of other products. There are no indications that mining flint was directed from 'above'. The farming societies were still fairly egalitarian.

The Stein group and SOM culture stand out with another collective enterprise, the place of multiple burial. The Cerny culture buried their dead, or at least their important dead, in monumental graves, but in the Michelsberg and Chasséen cultures this custom seems to have waned. With the Stein group and the SOM the custom of providing the dead with elaborate monuments returned. They took the form of a collective grave, serving a community. In areas with suitable stone, such graves had a megalithic character, such as the *allées couvertes* in the Paris Basin (Fig. 5.16). In other areas they were made of wood. These grave monuments replaced the large enclosures as territorial cores of the farming communities.

The difference in the kind of territorial cores has a cultural background. It has no roots in the economic, farming, basis of the local populations, at least as far as is known. But what the collective enterprises show is that the populations felt the need to erect structures that stood out in the environment. They underlined their presence there.

Because common settlements are hardly known, it is impossible to give an estimate of the number of people constituting a community. Nevertheless, it is perhaps not too far-fetched to think of a higher population density than before. Communities in sparsely settled areas will not have needed territorial markers. The pioneering and middle phases of the LBK, for instance, lacked them altogether.

The increase in population will also have triggered the settling of areas outside the traditional agricultural centres. We can see the first signs of this movement dur-

Fig. 5.15 Flint mine. In this case two bands of large nodules were exploited

Fig. 5.16 The megalithic monument La Pierre Turquaise near Saint-Martin-du-Tertre, (Dept. Val-d'Oise, France)

ing the Villeneuve-Saint-Germain and Cerny cultures, but the real spreading out took place in the Michelsberg-Chasséen horizon. By then the 'agriculturalisation' of the loess region was completed, at least for years to come. One restraint was still felt, however. Settlement did not occur in the middle of large plateaus without surface water. The well, and certainly the deep well was not yet part of daily life, if it existed at all.

Of course, the newly occupied land was not empty. People, living off hunting, fishing and gathering, lived here of old. How the process of 'agriculturalisation' did affect them is not yet very clear, but in one way or another the hunter-gatherer way of life changed into a farming life. It was possibly a combination of a cultural transformation and absorption of people into the farming communities, although it need not always have been a peaceful process.

And finally, the people of the western loess region did not live isolated on a cultural island. Cultural aspects such as enclosures and flintmines occurred all over Western, Central and Northern Europe. This does more than suggest supraregional contacts. The spread of certain types of well-recognisable flint, derived from a well-known mine, shows that these contacts reached distances of hundreds of kilometres. Such distances may not have been covered by individual people, but reflect long-distance communication networks.

Chapter 6
The First Millennia of Agricultural Landscape

6.1 The Original Vegetation

The arrival of the first farmers took place in a period which in the nineteenth century was given the name Atlantic Period because climatic reconstruction has shown that its climate had been maritime (influenced by the Atlantic ocean). It was the warmest period after the end of the last Ice Age, at least before the present global warming-up. In summer temperatures were c. 2°C higher than the average temperature of the twentieth century, and winters were the same or slightly milder. Summers may have been slightly drier and winters wetter (Fig. 6.1).

The natural vegetation associated with this climate was deciduous forest. The study of pollen grains, preserved in peat, and, rarely, ancient soils, revealed the main trees to be lime, elm, ash, oak, and in the eastern part of the loess region alder. Hazel, a shrub, completes the list of the main woody plants. Alder, in this case *Alnus glutinosa*, is a tree of wetlands and its stands must have covered permanently wet parts in valleys. The other species demand drier soils. In the past the four trees were thrown together to form a vegetation which went under the name *Quercetum mixtum*, mixed oak forest. But in reality such a forest never existed. These trees do not have the same requirements for optimal growing conditions. Moreover, the four trees in the pollen records comprise more than four true species. The reason is that related species share identical, or almost identical, pollen types.

'Lime' consists of two lime species, large-leaved lime (*Tilia platyphyllos*) and small-leaved lime (*Tilia cordata*). Large-leaved lime thrives on loamy soils rich or moderately rich in nutrients. Small-leaved lime is less demanding where soil quality is concerned, although it does not grow on poor soils. Both avoid waterlogged grounds. What the two have in common is that their seedlings do not require much daylight to survive. If the canopy of a lime forest happens to offer an opening, the seedlings are ready to shoot up, and this shooting up goes quickly. The trees reach great heights and the canopies cast much shade. The effect is that lime forests rejuvenate themselves, in the mean time eliminating tree species requiring more light.

A tree with light-demanding seedlings is oak. Again, the oak from the pollen records represents two species, pedunculate oak (*Quercus robur*) and sessile oak (*Quercus petraea*). The first grows on all kinds of soils, except for completely waterlogged ones. The second prefers loamy soils, though not calcareous, and, also not

C.C. Bakels, *The Western European Loess Belt*, DOI 10.1007/978-1-4020-9840-6_6,

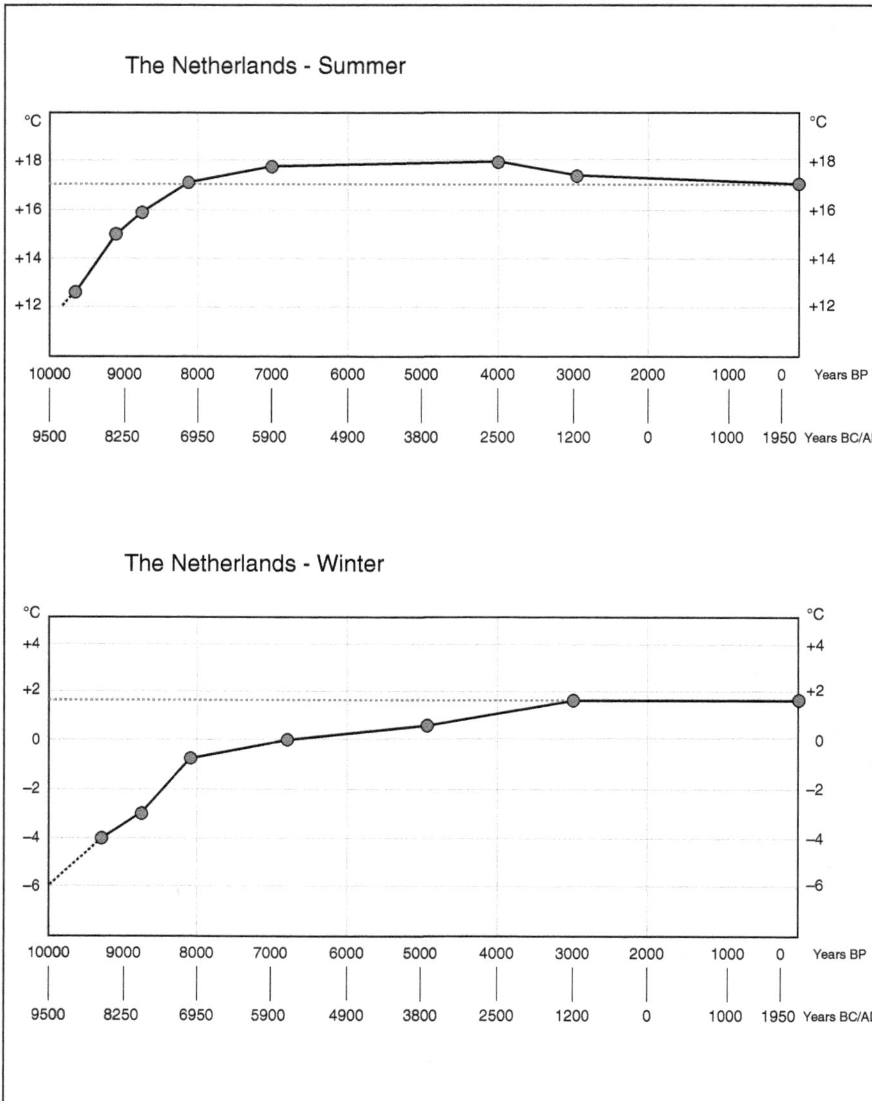

Fig. 6.1 Mean temperatures during the Holocene in the Netherlands

water-logged. Both species may have occurred in the loess region, though not in exactly the same stands. As mentioned, their seedlings require light to survive. They are also slower in their growth than seedlings of lime.

Elm pollen may have been released by three species. Smooth-leaved elm (*Ulmus minor*) was presumably the most common. The tree requires calcareous, nutrient-rich soils and tolerates temporary flooding. The second species is wych elm (*Ulmus glabra*), which can be found on relatively steep slopes. The last is fluttering elm

(*Ulmus laevis*), which resembles smooth-leaved elm in its requirements, although it tolerates soils less rich in nutrients. It is found mostly on damp grounds in valleys. All three are half-shade trees. They need less light than oak and ash, but more than lime. Their seedlings tolerate shade for longer periods.

Ash pollen, finally, represent a single species, ash (*Fraxinus excelsior*). It is the most demanding of the four main 'trees'. Ash requires in general moist, but not waterlogged, soils, but a high content of calcium can compensate for moisture. The tree tolerates temporary flooding. Seedlings can survive for a long time in the shade, but growing into tall trees requires light. If an opening-up of the canopy gives ash a chance, the tree grows fast.

In view of the requirements of all trees described above, it comes as no surprise that reconstructions of the Atlantic Period forest stock the loess plateaus with limes. This view is supported by the fact that within one and the same small subregion, percentages of lime pollen are higher in samples taken close to the plateau than in samples taken at some distance, for instance in the middle of a wider valley (Fig. 6.2). Such lime forests present a mosaic of trees of different age classes. They are not close-canopy forests with trees of the same age, looking like a hall with columns (the trunks) supporting a roof (the canopy) as has sometimes been put forward.

On moist soils, at the foot of slopes and especially on valley bottoms, elm was presumably an important component of the forest, as long as the water table was not permanently so high as to lead to wetland conditions with stagnant water.

The wet parts were in the eastern part of the loess region and in the Belgian Hainaut already covered by alder carr. In northern France (the Paris Basin), alder seems not yet to have colonised such grounds. The alder pollen percentages are still low there. The tree is such a good pollen releaser that, if alder carr was present, it should have been detected. What was growing on waterlogged soils in the Paris Basin is unclear, but in any case such wetland areas were very small because the main parts of the valley bottoms did not know stagnant water conditions at all (see Chapter 2).

Ash was not very important, but the 'tree-species' oak and the shrub hazel (*Corylus avellana*) are present with high percentages in the pollen records. Both require light. Hazel can survive in not too dark forests, but its stands cannot rejuvenate there and they do not flower, thus producing no pollen. Hazel is actually a shrub of forest edges. The pollen records suggest that growing conditions for oak and hazel were mainly present on lower parts of slopes and on valley floors. Nevertheless, it cannot be excluded that small areas with non-closed canopies and clearances were present on the plateaus as well. Some authors, first and foremost F. Vera, have introduced the idea that the Atlantic forest was never a vast forest with large stretches of closed canopies. The presence of large mammals, with their browsing and grazing, would have prevented this state. It is still open to debate whether the uplands were covered with dense lime forests, or with a patchwork of thicker and thinner stands of trees, possibly even interspersed with a considerable number of clearances. Small clearances in upland forests cannot easily be detected by pollen analysis. The pollen released by the herbs growing there will not reach the upper air to be transported

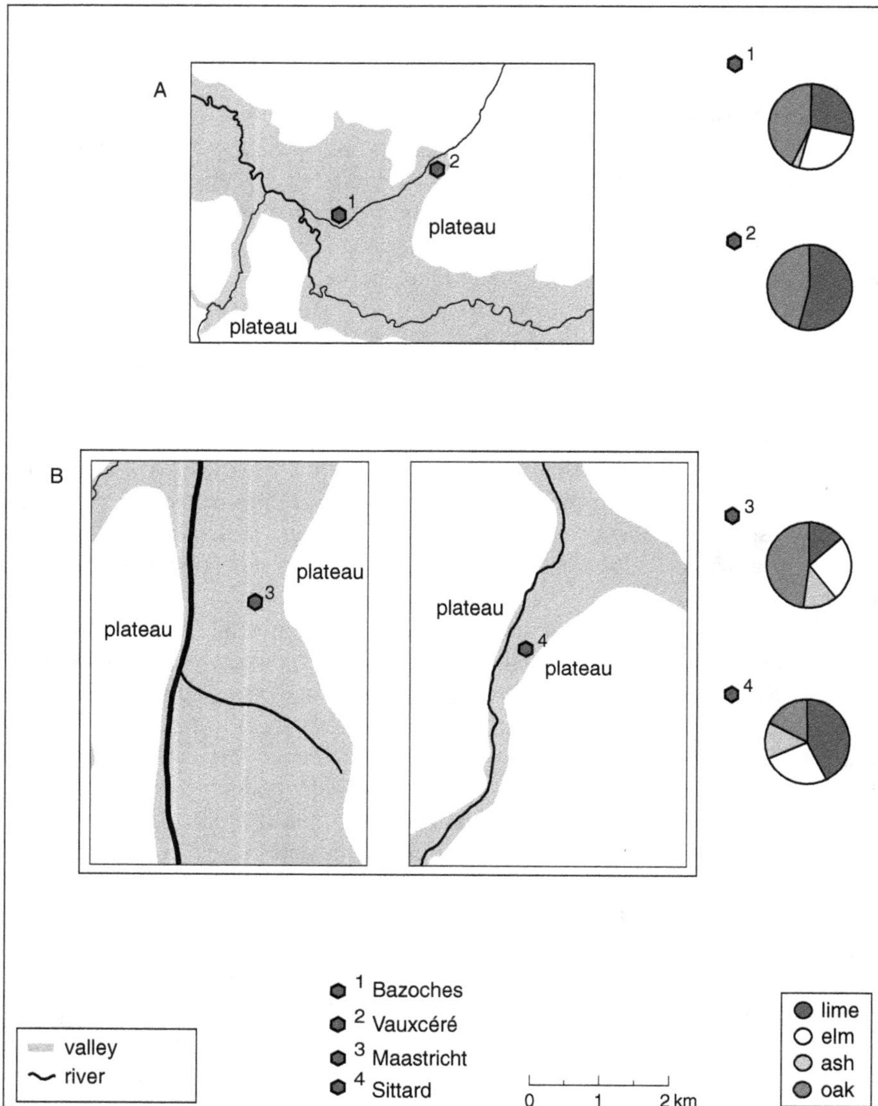

Fig. 6.2 Two instances of pollen data from, respectively, a wide and a narrow valley. **A**: The valley of the river Vesle (Dept. Aisne, France). **B**: The valley of the river Meuse (*left*) and its tributary Geleen (*right*), both in the Netherlands

to the waterlogged areas where they may land again and be preserved and detected later. They will be trapped by the taller surrounding vegetation. Therefore, the higher parts of the region may have carried a more varied plant life than was formerly supposed to have been present there. Nevertheless, notwithstanding this new view on the upland vegetation, the main diversity in the vegetation seems to have been present in the valleys of rivers and streams.

The elm forest of the valleys was certainly interspersed with stands of oak. A few herb pollen indicate the presence of open space. They include ribwort plantain (*Plantago lanceolata*), members of the goosefoot family (Chenopodiaceae) and wormwood (*Artemisia*). Such open spaces were bordered by a shrub vegetation, in which hazel may have played a major role. These open spaces may have had several causes, which need not have been independent. Firstly, a river valley offers a more dynamic environment than a plateau. Occasional flooding and changes in watercourses create diversity. Secondly, water attracts animals. With their trampling and browsing they maintain clearances. Special water-side clearances are, moreover, due to beavers. In the third place, hunters and gatherers may have cleared parts of the valley vegetation. There are no indications that humans cleared large stretches of vegetation by burning, either on the upland or elsewhere. The custom to maintain large areas free of tall vegetation to provide better grazing for their prey has not been detected in the loess region, but hunters and gatherers may very well have created some open space, although archaeological remains of their camps are scarce.

Taking all arguments together, the setting in which the first farmers arrived was one of mainly lime forest on the plateaus and a more diversified deciduous forest, interspersed with clearances, on lower slopes and valley floors.

6.2 The Impact of the Farming Communities on the Vegetation

The first farmers settled on the edge of plateaus or on gentle slopes, at least in most parts of the loess region. In the Paris Basin, they preferred to live on the higher parts of valley bottoms (see Section 3.6). The starting point may have been an already existing clearance. But the building of houses called for the felling of trees. Moreover, the only source of fuel was wood and cereals do not grow under trees. The livestock may have been fed with leaves and thin branches in addition to what the animals found for themselves. It was unavoidable that the original vegetation began to be affected by the activities of the new settlers. Moreover, the sites were permanent (see Sections 3.3 and 3.6) and caused continuous pressure on their immediate surroundings.

Changes in the composition of the vegetation are detected by the study of pollen and charcoal. Pollen diagrams show that the percentages of elm and lime decrease. For instance, the foundation of two LBK settlements opposite each other on either side of the rivulet Kleine Gete in Belgium was followed by a conspicuous decrease in elm. The inhabitants of the twin-settlements must have exploited the narrow valley to a large extent. The elm decline was followed by a decline of lime. Herb pollen, originating from open spaces on drier grounds, is still absent, but the rise in ash proves that the forest became lighter. The percentage of oak pollen rises as well, either because the absolute number of trees rose, or because the already existing trees did better and produced more pollen, or simply because the lower number of elm and lime pollen automatically made oak percentages rise. Anyhow, oak stands seem to have been spared large-scale damage, notwithstanding the fact that the main source of timber was oak. Refraining from causing damage to oak may have been a deliberate act, and cutting down oak may have been a restricted activity. It may even be that the value of acorns as food for pigs was playing a role. Hazel reacted in the same way as oak. Taking all together, the rise in ash, oak and hazel points towards an opening-up of the forest (Fig. 6.3).

A similar trend is seen everywhere where people chose to settle. The reverse is also seen. When a settlement was abandoned elm and lime percentages rose again. Nevertheless, in the long run dense forests disappeared in areas favourable to settlement.

The advance in vegetation of a forest-edge nature is best seen in the charcoal debris found in settlements. Such debris are the witnesses of hearth fires. The wood burnt in such domestic fires is commonly gathered in the vicinity of human dwellings. It reflects, to a certain extent, the woody vegetation of the near surroundings. Although charcoal spectra do not represent the entire vegetation of that time, they represent those kinds of wood that have good burning qualities. Charcoal, dated to a first phase of settlement, is characterised by a large proportion of forest tree wood, i.e. oak and elm, but also ash and some wood of the apple/cherry/sloe group (the wood of this group cannot always be readily distinguished). The latter group is obtained from forest edges. In later phases, proportionally more wood of this group is found. The share of hazel also increases. New members of forest-edge vegetation turn up, such as maple (*Acer campestre*). This implies that the forest edge gained importance. Some authors even suggest that this development was not entirely a result of more and larger clearances, but also of the deliberate promoting of such vegetations as semi-natural hedges to protect crops from domestic or wild animals.

The development is not restricted to the impact of the very first farming groups. Everywhere where new settlers arrived the effect on the vegetation was the same.

What lacks in the charcoal records is lime, and, to a lesser extent, alder. Alder grew presumably in spots that were not regularly visited to collect firewood. The absence of lime is conspicuous, but its wood is not a good fire wood (Fig. 6.4).

The impact of man on stands of trees can also be followed through the analysis of tree rings. A study of waterlogged oak trunks, found in wet sediments in the valleys of northern France, revealed that this tree was indeed valued, at least insofar as oak stands were given chances for re-growth. Oak, dating from before the arrival of the first farmers, presents a natural composition regarding age and trunk diameter. Oak, contemporaneous with the earliest settlers, is unfortunately still absent from the records, but trunks dating from a period between 4300 BC and 3600 BC are available and present a different picture. An important part came from trees that had known a rapid growth, because their radius has increased with c. 5 mm per year, whilst the remainder had gained only c. 0.7 mm per year. The fast growers must have grown up in clearances presumably related to settlements. The slow growers stood in more natural stands.

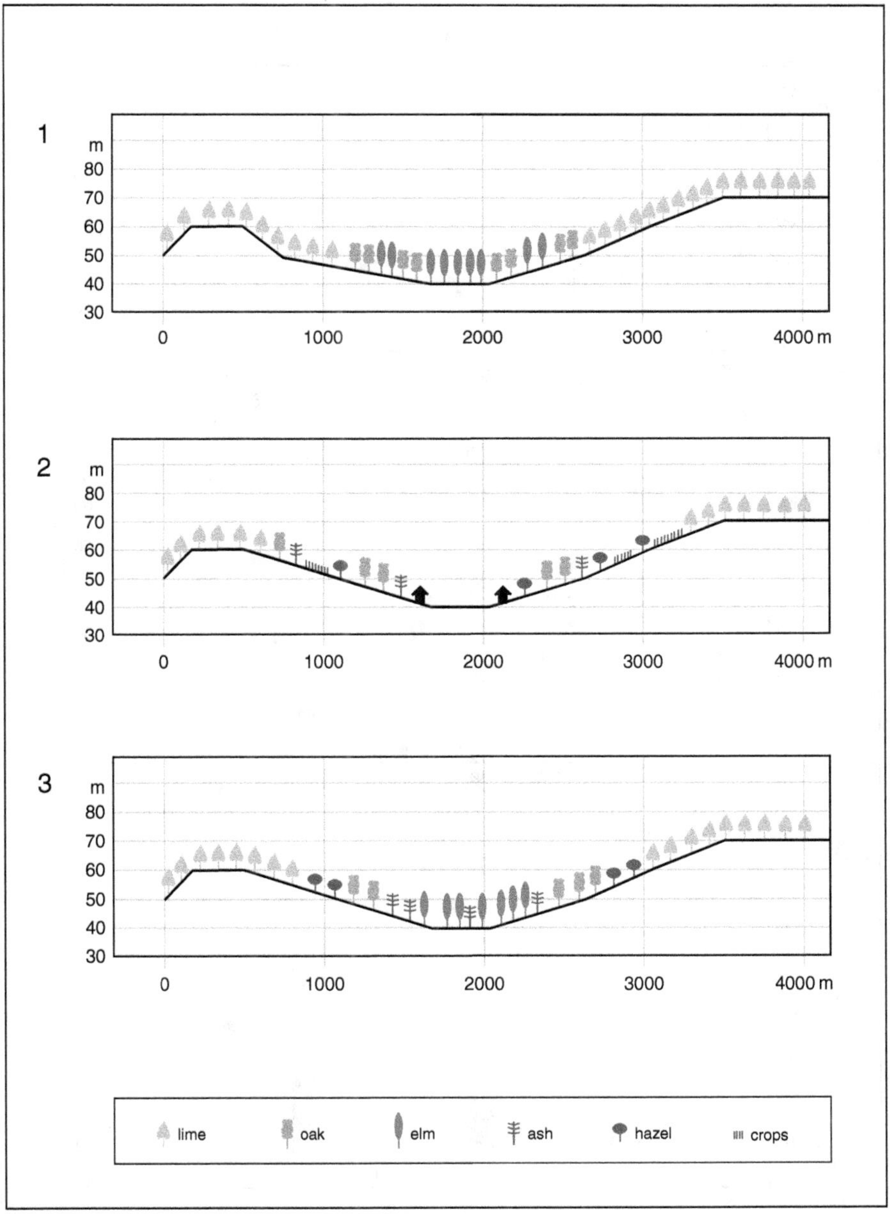

Fig. 6.3 Reconstruction of the vegetation around the stream Kleine Gete and the LBK settlements Overhespen and Wange (Belgium), before (**1**), during (**2**) and after (**3**) occupation. See also Fig. 3.4

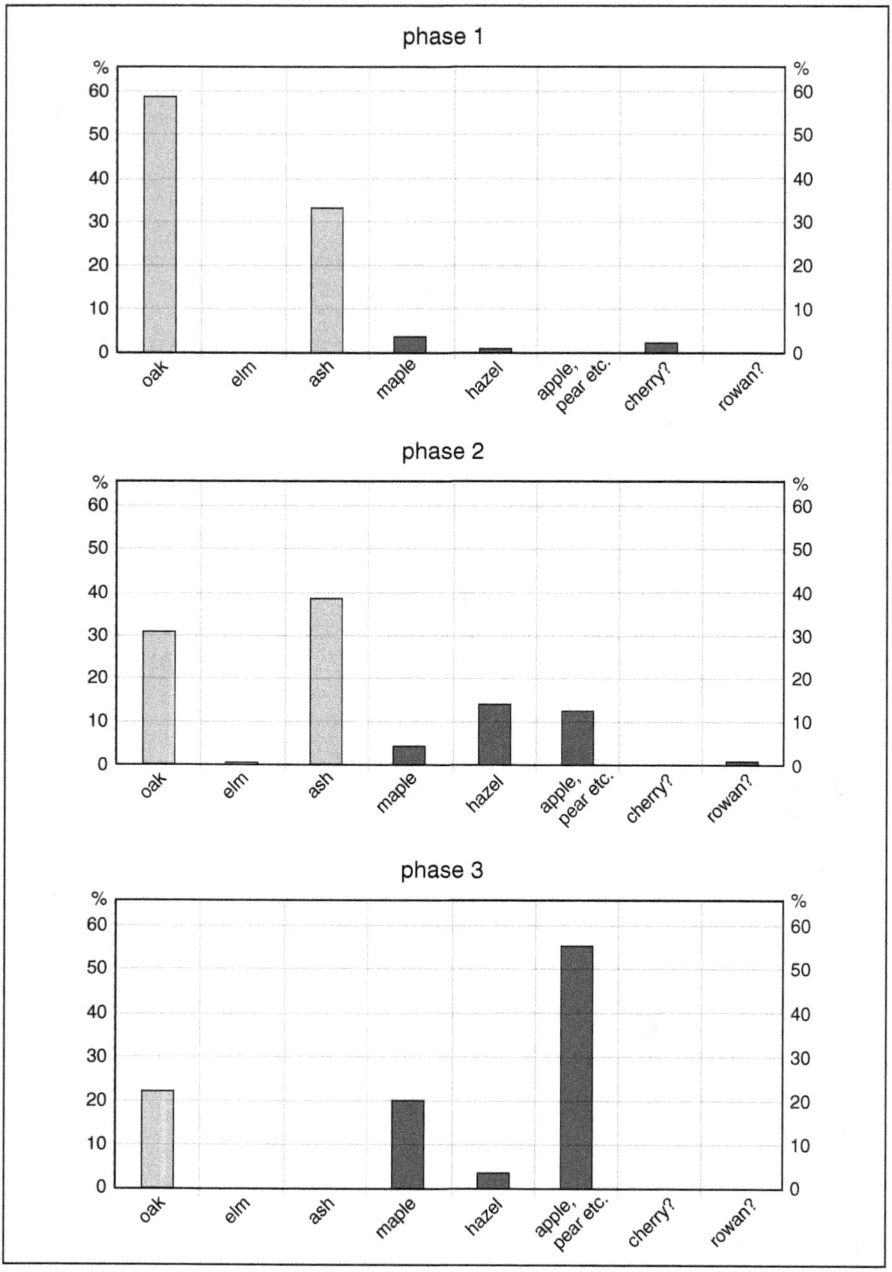

Fig. 6.4 The composition of charcoal retrieved from hearths in the LBK settlement Altwies, Luxemburg. Phase 1 is the first phase of occupation, etc

A similar study reveals that people belonging to the Chasséen culture felled oaks not just every year, but at certain intervals. In the site of Paris-Bercy, on the bank of the river Seine, four phases were detected. The first occurred around 3915 BC, the second around 3890 BC, the third around 3850 BC, and the last around 3795 BC (dates obtained through dendrochronology). The first phase concerned trees that had grown up, according to the width of their rings, in an open area, in a clearance dating from before the founding of the Chasséen settlement. Chasséen people obviously chose to settle in an area with a history of clearance, natural or man-made. The two following phases concerned trees from a forest environment with slow growths, presumably further away from the site. During the last phase, trees which were all of the same age were felled. These had known a very rapid growth during the 120 years of their life and their trunks had reached diameters of c. 50 cm. Such wood must have been obtained from a clearance that had been abandoned 120 years earlier, and allowed to revert to forest. The study shows that oak was required for special purposes, perhaps the renewal of structures. It reveals as well that secondary forest existed with trees of the same age.

What has become known about changes in the vegetation concerns presumably only the immediate environment of settlements. Seen are instances of secondary forest and forest-edge vegetation. In the beginning such changes were restricted to the edge of plateaus, slopes and the drier part of valley bottoms. As settlement proceeded to areas at larger distances from the first centres of human activity, the impact on the original forest will have spread from the edges of plateaus towards the interior. But, as put forward in Section 5.6, the very centres of large plateaus remained untouched.

The studies mentioned above demonstrate that the arrival of farming populations did affect the composition of the original forest. But human impact may not have been the only factor. For instance, vegetation is dependent on climate as well. During the period under review, 5300 BC–2650 BC, the climate did indeed change. The first part fell in the climatic period known as the Atlantic Period, as mentioned at the beginning of this chapter. During this period the climate underwent some minor oscillations. One of these has been mentioned in Section 4.2, where it was commented on in connection with a change in crop choice. But in general the oscillations are not considered to have had a real influence on the natural vegetation. The Atlantic Period ended around 3900 BC with a more important temporary change towards a colder climate. But this colder interlude seems to have had no measurable effect on the loess region. The climatic period that follows this colder period is called the Subboreal Period. Its climate has much in common with the Atlantic Period, although it is said to have been more continental with more pronounced oscillations. The continental aspect is, however, hardly detectable in the region under review and how far oscillations affected the general vegetation in the loess region is difficult to assess, because human interference was relatively strong near those places where climate can be studied with pollen analysis. This fact may have masked effects caused by climate.

Another factor affecting vegetation is the quality of the soil. Soil deterioration, especially a gradual loss of nutrients and natural acidification, causes changes in the natural vegetation. Again, such effects have not been recorded for the period under review. But one aspect of human impact will be considered next.

6.3 Erosion

Over the millennia, the surface occupied by open space is considered as slowly but constantly increasing, albeit with ups and downs. Some of this process is visible in pollen diagrams. Weed pollen and even occasionally grains of cereal pollen hint at the existence of fields. The presence of fields, though of unknown size, is moreover proven by finds of their products, i.e. crops in human settlements. One of the properties of fields is that their surface lies bare during longer or shorter times. Bare soils may lead to erosion and true loess is very susceptible to erosion. Very gentle slopes are already unstable. It comes therefore as no surprise that erosion already took place during LBK times. However, this erosion was very restricted, and the worst erosion was not triggered by raising crops, but by trampling surfaces in and around farm yards. Every pit in a settlement shows some fill caused by erosion of the topsoil. Depressions in or close to the settlements were also filled in with displaced sediment. This was not only the case in settlements founded on the edges of plateaus, but also in valley bottom sites.

The process became more serious with time. The construction and maintenance of large installations, such as the causewayed enclosures of the Michelsberg and Chasséen cultures, triggered erosion as well. Their long stretches of palisades required large-scale cutting of wood and their surface was presumably devoid of a closed vegetation cover. Bands of loam, washed down from these areas, are seen in the sediments of adjacent valleys. Nevertheless, erosion was not yet a large-scale affair. The severe erosion of the loess region came much later (see Chapter 10).

Chapter 7
Towards a More Complex Society: 2650 BC–50 BC

7.1 The So-Called Metal Ages

The farming communities described in Chapters 3, 4, and 5 lived in an age when metal was still unknown. Vessels were made of clay, wood or bark. The last two were only preserved under waterlogged conditions and are therefore very little known from the loess region, but they must have been common. Implements with cutting edges were made of stone (especially flint), bone or antler. However, in the SOM culture mentioned in Chapter 5 the first objects appear made of a metal, namely copper. These objects are jewellery, such as beads, and are considered to represent luxury products, not intended for daily use in households and on farms. But, as a matter of fact, not much is known of SOM daily life, because this culture, and the contemporaneous Stein group, are mainly known through burials (see Chapter 5).

Even less is known of the populations that followed. In the southern part of the region, cultural groups going under local names such as the Groupe de Gord have been described, but their characteristics remain hazy. Elsewhere such local groups seem to have been absent. Rather soon they were followed by far from local cultures of which the first goes under the name of Beaker culture and the next under the name of Bell Beaker culture. Both names are derived from certain types of pottery. In the loess region, their story is also mainly a story of the dead. Grave monuments dominate the records. Rare clusters of pits, filled with domestic waste, are usually the only witnesses of settlements, although some house-plans are attributed to those cultures as well. Their implements were still made of stone, bone and antler. The metals of the time are gold, used for ornaments, and copper, used for jewellery and implements like daggers and axe blades.

Pure copper is soft and the casting of pure copper tends to produce porous metal which requires much hammering to obtain serviceable objects. Mixing in other minerals or metals, such as arsenic, tin or lead, improves the quality. Such additives are sometimes already present in the original copper ore. During a long period, from c. 2500 until 1800 BC–1700 BC, and all over Europe, people experimented with mixing metals. In the end a mixture of copper and tin, called bronze, became the established alloy. Bronze remained the main raw material for implements until c. 800 BC, when iron took over.

C.C. Bakels, *The Western European Loess Belt*, DOI 10.1007/978-1-4020-9840-6_7,

Iron melts at a higher temperature than copper and the extraction of the metal from the ore requires a higher level of technological skill using fire. The first iron objects turn up before 800 BC, but these served, just as the first copper objects, for making ornaments. The shift from bronze to iron as material for the manufacture of common implements was a gradual process.

It is customary to call the first stage of this metal-using age the Copper Age or Chalcolithic. Next comes the Bronze Age, which in turn is followed by the Iron Age. Still, these are only names to serve as subdivisions of one long continuum. The Beaker and Bell Beaker cultures belong to the Copper Age, although some archaeologists place them at the very end of the Stone Age. The cultures in the Bronze Age are simply called Early, Middle or Late Bronze Age. The Iron Age is split up into the Early, Middle and Late Iron Age, or, in France, the First and Second Iron Age. The gradual transition from Bronze Age to Iron Age is reflected by the fact that the Late Bronze Age and Early Iron Age are often lumped together under the name Hallstatt period. The Middle and Late Iron Age may be combined as well, and are referred to as the La Tène period. The people living in the region during the latter period have become known as the Celts. The time of the Metal Ages ends with the conquest of the region by the Roman army.

7.2 Crops

The crops of the early stages of the period are not very well known. Some Bell Beaker pits, excavated in the area around the Moselle river in France, revealed the presence of some hulled wheat, presumably emmer. A site, attributed to a local variety of the Groupe de Gord in the Somme area revealed emmer wheat, einkorn wheat, a naked wheat and both naked and hulled barley. Lakeside settlements in the Alps show that crop growing was a normal part of the activities of contemporaneous Copper Age societies, proof that the art of crop cultivation was not lost. The lack of information in the loess region is merely due to a lack of suitable archaeological deposits.

The Early Bronze Age, 1800 BC–1500 BC is better known, but the list of crops comprises, surprisingly, only cereals. The presence of emmer wheat, spelt wheat, multirowed naked barley and broomcorn millet has been established. It is assumed that pulses and oil seeds were grown as well, but so far there is no trace of them.

The centuries between 1500 BC and 1100 BC bring more species: einkorn wheat, emmer wheat, a naked wheat (species not specified), spelt wheat, multirowed naked barley, multirowed hulled barley, broomcorn millet, pea, lentil, and horsebean. In addition to these ten crop plants, rye has been mentioned. However, this cereal was only found in two sites and moreover with two grains only and should not be considered as a crop plant. Rye was an arable weed long before it became cultivated.

Plant remains dating from the Late Bronze Age, 1100 BC–800 BC, add foxtail millet, bitter vetch, linseed, poppy and gold of pleasure to the list. Linseed and poppy were possibly not a truly new addition to the range of crops grown, because they had already been cultivated in the region in a distant past (see Chapters 3, 4, and 5). Nevertheless, the crops may have been absent for a time and then reintroduced.

The wide range of products is still added to during the next stage, 800 BC–50 BC, when oat, hemp and at the very end, common vetch appear. We can conclude therefore that the Bronze and Iron Age farmers knew a much larger range of crop plants than their predecessors. Where the Michelsberg and Chasséen farmers, for instance, could choose between three or four wheats (depending on whether they grew tetraploid or hexaploid naked wheat, or both), two barleys, one millet, one pulse and two oil seeds, the farmers discussed in this chapter could choose in the end between four or five wheats, two barleys, oat, two millets, five pulses, and four oil seeds. The list has grown from 9 or 10 crop plants to 18 or 19 at the end of the Iron Age. Most of the increase took place after c. 1100 BC (Fig. 7.1). The observation that the increase sets in around 1100 BC and not at 1800 BC, when records begin to be available after a time gap of over a millennium, allows for some confidence that the rise is a true rise and not just the result of missing records in a steady increase in crop species since the introduction of the first crops. Nevertheless, the absence of pulses and oil seeds in the period between 1800 BC and 1100 BC requires some caution. A chance remains that some of the new crops were introduced earlier than one is aware of at present.

The first newcomer is spelt wheat, a hulled hexaploid wheat. It is considered to be a hybrid between a tetraploid wheat of the emmer/durum group and the diploid wild grass *Aegilops squarrosa. Aegilops squarrosa* is common in northern Iran and adjacent Transcaucasia, Transcaspia and Afghanistan, where it occupies a broad range of habitats. It is a successful coloniser of secondary, man-made habitats and a common weed in cereal fields. In this scenario, spelt wheat is considered to trace its origin back to a natural hybridisation in a tetraploid wheat field within the natural range of *Aegilops squarrosa*. From its Asian origin it spread, in one way or another, to Europe. Recently, however, a second possibility has been proposed: a hybridisation of emmer and bread wheat. If this is indeed the case, spelt wheat could have had a second place of origin, namely in Europe.

Broomcorn millet, already encountered sporadically in Michelsberg contexts but now one of the main cereals, is an old crop with an origin in Central Asia. It is thought to have been developed in a centre of domestication that was independent from the centre in the Near East. Likewise, foxtail millet does not seem to belong to the Near Eastern assemblage either. Its place of origin is not known, but must lie somewhere in Central Asia as well.

Oat started not as a crop, but as a weed in fields of wheat and barley. In a rather late stage of prehistory, oat was picked up and sown intentionally as a crop. The cultivar, *Avena sativa*, appears first in Europe and is found from the second millennium BC onwards.

The horsebean, also known as the celtic bean, stands out because its wild ancestor is still unknown. The beginnings of its domestication are also unclear. During the third millennium BC the bean appears, rather suddenly, as a main crop in the Mediterranean and in Central Europe.

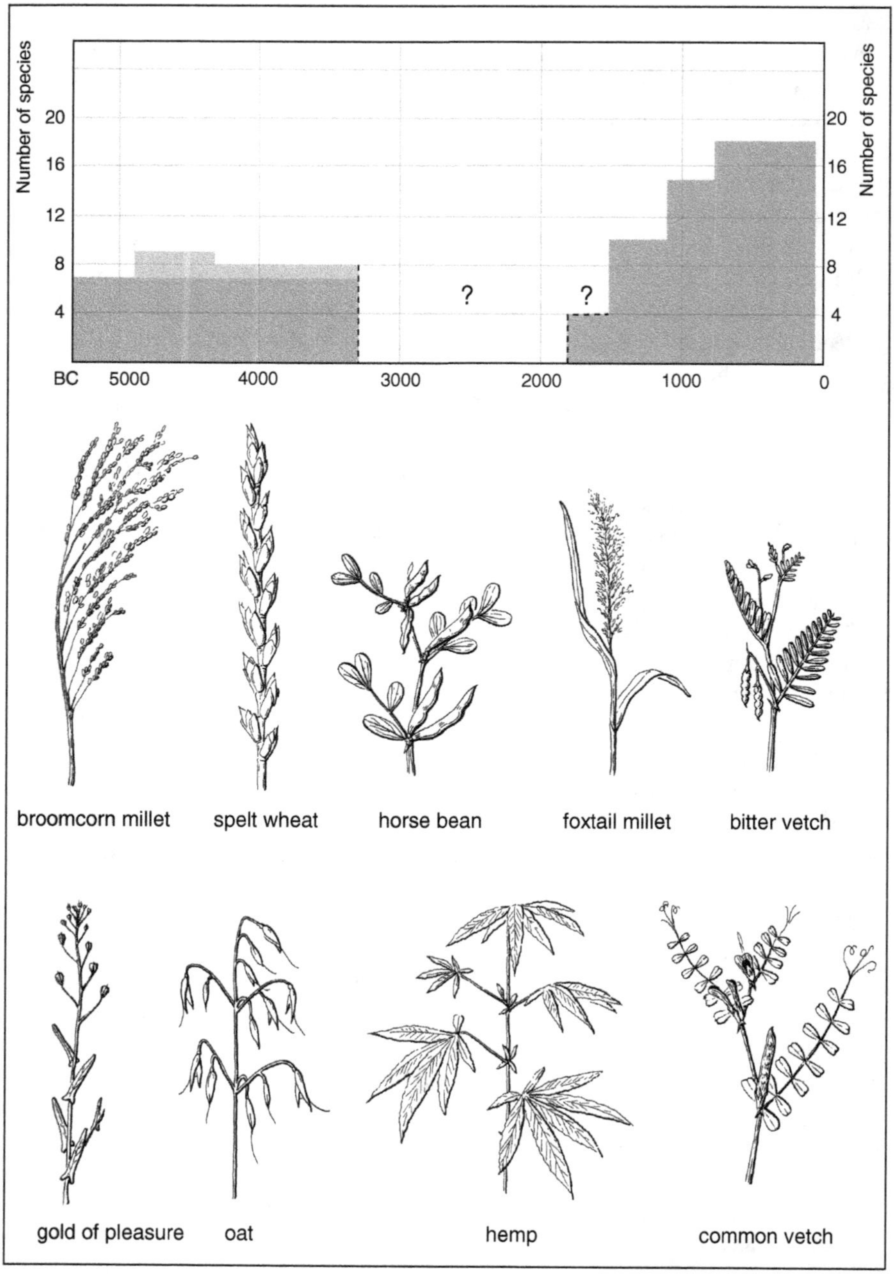

Fig. 7.1 The increase in crop species since the introduction of the first crops, and pictures of the nine newcomers

In contrast to the five newcomers mentioned so far, bitter vetch belongs to the well-known Near Eastern centre of domestication, though not exactly to its core area, which is the area bordering the valleys of the rivers Euphrates and Tigris and the place from where most early crop plants came (see also Chapter 3). Its origin is assumed to lie in Anatolia, Turkey. This vetch was cultivated of old in Greece and the Balkans, but reached the loess region west of the Rhine obviously not in the company of the other Near Eastern crop plants. It arrived much later. The reason for this tardiness is not well understood. It may have something to do with the use of this plant. Bitter vetch as such is toxic for humans and the pulse has to be leached in water to allow consumption without any grave health consequences. It is also toxic for monogastric animals, but not for ruminants, such as cattle and sheep, although it is not advisable to include more than 25% of bitter vetch in their daily feed ration.

Common vetch is another pulse that, for the same reasons, is not very attractive for human consumption. Today it is exclusively used as animal feed for cattle and sheep. Its origin lies probably in the Near East.

Another newcomer, gold of pleasure, has ancestors in the steppes of Eastern Europe and Asia. The wild plant has developed weedy forms that thrive in cereal and flax fields. The linicola form (the weed in flax) is a well-known example of plant mimicry, because the plant mimics the morphology of the flax plant rather well. It has developed into a weed so much adapted to flax, that the linicola form is now exclusively found in flax fields. Gold of pleasure seeds are rich in oil and one of the weedy gold of pleasure forms obviously developed into an appreciated oil seed crop, presumably somewhere in Eastern Europe.

The history of hemp can be traced back to China, where hemp was already cultivated around 2500 BC. Archaeobotanical records of the plant are mentioned for the European Iron Age, but in the loess region west of the Rhine serious hemp cultivation seems to have begun in the Roman Period. The seeds are used to extract oil and the fibres to make rope, string, thread, and cloth. A piece of string has been found in a first century BC context in France. Hemp seed is also used as animal feed, for instance for poultry, and is sometimes eaten by man, too. Its use as a drug is not easy to prove, but not unthinkable.

From the above-mentioned information it becomes clear that seven of the nine new crop plants do not belong to the assemblage of plants with origins in the Near East, where all plants cultivated so far came from, with the exception of the poppy. The range of crops has been widened by an influx of plants with roots in a more eastern, Central Asian, direction. Some have even been developed in Europe. The background of the steady influx of new crops may be connected to the broadening of the world in which society functioned. The fact that copper and tin are only to be found in distinct localities in Europe, which do not necessarily lie within reach of the average community, may have been one of the stimuli to widen social networks in which not only metal, but also other products were encountered, and adopted.

Not every crop developed into a main crop. The hulled wheats emmer and spelt were staple foods. Einkorn wheat lost importance, although it still appears in places as a true, separate, crop. Naked wheat was a main crop as well, but in most cases it is not known whether this wheat is the hexaploid bread wheat or the tetraploid macaroni wheat. In those cases where rachis remains were present (the only fully identifiable remains), the wheat could be identified as bread wheat. Multirowed barley was as important as wheat, but the preferred variety changed from naked to hulled. This change is observed all over Europe. It is surprising that, whilst naked wheat slowly but steadily replaces hulled wheat, in the case of barley the trend is reversed. Where easier threshing is looked for in wheats, a more difficult threshing seems not to matter for barley. One of the possible reasons is that the objective of barley growing changed. It may have been more and more destined to feed livestock. Better yields, or better resistance to diseases, may be others.

Broomcorn millet was a staple crop as well, but foxtail millet was never more than a very minor crop. Oat was in this period minor too. As far as the pulses are concerned, pea and horsebean were common products. Lentil was especially common in the southern part of the loess region and bitter vetch shared this pattern of distribution. Bitter vetch is hardly found in the German Rhineland, for instance. The culture of both crops may have been inhibited by climatic factors, because they may have reached the northern limit of their ecological tolerance in the region. Common vetch appears, as indicated above, only at the very end of the period and its importance is hard to assess.

The importance of the oil seeds (and flax and hemp also as fibre plants), is difficult to assess, because they suffer from underrepresentation in the records. But as they are regularly present in settlements, they must definitely have had a share in the crops commonly grown.

There remains the question whether all crops were destined for humans. The vetches, and possibly part of the hulled barley, may have been grown as animal feed.

One type of plant has not been mentioned so far in this book: kitchen herbs. In earlier periods they are obviously absent, but in the last centuries BC they turn up sporadically. Fennel seed has been reported from a settlement dated between 450 BC and 250 BC, and celery seed was found in a site from 250 BC to 50 BC. Nevertheless, this category of plants will be dealt with in Chapter 8 as they belong to the world of the Romans, but it is important to note that such species were already known earlier. Whether they were imported as seeds, for flavour or as medicine, or grown locally, is not yet clear.

7.3 Crop Cultivation

It may be safely assumed that after 3500 years of tilling the land, hardly any primeval forest was left (see Chapter 6). Only in the centre of vast plateaus, far from open water, may remnants of this forest have persisted. How much land was covered with secondary forest depends on the system of crop cultivation and the population density. The two factors are not independent.

The settlement pattern of the farming population, described in Chapters 3 and 4, and, as far as is known, the settlement pattern of the population described in

Chapter 5 as well, suggests a land use in which societies operated from long-living hamlets and tilled more or less continuously garden-size plots in their near surroundings. This kind of settlement and land use spread from the edge of plateaus or valley floors to more 'upland' settings, provided there was open water available. It is not known what happened between that period and the one discussed in the present chapter. What is clear is that from about 1800 BC the settlement patttern had undergone a substantial change.

Farmers no longer lived in small clusters of farms but on single farmsteads. Moreover, the duration of the occupation was short. It is said that the house lasted for one or two generations, and with a generation set at c. 30 years, this would mean an occupation between 30 and 60 years. The next generation rebuilt the farm in another location, though still within their own territory, and marked as such by their places of burial (see Section 7.6). On the basis of settlement evidence, it may be assumed that the land use system was mobile. After several years fields were given over to a long fallow period, and when the near surroundings of the farm could not provide new space for fields anymore, the complete farmstead was moved to a new place. This system of land use has been described by some authors as a kind of shifting cultivation. Some archaeologists ascribe the new settlement system to the limited durability of the buildings, but why these should have been less durable than the buildings of the Stone Age remains unexplained. Why new structures could not have been erected next to the old, decrepit, ones is also unexplained. Others ascribe the wandering to recurring exhaustion of soils. After several millennia of agriculture, with some but presumably not sufficient manuring, exhaustion of soils, tilled as garden-like plots near the house, may indeed have been a factor. But the distance between the old and new farm buildings is not all that large. There must therefore have been another factor guiding the behaviour of the farming population, probably a cultural (religious?) one.

Structures were rebuilt at distances of only several hundreds of metres and this within territories of 30–100 ha. At least, such was the case during most of the period. The distance between old and new places seems to have been decreasing over time. This was the case in the French Lorraine were settlement patterns have been studied intensively. As a matter of fact, in the last centuries BC rebuilding within a restricted space became customary again. But the small agglomerations of farmsteads did not recur, except in the north-eastern part of the region (see Section 7.6).

The change in land use and distribution of the farming population over the landscape must have had its repercussions on the vegetation cover. On new terrain or after a long fallow, light secondary forest and brush was the dominant type of vegetation to be cleared when new fields had to be laid out. The tool with which to achieve this was, as everywhere and always, the axe. But metal, first bronze and later iron, was now the material for the blade. In the course of time the shape of the axe underwent several changes, and the way of hafting also changed. Some of these changes were of a technical nature, but others seem to follow the prevailing fashion (Fig. 7.2).

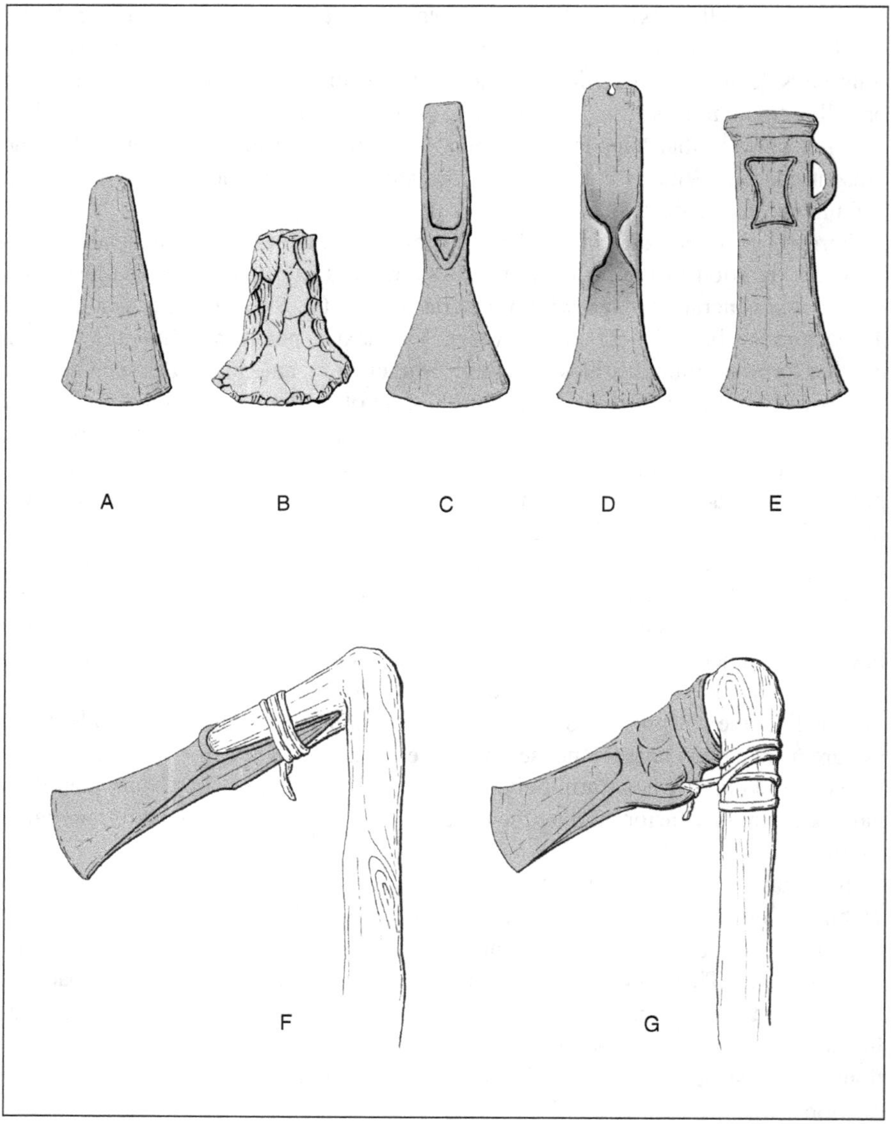

Fig. 7.2 Bronze axe blades (**A**, **C**, **D**, **E**) and the way in which they were hafted (**F**, **G**); large parts of the string used to attach the blades to the hafts have been omitted for clearness' sake. The blades are arranged following development in shape. **A** is the oldest. Blade **B** is of flint and an imitation of A made in the time when bronze was still scarce

The soil of new or existing plots was bound to have been loosened with an ard, which was by now an established implement. When iron became the prevailing metal, the share was fitted out with an iron sheath (Fig. 7.3). The sheaths show differences in their general shape and width, indicating that ards came in different

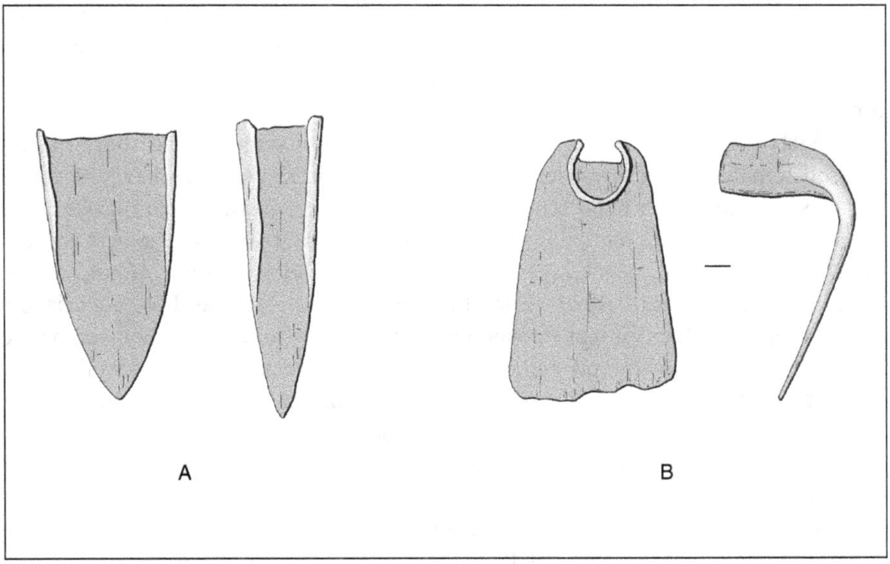

Fig. 7.3 Iron parts of tools. **A**: tips of ards, **B**: blade of a hoe

models, presumably in response to the type of soil to be worked. Next to the ard, the hoe was used to break up soils. Hoe blades made of iron are found in excavations, but hoes made of bronze are unknown (Fig. 7.3). This metal may have been unsuitable for this kind of implement. Still, not much is known about the farmers' tools. Most of these must have been made of wood, which has not been preserved. From regions that are more favoured in this respect, it is known that, a simple wooden harrow at least was part of the tool kit.

As remarked above, the range of crops to be sown on fields has been gradually widening. Some were main staples, others minor crops. The question is whether every field was sown with any seed, or whether some fields were allocated to staple cereals and others to minor crops requiring special care. Another question is whether all crops were sown separately. Some may have been sown as a mixture, i.e. as a maslin. The word maslin is used for a mixture of two crops, sometimes three, sown together. Mixed sowing is resorted to avoid risks. For instance, if one plant is less frost resistant than another, but more profitable when successful, a mixture will provide optimal results if frost occasionally poses a problem. A variant form is the resowing with a summer crop of patches in fields where the winter crop has failed, but in this case the final product is not a maslin in the strict sense of the word. Maslins are known from historical sources. At the start the ratio of two crop plants in the sowing seed varies from 1:1 to 3:1. The final product is, of course, more variable in composition. Maslins can be combinations of cereals or combinations of pulses. The written sources of historical times do not mention combinations of a cereal and a pulse, or combinations of other kinds of plants, at least not in this region. Of course, this does not prove that this combination did not exist in pre- and protohistorical times. But as such kinds of maslins obviously were not a success in the not-too-distant past, they may not have been a success either in much earlier times.

To start with the second question, this may be answered by the study of carbonised remains of the harvested product. Unfortunately, true remnants of single harvests are rare. Most plant remains retrieved during excavations are mixtures of several kinds of waste, including the remains of several harvests. The best sources of information are the black layers, consisting of carbonised grains, found on the bottom of underground silos (see for this kind of silos Section 3.5). Such layers seem to represent the original stocked product, but how they got carbonised on the bottom of a deep, narrow pit without any circulation of air is not yet well understood. Nevertheless the black layers do exist, although they are extremely rare. Most of them were detected in the southern part of the region and belong to the younger stages of the period. Almost as good are burnt-down granaries. If their contents had been smothered by the collapse of the structure, the original contents may not have been disturbed too much. Of course, if several products were stored separately in the same structure, some mixing-up may have occurred during the disaster, but in general a careful excavation can detect this. Both the bottom layers of silos and destroyed granaries are considered as a category A source of information.

A third category of finds consists of concentrations of grains that have obviously been thrown away as one single lot during one single event. They represent harvested products damaged by a fire elsewhere and subsequently discarded as waste in a pit. As a source of information on maslins or monocrops they are second choice, category B (Tables 7.1, 7.2 and 7.3).

The oldest suitable find dates from the period 1500 BC–1100 BC. The find poses a problem, in that the lot consists of a mixture of 82% hulled barley and 18% naked barley. This mixture may represent a maslin of two barleys with different properties. It is quite possible that through such maslin-growing farmers learned to prefer hulled barley to naked barley, a process leading to a virtual abandonment of the latter. But it is also possible that the farmer just grew a monocrop of barley and did not make any distinction between the two varieties.

More instances of suitable finds are found in the period 1100 BC–800 BC. Category A again comprises a mixture of hulled and naked barley. Other lots show mixtures of hulled barley/oat and hulled barley/emmer wheat/broomcorn millet. Barley/oat is known to have been grown as intentional maslin in historical times. The hulled barley/emmer wheat/broomcorn millet is a stranger combination. The three different crops were perhaps stored separately on the bottom of the silo, but how they became mixed later when they got burnt in situ remains to be explained. Another possibility is that the lot represents a maslin in a very wide sense of the word. Bare patches in a field with autumn-sown cereals, such as emmer and winter barley, may have been resown in spring with millet, and the result harvested

Table 7.1 Category A sources of information on the occurrence of maslins and monocrops. The crop plants are represented by their share (%) in the respective finds. The oldest period has no category A finds

Monocrop or maslin									
Site	Naked barley	Hulled barley	Emmer wheat	Spelt wheat	Oat	Millet	Einkorn	Horse-bean	Interpretation
Category A									
Period 1500 BC–1100 BC									
Period 1100 BC–800 BC									
Cuiry-lès-Chaudardes 807	–	54	–	–	46	–	–	–	Maslin
Cuiry-lès Chaudardes 830	–	20	32	–	–	48	–	–	Multiple storage or 'maslin'
Frouard HP 2091	50	50	–	–	–	–	–	–	Maslin or monocrop
Period 800 BC –50 BC									
Gondreville 4214	–	93	–	2	–	–	5	–	Monocrop
Gondreville 4219	–	90	–	9	–	–	1	–	Monocrop
Frouard HP	–	–	–	–	–	–	–	91	Monocrop + 9.1% cereals indet.
Compiègne	–	86	–	12	2	–	–	–	Monocrop
Menneville	–	–	50	50	–	–	–	–	Maslin
Jaux	–	–	100	–	–	–	–	–	Monocrop
Louvres Le-Vieux-Moulin 71	–	69	31	–	–	–	–	–	Maslin
Neerharen-Rekem 123	–	47	53	–	–	–	–	–	Maslin
Neerharen-Rekem 132	–	6	85	–	9	–	–	–	Monocrop
Maisnil 7	–	12	69	–	19	–	–	–	Multiple storage or maslin
Maisnil 17	–	–	100	–	–	–	–	–	Monocrop
Acy-Romance 3596	–	100	–	–	–	–	–	–	Monocrop
Forest-Monthiers	–	–	100	–	–	–	–	–	Monocrop

Table 7.2 Category B sources of information on the occurrence of maslins and monocrops. The crop plants are represented by their share (%) in the respective finds

Monocrop or maslin										
Site	Naked barley	Hulled barley	Emmer wheat	Spelt wheat	Oat	Millet	Einkorn	Horse-bean	Pea	Interpretation
Category B										
Period 1500 BC–1100 BC										
Crévéchamps	18	82	–	–	–	–	–	–	–	Maslin or 'monocrop'
Period 1100 BC–800 BC										
Ay-sur-Moselle	–	–	100	–	–	–	–	–	–	Monocrop
Cuiry-lès-Chaudardes 826	–	21	32	–	–	47	–	–	–	Multiple storage or 'maslin'
Cuiry-lès-Chaudardes 836	–	5	83	–	–	12	–	–	–	Multiple storage or 'maslin'
Period 800 BC–50 BC										
Tagnon 19	–	2	29	–	–	69	–	–	–	Multiple storage or 'maslin'
Tagnon 22	–	61	36	–	–	3	–	–	–	Maslin
Tagnon 23	–	53	44	–	–	3	–	–	–	Maslin
Tagnon 32	–	50	44	–	–	6	–	–	–	Maslin
Cairon 1 F2	–	–	–	–	–	–	–	13	87	Maslin
Cairon 1 F4	–	–	–	–	–	–	–	20	80	Maslin
Tahon 1	–	–	–	–	–	–	–	100	–	Monocrop
Tahon 3	–	100	–	–	–	–	–	–	–	Monocrop
Tahon 4	–	–	–	–	–	–	–	–	100	Monocrop
Tahon 6	–	–	–	100	–	–	–	–	–	Monocrop
Tahon 8	–	30	70	–	–	–	–	–	–	Maslin

Table 7.2 (continued)

Monocrop or maslin										
Site	Naked barley	Hulled barley	Emmer wheat	Spelt wheat	Oat	Millet	Einkorn	Horse-bean	Pea	Interpretation
Acy-Romance 4511	–	100	–	–	–	–	–	–	–	Monocrop
Acy-Romance 4512	–	100	–	–	–	–	–	–	–	Monocrop
Acy-Romance 4524	–	–	–	85	–	–	15	–	–	Maslin
Bétheny 262	–	100	–	–	–	–	–	–	–	Monocrop
Bétheny 267	–	100	–	–	–	–	–	–	–	Monocrop
Bétheny 276	–	15	80	–	–	–	5	–	–	Maslin
Bétheny 260	–	82	7	–	–	–	11	–	–	Maslin
Bétheny 298	–	–	100	–	–	–	–	–	–	Monocrop
Bétheny 569	–	100	–	–	–	–	–	–	–	Monocrop
Bétheny 595	–	100	–	–	–	–	–	–	–	Monocrop
Bétheny 633	–	16	47	–	–	–	37	–	–	Maslin
Bétheny 673	–	100	–	–	–	–	–	–	–	Monocrop
Louvres 34	–	100	–	–	–	–	–	–	–	Monocrop
Louvres 26	–	–	100	–	–	–	–	–	–	Monocrop
Louvres 30	–	100	–	–	–	–	–	–	–	Monocrop
Louvres 40	–	100	–	–	–	–	–	–	–	Monocrop

Table 7.3 The occurrence of monocrops and maslins and their composition

Occurrence of monocrops and maslins	Multiple storage + 'maslin'	Maslin	Monocrop
Category A			
1500 BC–1100 BC	0	0	0
1100 BC–800 BC	1	1	1
800 BC–50 BC	1	3	9
Category B			
1500 BC–1100 BC	0	0	1
1100 BC–800 BC	2	0	1
800 BC–50 BC	1	10	16

Composition of monocrops and maslins	1500 BC–1100 BC	1100 BC–800 BC	800 BC–500 BC
Category A monocrops			
Naked barley + hulled barley	–	1	–
Hulled barley	–	–	4
Emmer wheat	–	–	4
Horse bean	–	–	1
Category B monocrops			
Naked barley + hulled barley	1	–	–
Hulled barley	–	–	11
Emmer wheat	–	1	2
Spelt wheat	–	–	1
Horse bean	–	–	1
Pea	–	–	1
Category A maslins			
Hulled barley-emmer wheat	–	–	2
Hulled barley-oats	–	1	–
Emmer wheat-spelt wheat	–	–	1
Category B maslins			
Hulled barley-emmer wheat	–	–	6
Hulled barley-einkorn wheat	–	–	1
Spelt wheat-einkorn wheat	–	–	1
Horse bean-pea	–	–	2

together. A third hypothesis is that the millet with its tiny grains was added to a stored hulled barley/emmer maslin to fill the empty spaces between these cereals, stored as usual covered by their husks and therefore rather coarse products, in order to reduce the volume of air and thus promoting conservation under anaerobic conditions. Category B finds of this period (and on the same site) show this kind of mixture as well. Another site revealed a monocrop of emmer wheat.

After 800 BC suitable finds become more numerous. Monocrops are present consisting of hulled barley, emmer wheat, spelt wheat, horsebean, and pea. The most common mixture is hulled barley/emmer wheat, which may represent a true maslin. Emmer wheat/spelt wheat, spelt wheat/einkorn wheat, hulled barley/einkorn wheat, and horsebean/pea are found as well, and may be maslins too. An emmer wheat/broomcorn millet mixture poses the problem discussed above, as emmer wheat is commonly grown as a winter cereal, whilst broomcorn millet can only be sown in spring. Monocrops prevail, but if the interpretation of the mixtures as maslins is correct, maslin growing seems to have been rather common practice.

It is worth noting that the crops found as stored products represent only part of the total range of plants grown. Most of the stored products are hulled cereals. Some of the remaining plants may have been of very minor importance. Species that are seldom found, such as foxtail millet and gold of pleasure, are examples of this kind. But others, like naked wheat, are common in household waste and cannot, therefore, have been minor crops. They may have been stored in a different way, which did not provide an opportunity to preservation as concentrations of seeds.

The first question raised was whether all crops had the same status. A clear answer cannot be given. Hulled barley, hulled wheats, broomcorn millet and, to a lesser extent, horsebean and pea, are found in bulk and should be considered as having been staples. Whether they were the product of a different farming regime than the others is still an open question. A comparison of field weeds, trapped in the bulk finds, with field weeds found in the farmyard and household waste containing the other crops, revealed no differences in weed flora.

A detailed study of weeds, carried out in the Moselle area, showed, however, that the weed flora composition as such underwent a change through time. A shift is observed in the ratio annuals : perennials. The frequency (occurrence in individual samples with respect to the total number of samples) of perennials decreases. As perennials do not easily survive an intensive use of a field, the picture emerges that extensive farming regimes decreased in importance. Until 1500 BC a co-occurrence of an intensive small-scale cultivation regime and a more extensive cultivation regime may have existed, but from at least 1100 BC onwards intensive cultivation became more dominant (Fig. 7.4). By extensive is understood either short-lived or looked after with less care, or a combination of both. Under an intensive regime, a piece of land may be in use for a long stretch of time and/or tilled with care. The development goes hand in hand with the increase in the number of crop plants. The new plants may have been cultivated under the intensive regime, whereas the 'old' cereals may have been grown under a more extensive regime, but, as pointed out above, an analysis of their weeds revealed no clear-cut difference.

Most weed species prefer humid to dry, and moderate to highly fertile soils. Indicators of highly fertile soils are, for instance, *Capsella bursa-pastoris*, *Urtica urens*, *Solanum nigrum* and *Sisymbrium officinale*. Their presence is due to improvement of the soil by addition of fertilisers such as dung or other kinds of manure. Such species are especially abundant from 1100 BC onwards.

Unfortunately, the parcelling out of fields is not archaeologically detected until at least about 400 BC–300 BC. From the last centuries BC, ditch systems enclosing rectangular pieces of land show that parcels near the farmhouse are smaller than those at some distance, suggesting differences in land use. Despite the indications of an increasing intensification of land use, some variation in cultivation regimes may still be expected, with a more intensive cultivation near the house, and a less

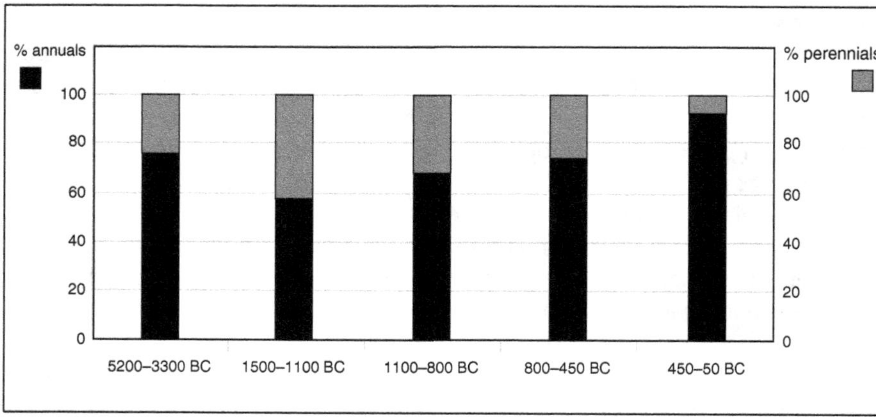

Fig. 7.4 The ratio annuals: perennials through time in the Moselle area

intensive cultivation further away. Unfortunately, the data are still too scanty to provide us with more detailed information.

Not only the ratio annual weeds : perennial weeds has shifted, but also the absolute number of weed species increased with time in the Moselle area study. The increase may reflect an increasing variation in agricultural methods, such as growing a wider range of crop plants, new types of crops, or a differentiation in manuring practices. But an increase in weed species may also reflect a change in harvesting methods. It must be kept in mind that the data available for weed analysis are provided by excavations in and around farmbuildings. They reflect that part of the weed flora that arrives with the harvested product. The farmers, dealt with in Chapters 3, 4, and 5, brought only relatively tall, or climbing, species to their yards, which led to the conclusion that they harvested their crops, at least their cereals, halfway the stalk. If crop plants are cut closer to the ground, or are uprooted, smaller weeds are collected too. The result is a longer list of species. Small plants like *Anagallis arvensis*, *Capsella bursa-pastoris*, *Spergula arvensis* and *Thlaspi arvense* are indeed more common in the finds than in earlier times.

An analysis of the height of weeds, present in stored harvests of category A finds used in the monocrop-maslin study, shows that five of the eleven lots did not contain small weeds, i.e. weeds that reach growing heights of 60 cm at most (Table 7.4). One lot did not contain any weeds at all. This is the lot of horsebean, which was probably stored already shelled. Two of the lots concern monocrops of emmer wheat, one a monocrop of hulled barley, and one a maslin of hulled barley and emmer wheat. But a monocrop found in the same settlement as the maslin contained a low-growing weed, *Capsella bursa-pastoris*, indicating that the practice of harvesting high or low was not mutually exclusive, and was certainly not a fixed procedure followed by specific farmers or farming communities. It seems that the height of harvesting cereals was variable during the period under review.

The methods of harvesting other crops remains unknown. A category B monocrop of horsebean was as clean as the category A lot, whilst a monocrop of

Table 7.4 The weeds in stored harvests of category A. 1: Gondreville 4214; 2: Gondreville 4219; 3: Frouard; 4: Menneville; 5: Jaux; 6: Louvres 71; 7: Neerharen 123; 8: Neerharen 132; 9: Maisnil 17; 10: Acy-Romance 3596; 11: Forest-Monthiers

Weeds in stored harvests												
	Height in cm	1 h b	2 h b	3 n/h b	4 e/s	5 e	6 h b /e	7 h b /e	8 e	9 e	10 h b	11 e
Tall and climbing weeds												
Agrostemma githago	20–100	■	■	–	■	–	■	–	–	–	–	–
Agrostis sp.	10–120	■	–	–	–	–	–	–	–	–	–	–
Alchemilla sp. (Aphanes?)	10–70 (2–20)	–	–	–	■	–	–	–	–	–	–	–
Artemisia vulgaris	30–120	–	–	–	■	–	–	–	–	■	–	–
Atriplex sp.	20–90	–	–	–	■	■	–	■	–	–	–	–
Avena fatua	60–120	–	–	–	■	■	–	–	–	–	■	–
Bromus arvensis	30–100	–	–	–	■	–	–	–	–	–	–	–
Bromus secalinus type	40–100	■	–	■	■	■	■	■	–	–	–	■
Bromus sterilis/tectorum	15–120	■	–	■	–	–	–	–	–	–	–	–
Buglossoides arvensis	10–70	■	–	–	–	–	–	–	–	–	–	–
Chenopodium album	15–150	–	–	■	■	■	■	■	–	–	–	–
Chenopodium hybridum	30–90	–	–	–	■	–	–	–	–	–	–	–
Convolvulus arvensis	20–100	–	■	–	–	–	–	–	–	–	–	–
Daucus carota	10–80	–	–	–	■	–	–	–	–	–	–	–
Echinochloa crus-galli	10–120	–	–	–	–	–	–	■	■	–	–	–
Fallopia convolvulus	10–100	–	–	–	■	■	–	■	–	–	–	■
Galium aparine	15–100	■	■	■	–	■	–	■	–	–	–	–
Hypericum perforatum	20–80	–	–	–	■	–	–	–	–	–	–	–
Lapsana communis	30–120	■	■	–	–	–	■	–	–	–	–	–

Table 7.4 (continued)

Weeds in stored harvests	Height in cm	1 h b	2 h b	3 n/h b	4 e/s	5 e	6 h b /e	7 h b /e	8 e	9 e	10 h b	11 e
Malva sylvestris	30–120	■	–	–	■	–	–	–	–	–	–	–
Persicaria lapathifolia	10–120	–	–	–	■	–	–	–	–	–	–	■
Phleum pratense	10–150	–	–	–	■	■	–	–	–	–	–	–
Silene dioica/latifolia	30–100	–	–	–	■	–	–	–	–	–	–	–
Solanum nigrum	5–60	–	–	–	■	–	–	–	–	–	–	–
Vicia hirsuta/tetrasperma	15–70	■	■	–	■	■	–	–	–	–	–	■
Low-growing weeds												–
Adonis sp.	15–50	■	–	–	–	–	–	–	–	–	–	–
Anagallis arvensis	6–20	–	–	–	■	–	–	–	–	–	–	–
Capsella bursa-pastoris	5–50	–	–	–	–	–	–	–	■	–	–	–
Galium spurium	10–40	■	■	–	–	–	–	–	–	–	–	–
Hieracium subgen. Pilosella	2–30	–	–	–	■	–	–	–	–	–	–	–
Orlaya grandiflora	10–50	–	■	–	–	–	–	–	–	–	–	–
Picris hieracioides	3–50	–	–	–	■	–	–	–	–	–	–	–
Plantago lanceolata	5–45	–	–	–	–	–	■	–	–	–	–	–
Poa annua	15–50	–	–	–	–	–	■	–	–	–	–	–
Polygonum aviculare	5– . . .	–	–	–	–	■	–	–	–	–	–	–
Stachys arvensis/sylvatica	7–30	–	–	–	■	–	–	–	–	–	–	–
Thlaspi arvense	15–50	–	–	–	■	–	–	–	–	–	–	–
Veronica hederifolia	5–30	–	■	–	–	–	–	–	–	–	–	–

h b: hulled barley; n b: naked barley; e: emmer wheat; s: spelt wheat. The harvests of Cuiry-lès-Chaudardes 807 and Compiègne, and the Frouard horse bean crop did not contain any weeds. Instances of multiple storage were omitted.

pea showed low growing plants. But, as the latter was a B category find, and slightly contaminated by some emmer wheat, it is not clear whether the weeds came with the peas or not. Lots of other crops, suitable for this kind of analysis, are absent from the records.

Excavations rarely yield implements used for harvesting. Until 1500 BC sickles were still being made of flint set into wooden handles, although their precise form is not known. Bronze replaced flint as material for this kind of tool at a much later period than in the case of axes. One of the reasons for this may be that the sickle had a lower social status than the axe. Only when the metal became more widespread and common, were sickle blades also made of bronze. The blade is a curved piece of metal, strengthened at its back opposite the cutting edge, by two thicker parallel ridges. It is 11–15 cm long and 2.5–3.5 cm wide. The implements are uniform in style. Sickle blades lack the variety of forms shown by axe blades. Some difference is present at the short side where the haft was attached. As indicated, finds are relatively scarce, although the sickle must have been a common tool. But bronze is a recyclable material, and, moreover, subject to corrosion. Well-preserved specimens have turned up as part of hoards or offerings. They show traces of resharpening and some of them are extremely worn. These were obviously resharpened until the strengthening ribs were reached (Fig. 7.5).

The fact that during a long period only one type of cutting implement was used suggests a rather uniform way of harvesting in the case of crops that were harvested by cutting. Whether cutting high or low, it was done with the same implement, at least during the Bronze Age, until 800 BC.

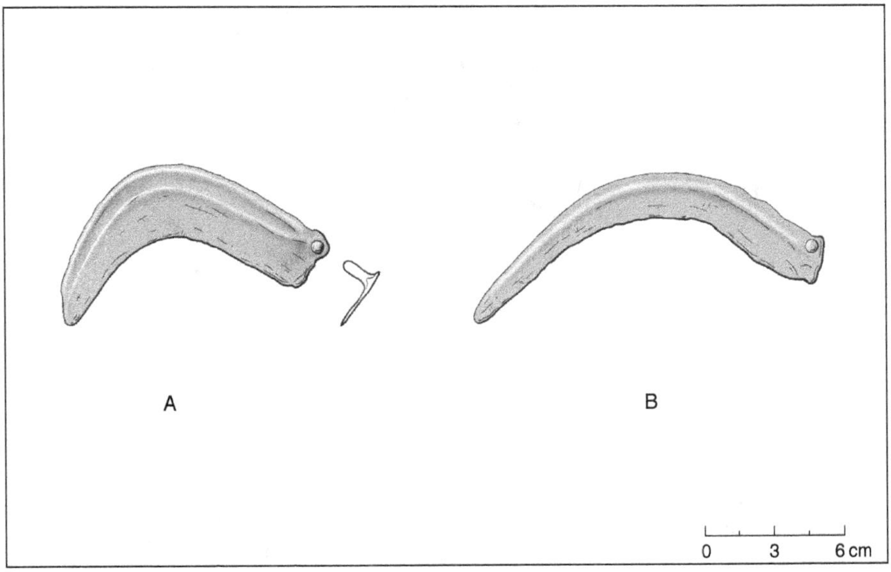

Fig. 7.5 Bronze sickle blades; the protrusion serves to fix the haft to the blade. **A**: complete specimen, **B**: well-worn blade

Of course, other types of harvesting are feasible and not every crop is fit to be harvested by cutting. Flax, for instance, is notoriously difficult to cut and is mainly harvested by uprooting, a practice that does not require tools. Pulses are commonly uprooted as well, or their pods are collected by hand-picking. Simple tools, such as the wooden sticks (mesorias) described in Section 5.3, will not survive. The same is true for the handheld wooden comb, which is a common tool used to strip the seeds of broomcorn millet off their stalk. Also knives could have been used for harvesting, but they are such multi-purpose tools, that they do not stand out in the archaeological record as such.

From 800 BC onwards, iron replaced bronze as the material for sickle blades (Fig. 7.6). Iron is much susceptible to corrosion and iron tools have been rarely preserved. The curve of the blade resembles the earlier bronze one. They were attached to a perishable haft by means of either a socket or a tang. In addition to the short sickle blade a much longer blade, with lengths up to 50 cm, has been discovered. It is interpreted as a scythe (Fig. 7.6). Comparable blades, still attached to their hafts, have been found in contemporaneous sites in Switzerland. The hafts have lengths of

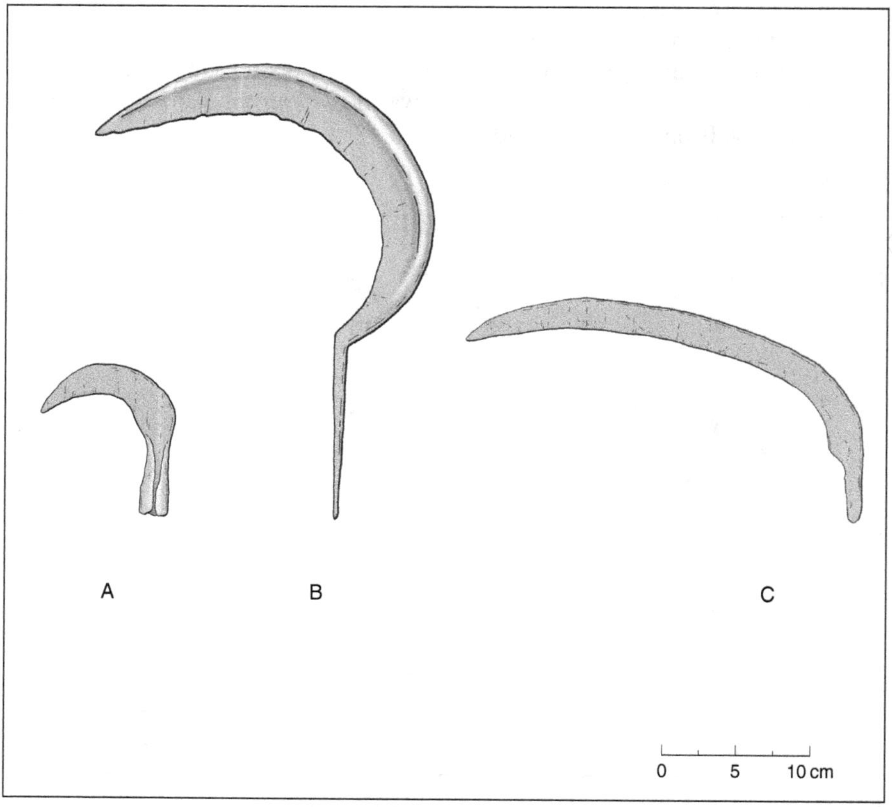

Fig. 7.6 Two iron sickle blades (**A** and **B**) and a scythe (**C**)

60–70 cm. The implement may have been the tool for the low-on-the-stalk mode of cutting cereals, when apart from the grain, the straw was a desired product. However, the swing of the scythe strikes the stalk with more force, and with a different kind of blow, than the sickle. When cereals ripen unevenly, or have rather brittle ears like emmer wheat, harvesting with a scythe causes more losses. As a matter of fact, the scythe became a harvesting tool for cereals only late in historical times. The implement was perhaps used to collect straw after hand-picking of the ears, or to cut grass. Hay is a product that has not been mentioned so far, because traces of hay are mostly lacking in the archaeological records. Nevertheless, the collection of hay may have started around this time (see Section 10.1).

The range of harvesting tools has been further widened by the invention of a true machine. This harvester was a large box, open at the top and at the front end. Horizontal teeth, attached to its bottom, protruded forward. The whole was mounted on two wheels, and provided with two shafts at the back to which an animal could be harnessed with is head in the direction of the box. The animal pushed the harvester through the cereal field. The teeth in front gathered the ears and stripped them off, after which they fell into the box. It is not quite known when this harvester was first put into practice. It is known from Roman written sources, especially Pliny and Palladius, and from pictures in relief on stone, dating from the first and second centuries AD (Fig. 7.7). The Romans knew the harvester as something from 'Gallia' and called it 'vallus'. According to them it was a machine to be used on more or less level ground and on large fields, and then only if the straw was not required too. Such conditions are considered to have been met on the large farms characteristic of

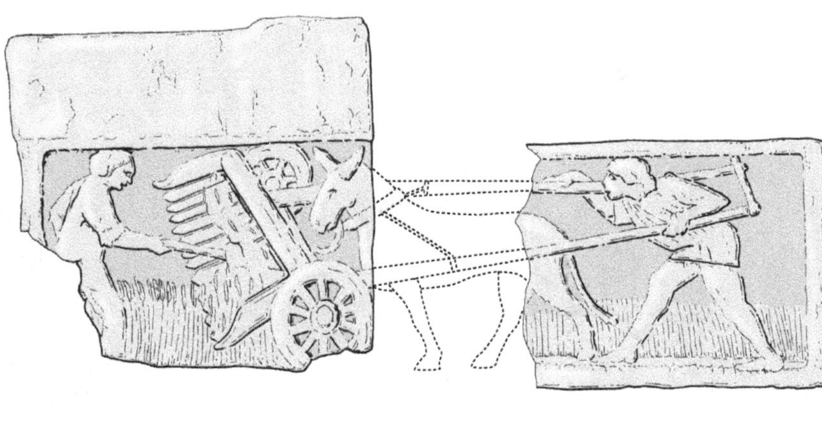

Fig. 7.7 The vallus. The picture is composed of two fragments of sculptures found at different places

the period after the Roman conquest (see Section 6.2). It is questionable whether they existed in pre-Roman times. Nevertheless, the harvester was not a Roman invention. It must have had its roots in an indigenous tradition. In fact, the harvester is a mechanised version of the handheld comb used in harvesting cereals like millet.

The treatment of the crops after the harvest remains unknown. Special threshing floors outside or within the farmyard have not yet been detected. The harvest is presumed to have been processed near or in the farmbuildings. A wagon, drawn by cattle, especially oxen, may have been used to transport the more bulky crops. Implements for threshing and further processing have disappeared from the records.

Storage took place either underground in silos, or aboveground in granaries or large containers. The silos are of the same nature as those described for the earlier periods. They are a continuation of an ancient, long-standing tradition. Most of the products stored in them were semi-threshed hulled cereals, that is the grain without straw, but still encased by chaff. Some remnants of the original fill of stored grain are almost weed free, others are not. The difference in weed contamination may be the result of a different type of harvesting, or the result of the presence or absence of cleaning procedures before storage. Weeds can to a large extent be removed by sieving.

Storage aboveground in detached granaries is a new development. In previous periods grain is considered to have been stored in lofts in the main building, with, perhaps, an exception in the Rössen culture (Chapters 3 and 4). The most simple and common granary is a building erected on four sturdy posts, set in a square,

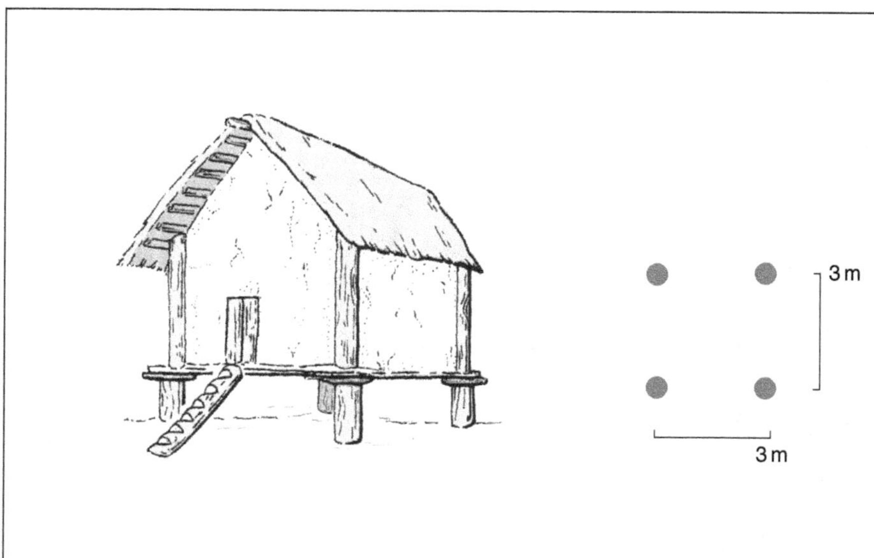

Fig. 7.8 A possible reconstruction of a four-poster granary and what is found in an excavation: four postholes set in a square

and deeply dug into the subsoil. The four deep foundation pits are the only features left. Sometimes they show a trace of the original post (see Section 2.5). The superstructure is not known, but on the basis of historical and ethnographical parallels, the granary is seen as a structure, placed on stilts with a floor well above ground (Fig. 7.8). Common sizes are 3 × 3 m with an unknown height. Larger sized structures were rectangular and founded on six, eight or nine posts. But these are more in the nature of outhouses.

The granary was certainly not, at least not commonly, used for bulk storage of grain. In the few instances where the burnt contents of granaries, destroyed by fire, could be studied, the conclusion was that they held several products stored apart, in separate heaps, sacks, baskets or chests. This is confirmed by finds in similar granaries outside the loess region. The detached granary was a common phenomenon in Western Europe by this time (Fig. 7.9). It is even feasible that the granaries were also used to store other things beside crops.

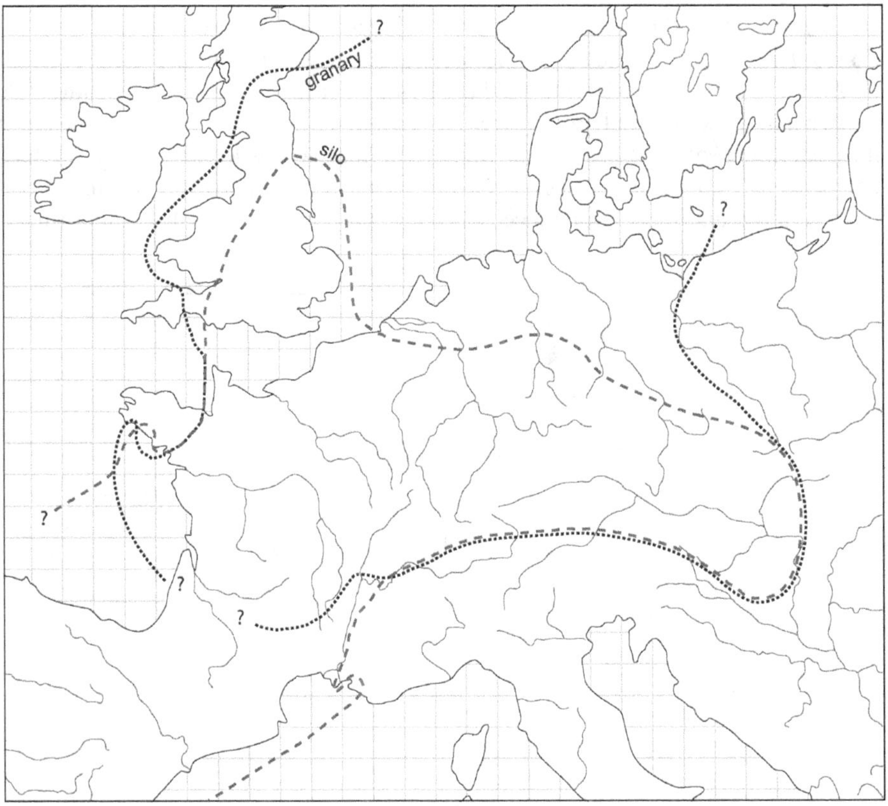

Fig. 7.9 The occurrence of granaries and silos in the Metal Ages of central and western Europe. Granaries are a more northern and silos a more south to south-eastern phenomenon, but the areas show considerable overlap, including the loess region

When exactly the granary became part of the crop handling system is not yet known, because buildings of the period 2650 BC–1500 BC are practically unknown. But from 1500 BC onwards they became common.

Most large containers for storing crops will have been made of perishable materials, as is known from historical and ethnographical sources. However, one type of container was of durable material: the large ceramic vessel. The vessels found have capacities of 40–120 L (Fig. 7.10). They functioned outside and inside buildings and were usually partly inserted in the ground.

Storage underground is a long-term storage, because a silo cannot be opened and closed again and again. Once air enters the pit, the grain will start to decay (see Section 3.5). Storage aboveground can be both long-term and short-term storage. As mentioned earlier, underground storage was mainly used for hulled cereals, although some pulses were kept in this way, too. The other products seem to have been stored aboveground, in the granaries and/or vessels. Some possible chains of operation are depicted in Fig. 7.11.

During most of the period the time-honoured saddle quern was used for grinding grain into a coarse semolina or a flour. Its size, and especially the size of the upper stone in relation to the size of the lower stone varied with cultures and times, but the principle had remained the same since the querns of the first farmers.

A minor disadvantage of the saddle quern is that the lower stone tends to slide away when working on a hard surface. To counteract this tendency, quern makers in the eastern part of the region invented a lower stone with a lengthwise keel, which could be set into the floor. Such keels are seen from c. 1000 BC onwards. Keels became more and more pronounced until around 800 BC specialised quern mak-

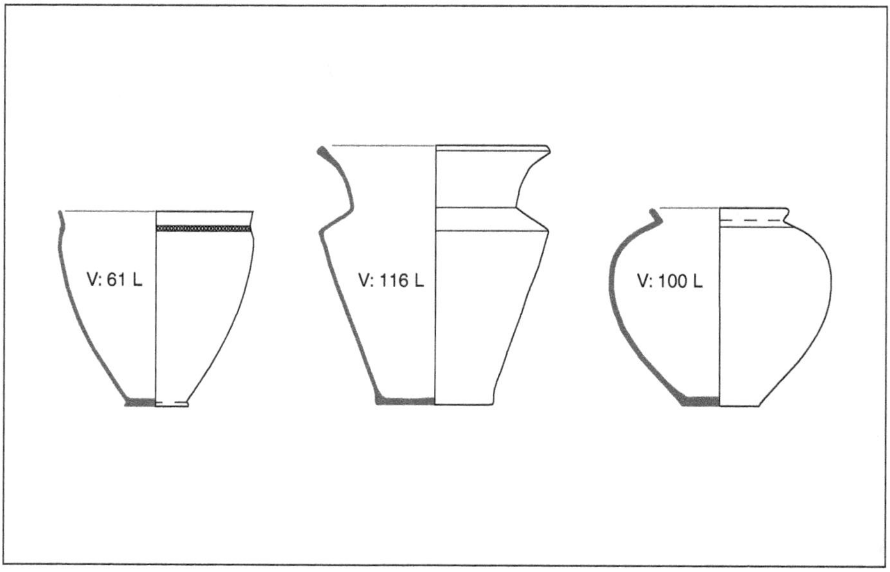

Fig. 7.10 Ceramic containers for storing crops with a capacity of respectively 61, 116 and 100 L

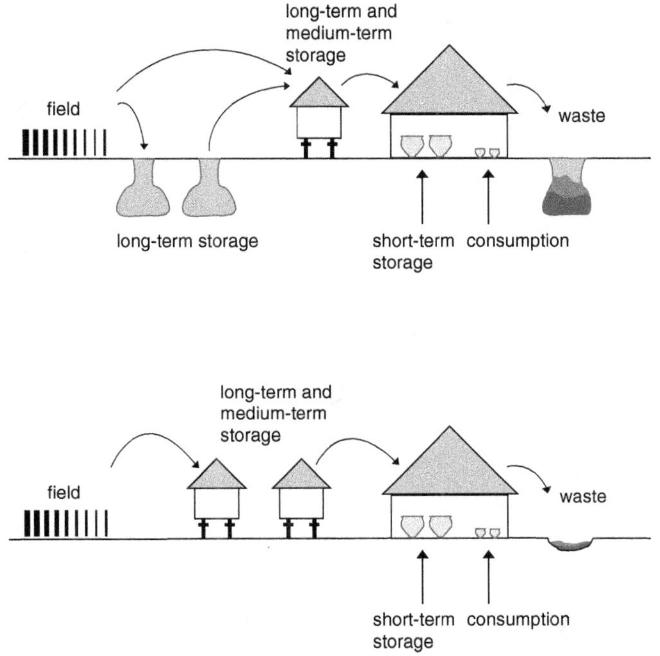

Fig. 7.11 Two possible chains of operation. **A** with silos and **B** without. Waste was dumped either in an abandoned silo or in a pit

ers in the area of Mayen-Koblenz (Eifel mountains, Germany) developed the 'hat of Napoleon', a lower quern stone with concave sides and a width-height ratio of 1 : 1. Its name is derived from the favourite hat of the French emperor (Fig. 7.12). The Mayen-Koblenz stone quarries provided querns of good-quality tephrite (a kind of lava), that were much sought after. The area covers only six square kilometres, but is known as the largest production centre of querns in Western Europe. Its produce, consisting of ready-made implements, were exported as far as the Meuse and Moselle areas, but not farther to the west. In the Paris Basin, for instance, the manufacture of saddle querns was a local, presumably entirely domestic affair.

From the second century onwards the era of the saddle quern came to an end. It was replaced by the rotary quern, a type of quern developed in the Mediterranean region. The rotary quern consists of a circular lower and upper stone. The upper

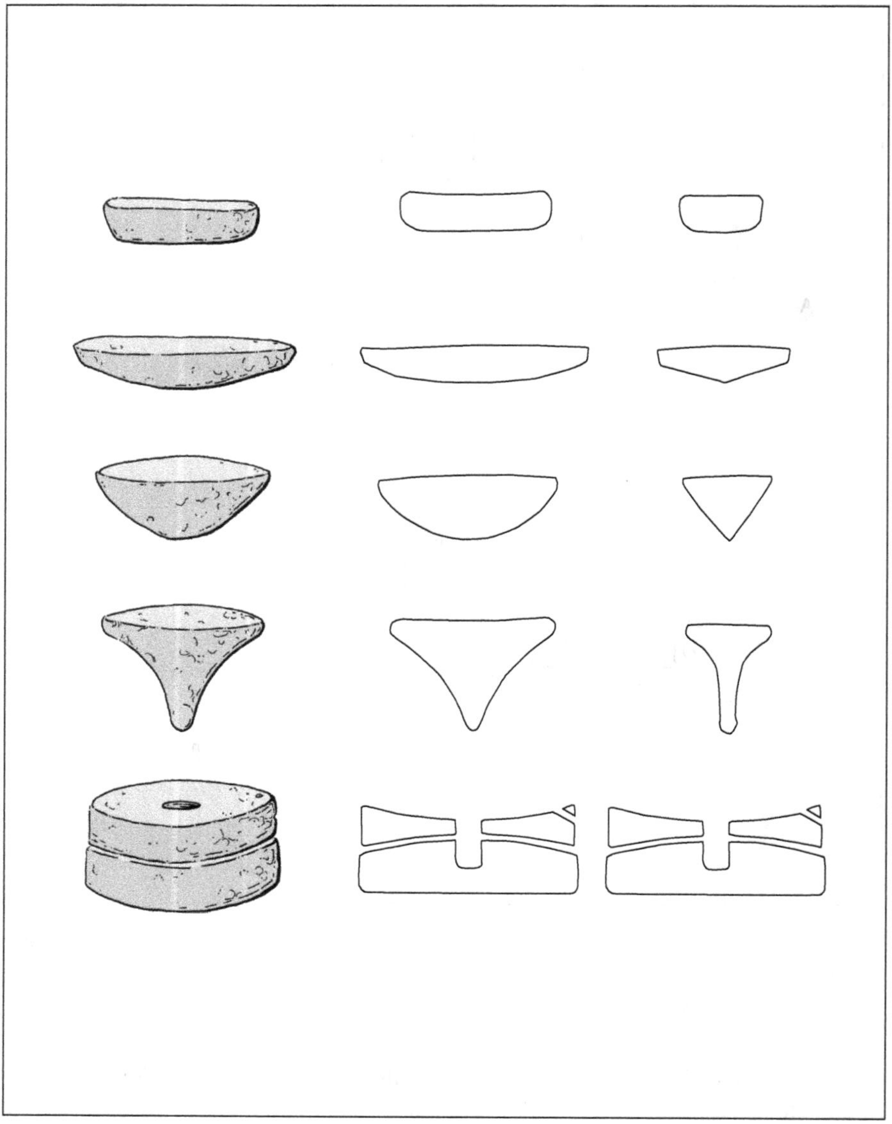

Fig. 7.12 The development of the quern, form top to bottom. The fourth type is the so-called hat of Napoleon

surface of the lower stone is convex and the underside of the upper stone concave. They are held together by a wooden spindle inserted into a central hole. The upper stone has a second hole at its rim to accommodate a handle (Fig. 7.12). The quern can be worked with a rotating or swinging movement. Grain is poured in through the hole in the centre, and flour escapes from between the stones. The diameter of the implement was c. 35 cm at the beginning, to increase to c. 40 cm at the end of the period.

The main improvement gained by the adoption of the rotary quern lies in its effectiveness in producing flour. Experiments have shown that working with a saddle quern results in the production of c. 0.6 kg flour per hour, whilst milling with a rotary quern using a swinging movement produces 2 kg per hour. A rotation movement results in 6 kg per hour.

The Mayen-Koblenz quern makers switched to the production of this new type of quern. Elsewhere the domestic production of querns was abandoned and making rotary querns was from then on the concern of specialised workshops. But making flour remained a task to be done at home.

7.4 Livestock and Animal Husbandry

As before, the core of the livestock consisted of cattle, pig, sheep and goat. But the composition did not remain the same. Two animals were added to the traditional four: the horse and the chicken.

The horse was first. There are a number of theories regarding the domestication of the horse. Of old, horses have played a role in the life of humans. They appear already in palaeolithic cave art, but these horses were truly wild and were hunted for meat. Wild horses were widely distributed throughout Eurasia, especially so on the steppes of the Ice Age and its aftermath. The number of their remains falls steadily with the disappearance of steppe conditions. In the recent past only one putative wild population, the Przewalski horse, had remained.

How, when and where the horse became domesticated is not yet clear. One hypothesis is that the centre of domestication lies in the southern Ukraine and Kazakhstan, where remains are frequently found in archaeological sites from 4500 BC onwards. Some teeth show wear marks that could have resulted from the friction of a bit against the molars, indicating captive, but not necessarily domesticated, animals. The horses with bit wear could have been ridden, but may also have been led with a leash attached to a bit. Skeletal changes, which would provide secure proof that the horses were actually bred in captivity, and not merely tamed, are lacking.

DNA analysis has shown that the southern Ukraine and Kazakhstan cannot have been the only centre of origin. Numerous lineages have been discovered and there must have been multiple origins, whose exact location is so far unknown. The oldest reliable archaeological finds of domestic horses date from only c. 2000 BC, when teams of two horses were found together with spoke-wheeled chariots as part of grave gifts in Kazakhstan. From that period onwards the horse and the horse-drawn wagon or chariot spread fast into other areas, including Europe. In the loess region west of the Rhine, the horse as a domestic animal appears in the late Bronze Age, from 1100 BC onwards. Finds become more numerous from 450 BC onwards.

The bird, generally considered to represent the wild ancestor of the domestic chicken is the Red Jungle Fowl, also known as Bankiva fowl, from South-East Asia.

The history of its domestication and dispersal is not very well understood, but the available evidence indicates that it was first domesticated in Burma. The first domestic chickens appear in the region east of the Rhine around 600 BC; and it is quite possible that the birds arrived in the loess area west of the Rhine at the same time. After 450 BC the chicken turns up at several sites. It is possible that the duck and the goose were kept as well, but their bones give no clues whatsoever to the status of these birds. They may represent wild fowl, as these species also occurred in the natural environment.

Finds of bones dating to the period 2650 BC–1100 BC are too scarce to allow for descriptions of the livestock that go beyond the species list. From 1100 BC onwards, they become progressively more numerous.

Finds from the period 1100 BC–450 BC reveal sites with a dominance of cattle, followed by pigs and some sheep/goats, sites with a dominance of pigs, followed by cattle and a few sheep/goats, and sites with a high proportion of sheep/goats, followed by pigs and cattle. There is no obvious relationship with subperiods or subregions. A more precise picture of the composition of livestocks and their possible local differences must await future research. Nevertheless, some characteristics are clear. Oxen formed definitively part of the livestock. Moreover, all domestic animals were small. The height at the shoulders of cows and bulls reached 106 cm and 113 cm respectively. Oxen were larger and could reach 122 cm. Pigs were not very small, 77 cm, but ewes reached only c. 56 cm and rams some 64 cm. Goats, although present everywhere, were not numerous enough in the finds to produce estimations of shoulder height. The conclusion is that animals were again smaller than before (see Section 5.4). Some archaeologists think that the trend must be ascribed to the sheer effect of captivity, including curtailing of movement and far from optimal feeding. Be this as it may, size seems not to have been very important to this kind of farming society. Numbers may have been of far greater importance. Wealth may have been (partly?) expressed in terms of numbers of livestock.

The age at slaughter differs per species. Half of the cattle seems to have been slaughtered young, that is before an age of 6.5 years. There is no preference for the age class of 0–2 years. The other half lived to ages of 6.5 to over 11.5 years. Many members of the older age classes are oxen and cows. This implies that oxen were raised for labour and not only for meat. Cows may have been kept longer for reproduction, but possibly also for their milk. Evidence in the form of utensils for milking and cheese making is not available, but this could be due to the fact that such utensils are commonly made of perishable materials.

Pigs were mainly slaughtered young, at least before the age when fast growth came to a halt. Only few animals attained ages beyond their third year. Pigs were obviously kept for their meat.

Sheep/goats were also slaughtered young, before their third year, which seems rather strange, because it would have been expected that both sheep and goats would have been kept for their milk, and, at least sheep, for their wool. In the age classes older than two years, ewes are much more numerous than rams. Ewes were obviously kept for reproduction, but they may have given some milk as well. The presence of spindle whorls and loom weights, both made of ceramic material, attest to

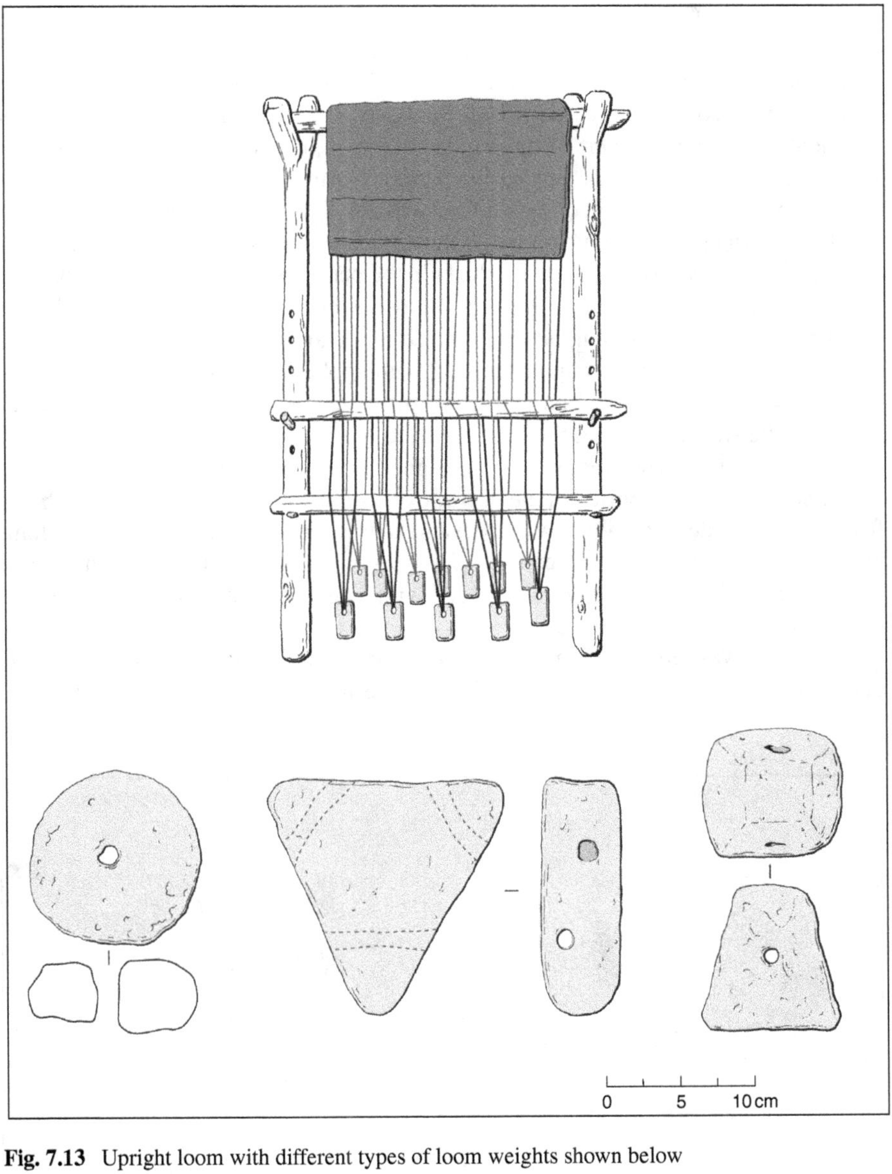

Fig. 7.13 Upright loom with different types of loom weights shown below

spinning and weaving. Spindle whorls, fixed to the end of a spindle, provide extra momentum when twisting fibres by hand. Loom weights, tied to the lower end of the warp threads, hold the warp straight in upright looms (Fig. 7.13). Of course, vegetable fibres, such as flax fibres, are also spun and woven. But woollen fabrics, dating to this period, are known from elsewhere in Europe, and the supposition is not very credible that they were unknown in the loess region west of the Rhine. Felt is another product made from the hair of sheep and goats, but, in fact, felt is hardly mentioned from Central and Western European archaeological sources. But there are no reasons to assume that felt making was unknown.

The horse had in common with the other species that it was a small animal, at least as far as could be ascertained on the basis of scarce finds. But the horse is not mentioned above as part of the regular livestock, because this animal played a special role in the rural economy of the period. Ordinary traction was provided by oxen, drawing ards and wagons with disc wheels. The horse bones, discovered so far, do not turn up in combination with kitchen refuse, so it seems that the horse was not kept for its meat. A clue to its use early in the period are cheek pieces, belonging to bits, and made of antler. The 'bit' itself may have been of string or leather. When metal became more common, bits were made of metal and were of the snaffle type. After 900–800 BC luxury wagons appear, with spoked wheels. These wagons are known because they formed part of grave gifts, destined to accompany a special person to the world hereafter. Such graves have been found in the south-eastern part of the loess region. The wagons were designed to be drawn by horses, though the actual horses were not buried in the grave. Contemporaneous pictures, although not from the area under review, show these wagons drawn by a team harnessed to a long pole (Fig. 7.14). A wheel diameter of 0.70–0.95 m was the norm, as was the use of eight to ten spokes. Wheels were provided with iron tyres. The norm for the wheel gauge was 1.10–1.30 m.

The horse was, obviously, not an ordinary farm animal, but an animal to be ridden, or to draw a luxury or ceremonial vehicle. Stirrups were unknown at the time,

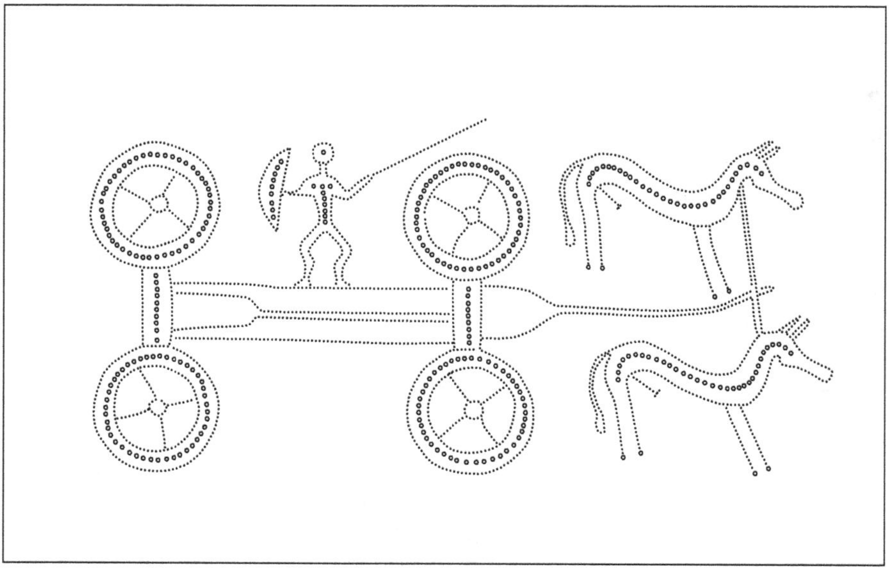

Fig. 7.14 A four-wheeled vehicle drawn by a pair of horses and a warrior riding on top. This image is depicted on the back of a bronze sofa discovered in a chieftain's grave at Eberdingen-Hochdorf, Germany

and riding would have been without them. Horseshoes were not known either. Saddles, if used at all, and harnesses have not been discovered, although some ornaments are interpreted as parts of the harness. But as wood and leather would have been the main materials used in their manufacture, saddles and harnesses can hardly be expected to have survived. Nevertheless, the history of the horse harness elsewhere in Europe shows, that in this period only the throat-and-girth harness was known. This consists of a girth circling the belly and the rear part of the ribs, and a strap crossing the withers diagonally and surrounding the throat. They meet at the spine, where the point of attachment to the wagon is located. It is not an optimal way to harness a horse, as it hampers breathing.

The number of excavated sites containing animal bones dating to the period after 450 BC is much larger and the conclusions concerning livestock become therefore more reliable. Between 450 BC and 250 BC the part of the region that is best known, northern and north-eastern France, shows more or less equal proportions of cattle, pigs and sheep/goats (Fig. 7.15). If there is a slight dominance of one of these categories, it concerns pigs. Animals were small. All settlements revealed

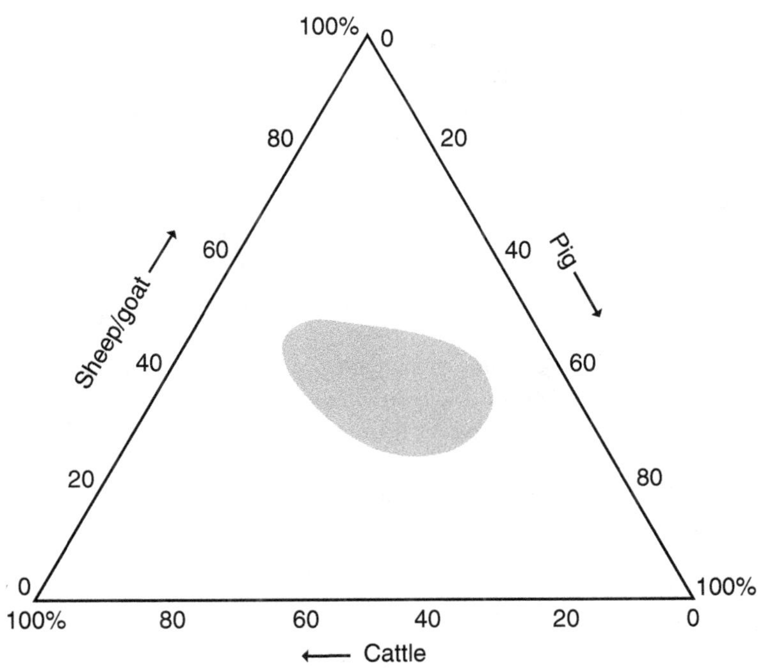

Fig. 7.15 Composition of the livestock in the period 450 BC –250 BC

remains of horses. In one exceptional case, Compiègne 'Le Fond Pernant' (Dept. Oise), remains of horses constituted 26% of all animal bones found. The remains were retrieved from special pits, in which they were found associated with entire vessels. Obviously, this site was not just one of the common domestic sites, but was somehow connected with rituals. In general, horse remains constitute less than 3% of the bones. It is, however, far from certain whether this low percentage reflects the true state of affairs. Most of the bones do not show traces of butchering and it seems that, normally, horse flesh was not consumed. Because domestic waste is the main source of information, horses may be underrepresented in the available set of data.

Some sites revealed chickens, sometimes in percentages up to 1.5%, but also chickens may be underrepresented, because their bones are more fragile than the bones of the larger animals.

The most numerous data concerning livestock are retrieved from farms dating from the period 250 BC–50 BC. During most of this period animals were still small. Cows reached 100–115 cm at the shoulders and bulls were hardly larger. The height of oxen was 120–130 cm. The size of pigs had fallen to 70–75 cm. The height at the shoulder of ewes and rams was 56 cm and 65 cm respectively. Horses measured 110–137 cm. All members of the livestock had a gracile appearance. Ewes are described to resemble roe deer, except for the horns, of course, because most ewes still had horns.

The composition of the livestock had changed compared to the previous period. Cattle steadily lost importance, whilst pigs, and in some sites sheep, had a larger share. In those settlements, where a sufficient number of bones could be identified, goats constituted a minority in the sheep/goats category. The sheep:goats ratio was c. 12:1.

The predominance of either pigs or sheep was not independent of the kind of settlement. During the period large, nucleated settlements appear, in addition to the single farms (see Section 5.5). Bones of sheep have a larger share in the bones retrieved from small sites, whilst pig bones dominate in the refuse of the large sites.

The age of cattle at the time of slaughter varies. In some farms most of the animals did not survive their second year, whilst the remainder lived over nine years. Such farms raised cattle for its meat. In others most of the animals reached ages of over nine years. The oldest animals have been observed in ritual places, where cattle have been sacrified. These sanctuaries revealed dozens and dozens of truly old animals. With the most probable age pyramid in mind, those animals are considered to have lived in large herds of over a hundred head.

Pigs were invariable slaughtered young, an indication that they were providers of meat. A minority was kept for reproduction. A skull of a boar with its canines sawn off during its life-time, found at Varennes-sur-Seine (Dept. Seine-et-Marne), indicates that, despite their small size and gracility, boars could be difficult to handle. The treatment of the canines will have reduced possible danger.

The age classes of sheep follow the pattern observed in cattle. On some farms young animals of up to two years were dominant in the bone spectra, whilst in others sheep bones belonged to animals of four to six years. Still older animals are rare. This suggests that sheep were raised for more than meat alone. Cheese and/or wool were obviously valued products. Perforated vessels, found as sherds in household litter, may have been used, lined with cloth, in a cheese-making process (Fig. 7.16). But, of course, they may have been used to sieve anything. Spindle whorls and loom weights point to spinning and weaving activities. Of the quality of the wool nothing definitive is known, but it would have been coarse, as wool from later periods was also coarser than it is today.

Horses were slaughtered as well, mostly as adults, and were obviously consumed after a working life. The consumption of horsemeat is a break with earlier customs. It may well be that horses had lost some of their special, luxury status.

A substantial part of the chickens did not attain a fully adult age. Apparently they were consumed young. Of eggs there is no trace. But the fowls were possibly not only raised for consumption. The earliest chickens were the property of important people. The birds, and especially the cocks, may have been kept for show, or for cockfighting contests.

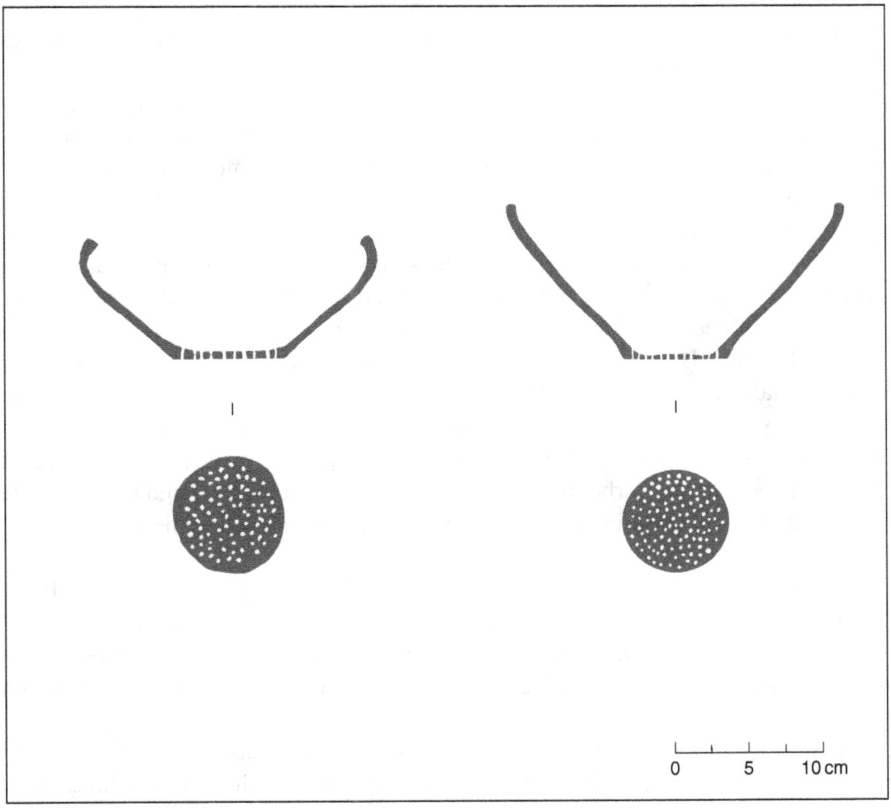

Fig. 7.16 Vessels with perforated bottoms, possibly used in cheese-making

Dogs were present on farms from the very beginning of agriculture, but as they are generally not considered to have formed part of the livestock, they have been ignored so far in this book. However, from 250 BC onwards, the evidence that young dogs were slaughtered for their meat is undeniable. Moreover, find concentrations of only paws (paws being the parts of the skeleton which remain attached to the skin after skinning), suggest that their skins were valued as such and tanned into leather. Such concentrations are especially found in the nucleated settlements. Thus, dogs may have been bred for their skin as well.

As mentioned above, farm animals were small, sometimes even extremely small according to modern standards. But, from c. 100 BC onwards, larger animals appear, in small numbers, in the records. Most of these are cattle and horses. As far as can be ascertained, they turn up principally in the southern part of the region. They include cows of 120 cm at the shoulders, bulls of 124 cm and horses of over 140 cm. They are found side by side with the customary small livestock, but not on every farm. Farms where they have been found also reveal foreign goods such as wine vessels (amphorae) from Italy. There is obviously an influx of larger sized livestock from elsewhere. That 'elsewhere' must be sought in a southerly direction. There are strong indications that the people living in the southern part of the loess region maintained relationships with the Mediterranean world, even Italy. One of the established commercial routes followed the rivers Rhone and Saône to the upper courses of the rivers Seine and Moselle (Fig. 7.17).

The next topic is, Where did the livestock feed? As mentioned in Chapter 6, primeval forest would not have covered most of the surface. Open spaces and secondary forest must have prevailed. This would have meant more food for animals. In primeval forests the growth of herbs and shrubs is restricted. After degradation of the forest, browsers and grazers find more to eat.

It is presumed that most of the livestock fed in the open air, and was kept in check by a herdsman, possibly aided by dogs. At least from 250 BC onwards several different breeds of dogs were present, and some of these may well have been shepherds' dogs. However, for special animals such as draught oxen, horses, or the large animals present after 100 BC, additional feed may have been needed. One of these is hay. Assemblages of seeds, which could be interpreted as the remnants of rough hay, are incidentally found in contexts dating from after 450 BC. Stalks of grasses do not survive carbonisation, but herb seeds from pastures and meadows do. The scythes, mentioned in Section 7.3, may have been used to collect this kind of fodder.

Other types of animal feed were probably hulled barley and the pulses bitter vetch and common vetch. Both vetches are, without treatment, toxic for humans and horses, but not for cattle and sheep (see Section 7.2). It is striking that common vetch appears when the new animal breeds arrive. There may be a connection between the two.

As will be explained in the next section, most of the animals were kept year round in the open air. In some periods, regions and establishments, however, stalling under a roof may have been practised.

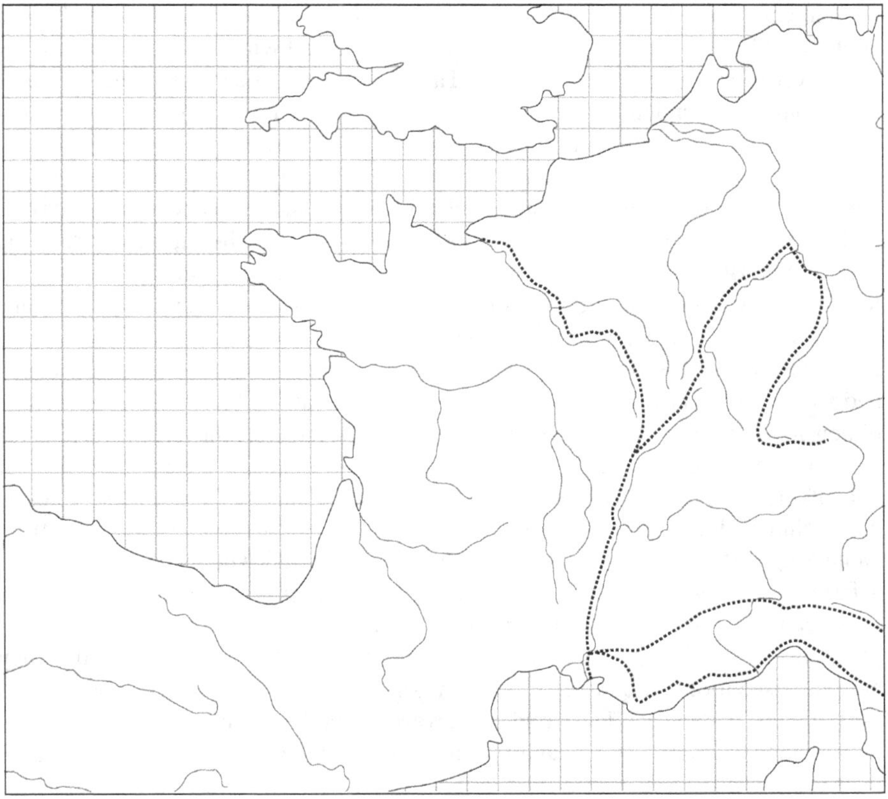

Fig. 7.17 Trading routes connecting the Mediterranean world with the loess region

7.5 Farmbuildings and Yards

The core of rural life was the single farm. It consisted of minimally one main building and one or several outhouses, such as granaries. Quite often, one or two underground silos were present as well, but this was not the case everywhere. Wells were rare in most regions, which is rather peculiar, as the phenomenon of the well was very familiar in Europe at the time. Most wells are reported from the Moselle area. Fences are hardly mentioned before 800 BC, but this absence is possibly not a true absence, because fences are commonly not inserted deeply into the ground and their faint traces, if preserved at all, are not easily recognised.

Archaeologists have trouble detecting settlements dating from the period 2650 BC–800 BC, a long stretch of time. That the region was not devoid of people is proven by burials. Their graves are well known. The most likely explanation is that, in general, buildings lacked deep foundations. What was left after abandonment was easy prey to soil erosion or later ploughing activities. Some archaeologists even suggest that the frames of the houses were not entrenched into the ground at all, but were erected on sills, laid directly on the surface. Indeed, from the end of the 2650 BC–800 BC period, buildings have been detected with walls erected directly on a layer of stone rubble (see below). Had the stone not been there, the buildings would have gone undetected. Nevertheless, constructing frames by erecting posts in postholes was also still practised, just as in previous times. Traces of these postholes have been found regularly. As far as is known, whatever their foundations, buildings consisted of a wooden frame supplemented by wattle and daub and a thatched roof.

Information dating from before 1800 BC is virtually absent, while from the period 1800 BC–1100 BC some buildings have been reported. During this time, the loess region was divided into two cultural subregions: an eastern and a western one.

The eastern part comprised the German Rhineland, Southern Limburg in the Netherlands, the Moselle area, French Lorraine and presumably the areas in between and to the east. The earliest houses are large, three-aisled buildings with average dimensions of 18 × 6 m. The central aisle is wider than the others. The short sides are semicircular, suggesting a hipped roof. Entrances were in the middle of one of the long walls, or both, whilst an entrance in one of the short sides is not excluded (Fig. 7.18). Such houses resemble the Bronze Age longhouses from Denmark, north-western Germany and the Netherlands, where the longhouse is a farmhouse in which two separate parts can be discerned, left and right of the entrance in the long wall. One half is interpreted as a stable, because in some house-plans the side aisles in this part show partitions perpendicular to the wall. They are reminiscent of cattle boxes. The other part lacks these boxes and is interpreted as the living part. In the loess region, however, such boxes are missing in the few house-plans excavated so far; and it is not certain that animals were stabled under their roof.

The successors to these three-aisled houses were two-aisled houses. The rounded ends disappeared, and the plans became rectangular. Houses also became shorter, some 12 × 5 m. In one settlement, Rosières-aux-Salins (Dept. Meurthe-et-Moselle), a vestige of a division into two different parts can still be seen. The position of roof-bearing posts suggests the existence of a loft (Fig. 7.18). Other sites show plainer plans.

The main house of the period was accompanied by four-poster and/or six-poster outhouses (Fig. 7.19a and b). The four-posters are interpreted as granaries. The six-posters may have been granaries as well, although they can have functioned as sheds or even byres. But if they were used for animal shelter, it must have been for only a part of the livestock. Two such rectangular six-poster outhouses, accompanying a two-aisled longhouse with rounded short sides, excavated at Hettange-Grande (Dept. Moselle), measured 4 × 2.9 m and 4 × 3.5 m respectively, surely not enough to house all animals belonging to such an establishment even if they were as small as described in Section 7.4. Comparable six-poster outhouses, with lengths of 4.1–5.8 m and widths of 2.9–4.5 m have been described as having probably the same age as the two-aisled houses of Rosières-aux-Salins mentioned above. Yards also had silos, although these are not reported as being present everywhere. Some of these pits encased a big piece of pottery, which was the true container for something to be stored (Fig. 7.20).

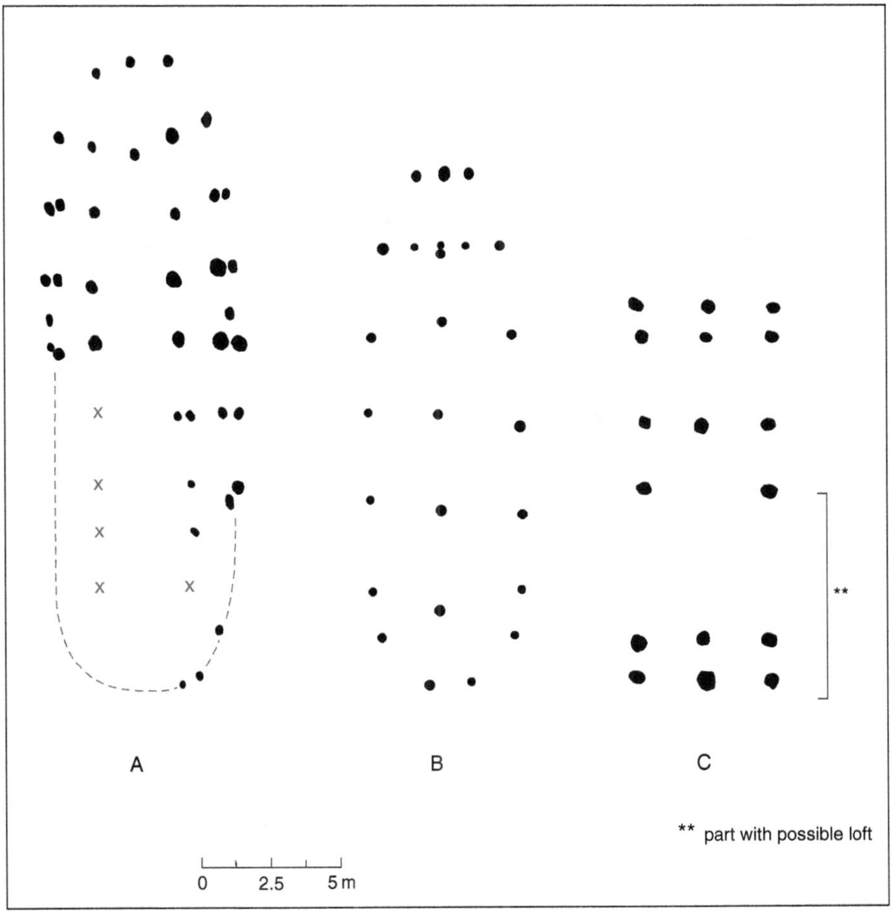

Fig. 7.18 Buildings from the period 1800 BC –1100 BC. *Left*: a three-aisled longhouse, excavated in Frouard (Dept. Meurthe-et-Moselle), *middle*: a two-aisled house from Hettange (Dept. Moselle) and *right*: a rectangular house with a possible loft, from Rosières-aux-Salins (Dept. Meurthe-et-Moselle), all in France

The farms of the western cultural subregion are even less well known. There is a debate going on whether some circular structures should be regarded as regular farmhouses. An example is a circular plan, excavated at Roeux (Dept. Pas-de-Calais). Its traces consist of ten posts of which two are paired, leaving a construction of essentially eight upright posts. These posts are spaced 2 m apart, except for the south-east where a spacing of 4 m is observed (Fig. 7.21). This is interpreted as the entrance. The best parallels for such constructions are to be found in southern England, where circular houses of similar size were the norm. Indeed, the material culture has much in common with that of the other side of the Channel. North-western France and southern England shared a cultural tradition in metalworking, pottery and burial customs during this period. Archaeologists speak of 'the Atlantic

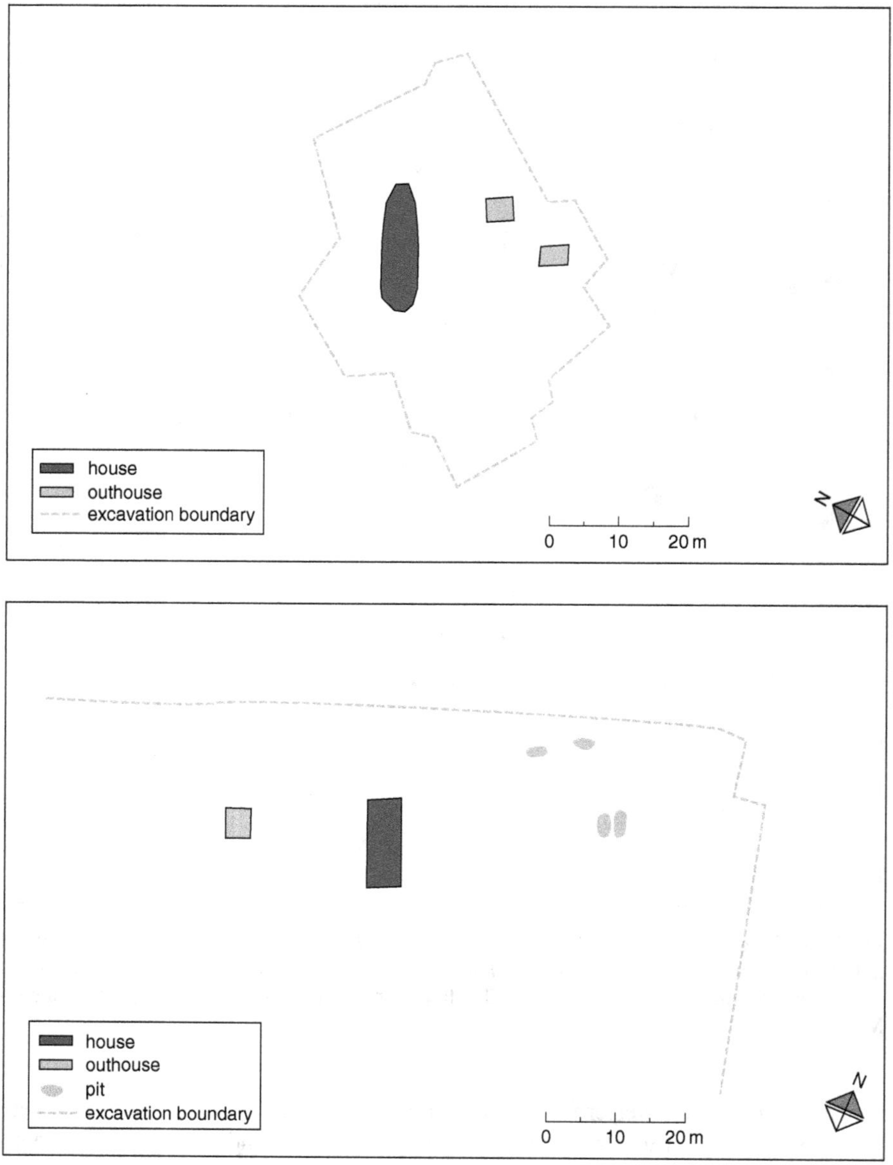

Fig. 7.19 **a** The yard of Hettange. **b** The yard of a farm at Rosières-aux-Salins

Tradition'. The sea was not an insurmountable barrier at all. A seaworthy ship, dating from c. 1300 BC, has been unearthed at Dover in England. It could have carried up to three tonnes of cargo. It seems, therefore, permissable to look at England for a model of rural settlement. The English sites show dispersed farms with round houses, rectangular outhouses and four-poster granaries in their yards. Some yards

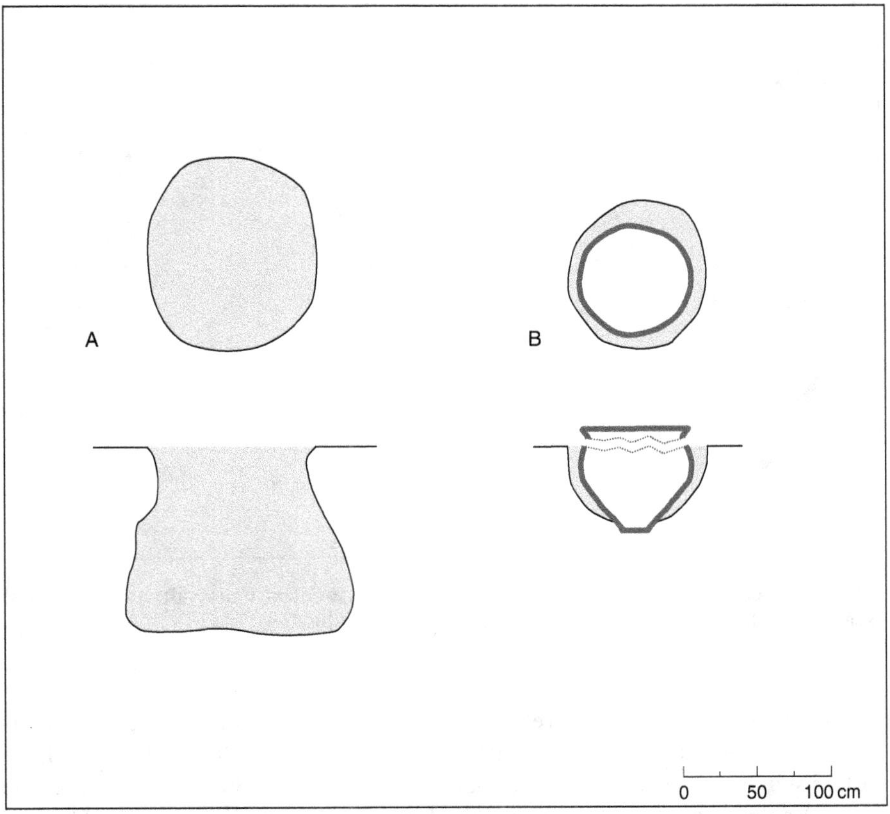

Fig. 7.20 Silo without (**A**) and with a ceramic container (**B**) shown from above and in section. The rim of the vessel was shattered by later ploughing

were set into enclosures with multiple entries, so-called causewayed camps. Such an enclosure, with a diameter of 130 m, has been found in north-western France at Etaples (Dept. Pas-de-Calais). It has a founding date of c. 1500 BC. Similar enclosures, especially found along the river Somme, are interpreted as possible locations of comparable establishments. Nevertheless, some caution is necessary. Circular structures are very common in the ceremonial sector of society, also in other parts of the loess region. It cannot be concluded, therefore, that every round structure is necessarily a farm.

The period of 1100 BC–800 BC provides more information. A two-aisled rectangular building, found in the north-eastern part of the loess region at Sittard-Hoogveld in the Netherlands, still resembles the two-aisled houses of the second part of the period described above for the Lorraine in France. Its dimensions are 9.3 × 3.9 m. The tradition may have lingered on in this part. But elsewhere the farmhouse has changed to a stereotypical, rectangular one-aisled building, measuring 9 × 5 m on average (Fig. 7.21). It is possible that these structures are slightly

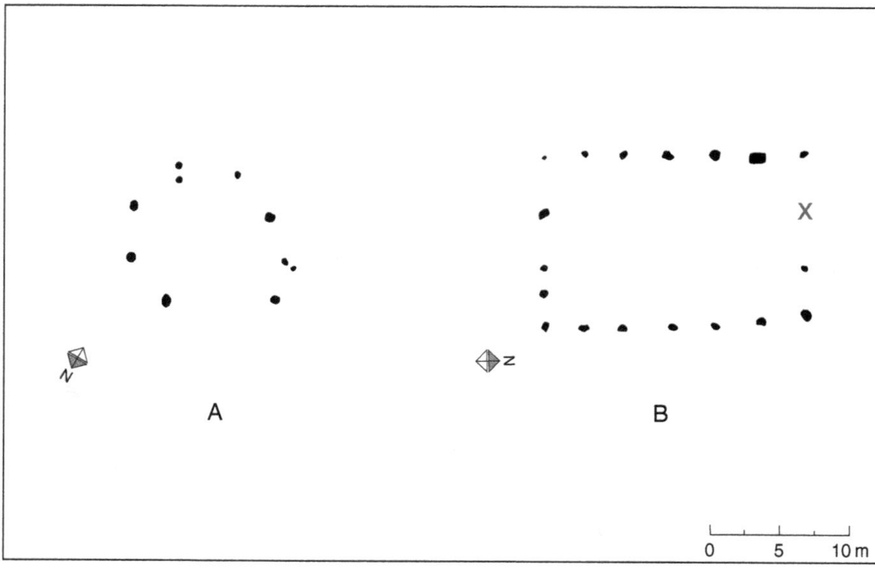

Fig. 7.21 **A**: circular house excavated at Roeux (Dept. Pas-de-Calais) France. **B**: rectangular house excavated on the terrain of the Aéroport de Lorraine (Dept. Moselle), France

younger than the Sittard-Hoogveld house, but another possiblity is that the difference must be attributed, again, to a different cultural sphere. During this period, the eastern subregion had split up into a southern part, the Rhin-Suisse-France orientale (RSFO) culture, and a northern part, the Urnfield Culture. However, the Urnfield Culture houses proposed for the German Rhineland, measuring some 4.5 × 3.5 m, one-aisled and with six uprights only, do not match the Sittard-Hoogveld house, although they belong to the same cultural sphere. They are almost identical to the six-poster outhouses of the period before and whether they actually represent the main farmhouse, a question to be considered. Still, no larger buildings have been detected so far. It looks as if the eastern sphere has split up into several provinces, each with its own building tradition. Nevertheless, it is quite possible that the differences are due to a lack of sufficient data and/or differences in subperiod. Notwithstanding the differences, all subregions still show the four-poster granary and the underground silo.

Some archaeologists think that in the western subregion houses were still round, but definitive proof is absent. Circular structures have been encountered, but it is questionable whether they represent ordinary dwellings. No other types of houses have been described so far.

The single farm was, however, not the only kind of site present during this period. At certain places in the southern part of the loess region, so-called hilltop settlements (*camps de hauteur*) appeared, presenting a more nucleated type of settlement surrounded by walls and ditches (see Section 7.6). One of these camps, Catenoy (Dept. Oise) revealed one-aisled rectangular buildings with walls erected on a foundation layer of stone rubble. They have a width of 4.5 m, but the determination of their length awaits continuation of the excavations. It would go too far to describe them as farm buildings, because the camps were certainly more than just an agglomeration of farms.

Buildings, similar to those found at Catenoy, have been excavated at a site dated to the very beginning of what is considered to be the next period, 800 BC–450 BC. The settlement in question, Choisy-au-Bac 'le Confluent' (Dept. Oise) is a kind of lowland version of the hilltop settlements described above. Founded on a triangular piece of land where two rivers meet, it was, at least during some time of its occupation, defended by a wall on the land side. Occasional, gentle floodings have covered successions of house generations. The walls of the dwellings were founded on a layer of rubble, consisting of pottery sherds, fragments of bone, and, in later phases, stone. There are no traces of roof-bearing posts in the centre of the structure. The first phase of occupation comprised at least four houses, with lengths of 4.8–6 m and widths of 3.6–4.8 m. Three ovens, outside the houses, were obviously destined for the working of bronze as attested by the associated waste. In the subsequent phases house-plans became slightly larger, 6–8 m long and 4–5 m wide, and also more numerous. Ovens for making pottery and working bronze or iron were outside. It is questionable whether this settlement represents the farming village of the period. Most probably not. Nevertheless, Choisy-au-Bac shows that house-building was not necessarily based on wooden skeletons of upright posts set into the ground. This explains perhaps why in some areas still no houses dating from the 800 BC–450 BC period have been found, at least not main buildings. What has been found is the granary, because this was still sitting on its sturdy posts, entrenched deeply into the subsoil.

Notwithstanding the above-mentioned remarks on houses without deep foundations, the tradition of erecting buildings with posts other than granaries subsisted in the southernmost, eastern and northern part of the loess region. Below, some examples will be presented.

A single two-aisled farmhouse, 15 × 8.5 m, rectangular with rounded corners, and accompanied by a four-poster granary of 3.5 × 3.5 m, excavated at Barbey 'La Hay Guyonne' (Dept. Seine-et-Marne), represents possibly the common rural site of the south. Underground silos are not reported and the yard comprised only some indifferent pits (Fig. 7.22).

Something similar, but now with a rectangular three-aisled house of some 12 × 7 m and a four-poster granary of 2.6 × 2.6 m, was found at Joey-aux-Arches 'La Machotte' (Dept. Moselle) (Fig. 7.23). Another complex consisting of a three-aisled house and surrounded by four granaries was detected in the same area, on the location of the regional Airport of Lorraine, south of Metz (Dept. Moselle).

Two instances in the German Rhineland revealed rectangular main buildings with two-aisles. In Hambach 502, the house measured 9 × 4 m. Its yard held an outhouse based on nine posts, a second one based on six posts, two four-poster granaries and two underground silos. The other, at Inden-Altdorf, measured 11 × 4.5 m. Its yard also had a nine-poster outhouse, a six-poster outhouse, one four-poster granary, but no silo (Fig. 7.24). The houses resemble the type found in Sittard-Hoogveld,

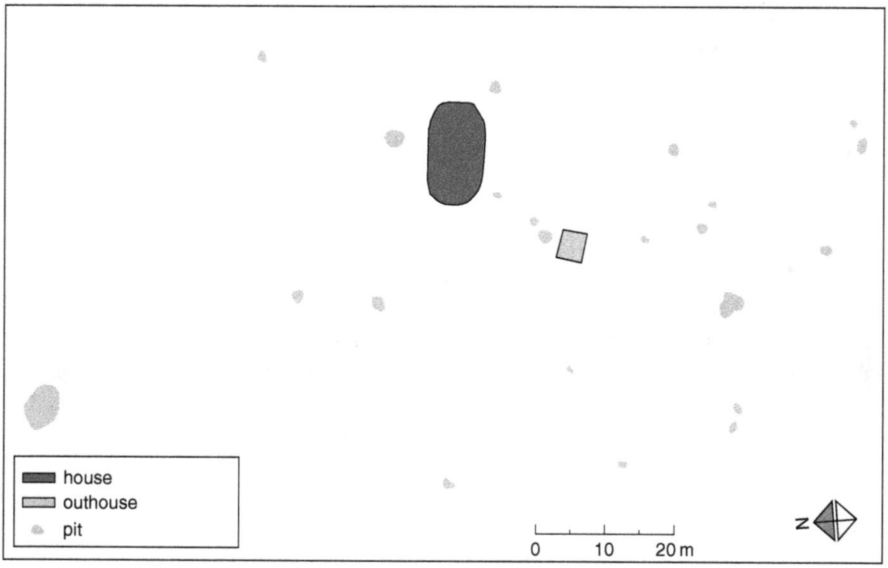

Fig. 7.22 The yard of a farm excavated at Barbey (Dept. Seine-et-Marne), France

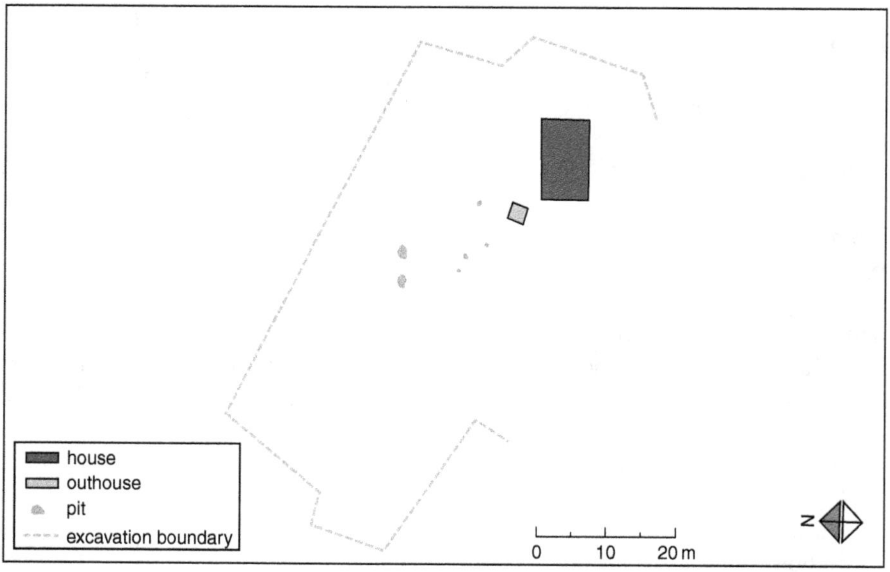

Fig. 7.23 The yard of a farm excavated at Joey-aux-Arches (Dept. Moselle), France

described above. The Sittard and Inden-Altdorf houses even share a cultural peculiarity. The north-western corner posthole of the latter was filled with a large number of sherds, belonging to the same pot. In Sittard-Hoogveld, the north-eastern corner posthole contained an unusual large number of sherds. Removing a corner post,

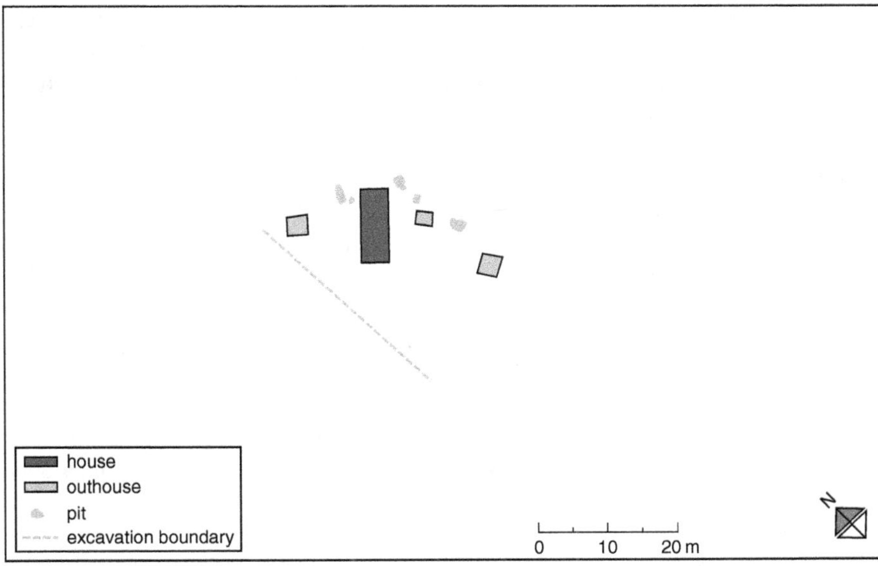

Fig. 7.24 The yard of a farm excavated at Inden-Altdorf, Germany

and filling the hole with sherds, may have been part of a ritual connected with the abandonment of the place.

It would have been satisfying if a yard with a rectangular two-aisled house, some outbuildings and one or two granaries had turned out to be the standard for the north-eastern part of the loess region, but this is not the case. In the very site of Sittard-Hoogveld, the period 800 BC–450 BC farmhouse is a three-aisled building. It measures 12 × 7.5 m and is accompanied by three outhouses. One is an eight-poster, 3.25 × 2.5, a second a six-poster, 3.5 × 2.5 m, whilst the third was too damaged to provide a length or width. A two-aisled house-plan excavated in nearby Geleen, measuring 12.6 × 4.8 m, revealed one straight and one semi-circular short side. It was flanked by a nine-poster outhouse and a four-poster granary. One of the postholes in the middle of a long side was found filled with a large number of sherds, possibly a ritual deposit, which connects this house with the German houses described above, though it differs in its general shape.

In view of these examples, the conclusion seems to be justified that the general plan of the farmhouse is far from uniform. One-aisled structures are absent from the yards best documented, and therefore presented here, but some authors maintain that even one-aisled houses existed as well. Nevertheless, the houses also had a lot in common. Frames were of wood, walls of wattle and daub, and roofs presumably thatched, although wooden shingles cannot be excluded. Shingles are known from areas east of the Rhine. In most cases, or perhaps all, the entrance was at one of the short sides. They had a length between 11 and 15 m. Traces of cattle boxes are absent and authors agree that the main houses were for people to live in, not to share with livestock. Every yard contained at least one granary. Larger outhouses and silos seem to have been less a standard facility on yards. In all examples wells are missing, notwithstanding the fact that well-digging was known in the period. The entire complex was an entity on its own, a single farm. There is no structured relationship to neighbouring farms.

At the end of the period 800 BC–450 BC, a new phenomenon turns up: large agglomerations of either granaries or silos, especially in the eastern and southern part of the region. Clusters of granaries number 15–30 structures, sometimes enclosed by a fence (Fig. 7.25). They are situated in the immediate vicinity of a settlement. Clusters of silos lie in the open. They can contain as many as 80 of these pits. In every case a few silos intersect, showing that not all can have been contemporaneous, but still, quite a few probably were. The clusters are sometimes situated next to settlements, but they can also lie at a distance of several hundred metres (Fig. 7.26).

The occurrence of the clusters corresponds with a relative climatic optimum occurring in the fifth-fourth century BC. In the centuries before then, starting around

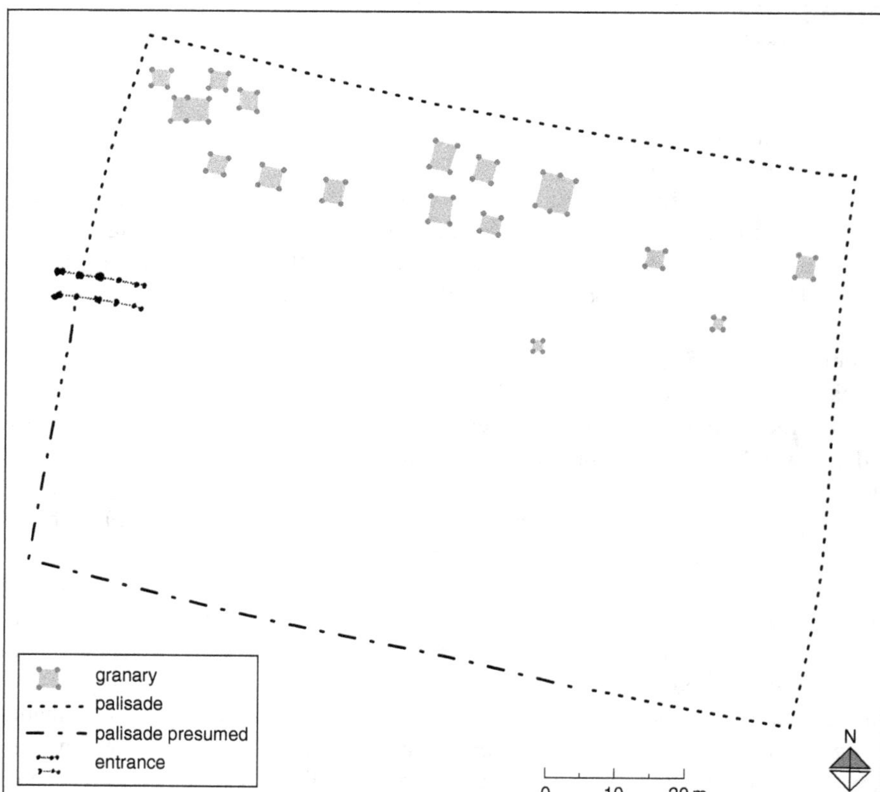

Fig. 7.25 Cluster of granaries in an enclosure. Example from Bucy-le-Long (Dept. Aisne), France

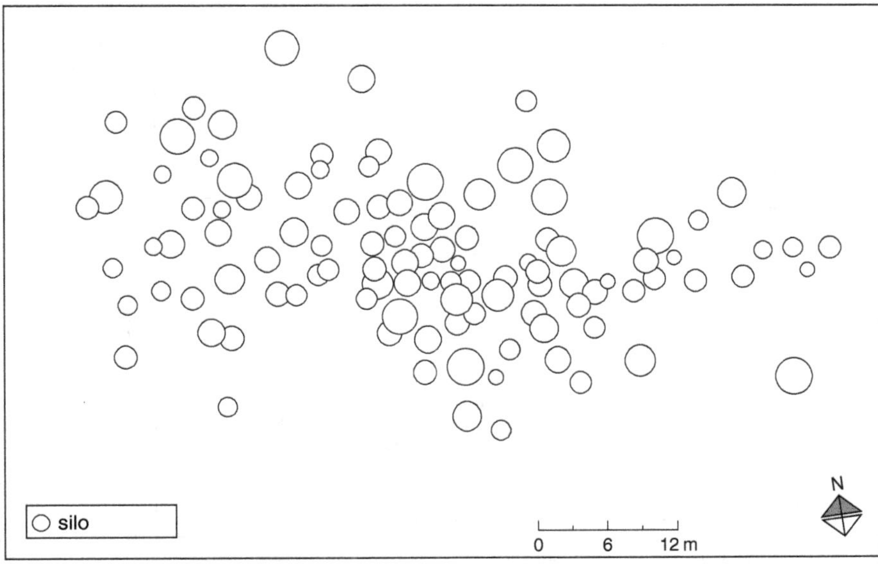

Fig. 7.26 Cluster of silos. Example from Soupir (Dept. Aisne), France. Overlapping silos show that not all were contemporaneous

800 BC, there was a dip in the general climate towards colder and wetter conditions. In the fifth century conditions became presumably much better for agriculture. It may be that cereal production was given such a boost that true surpluses were generated.

The custom of storing products in special areas disappeared during the first half of the next period, 450 BC–50 BC. Storage returns to individual farmyards. But there the resemblance to earlier periods ends. Changes at the level of the organisation of sites appear. In the north-east, change is directed towards a clustering of farms into small hamlets. In the east and south, farms present differences in size and wealth. Ditches defined the yards, or, better, the banks realised by digging the ditches. But banks do not survive ages of intensive land use, whilst a filled-in ditch does and can be found during excavations. Indeed, most ditches were obviously badly kept and soon filled with waste. In some cases the ditch seems to have served for drainage, but this was not the rule. The bank was the main part. It was sometimes supplemented by a palisade. Hedges may also have played a role.

At first, ditches were laid out along crooked lines, but in the course of time they became straighter and straighter (Fig. 7.27). The enclosure is considered to have served as a territorial marker, or to keep animals in or out of the premises, or both.

In northern France the surface of the enclosed yards falls into four classes. The smallest type covers less than 5000 m^2, and this applies to 45% of the 142 enclosures known sufficiently to be measured. The second class, which covers 5000–9000 m^2,

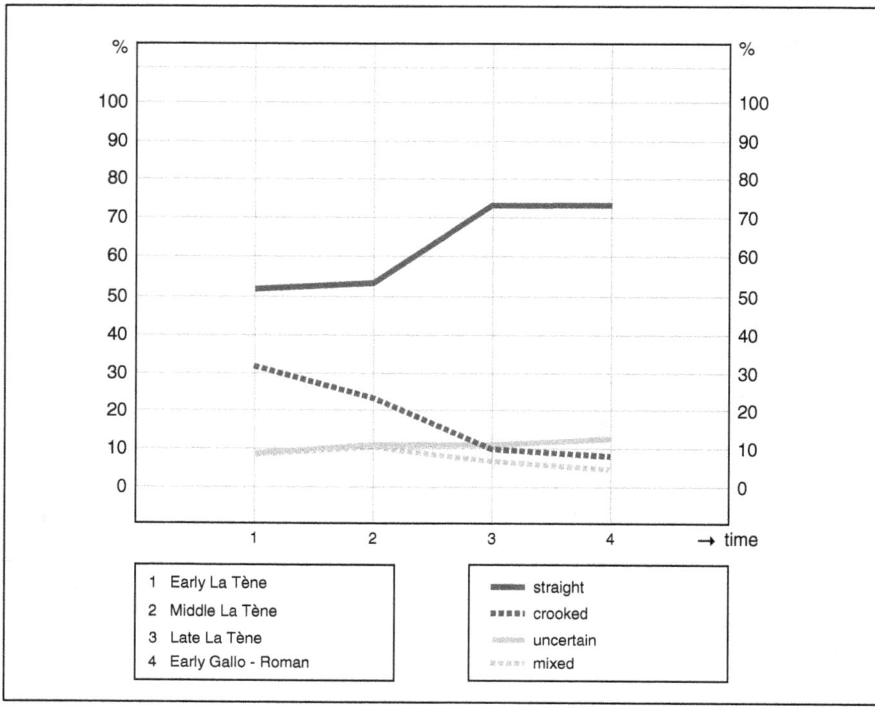

Fig. 7.27 The lay-out of ditches surrounding yards through time, showing that crooked lay-outs disappear

comprises 33% of the yards. The third, 12%, has surfaces between 9000–12000 m^2. And the 10% of enclosures belonging to the fourth class cover more than 12000 m^2. Classes are not tied to specific areas (Fig. 7.28).

French archaeologists could also distinguish four classes in farms. These four do not tally exactly with the four classes in enclosures, because the smallest farm (farm class 4) had no enclosure. It consists of a single building and some pits. Establishments of this class may not have been independent, but satellites of other farms. Or they belonged to the poor. Farm class 3 is the most common. It has a modest farmhouse, surrounded by several smaller structures (outhouses?), granaries and silos. Farm class 2 has more differentiation between the main house and other houses, which are considered to have been inhabited by lower class families. This is concluded on the basis of finds associated with these buildings. The inhabitants of the biggest house had more luxury goods to throw away. Outbuildings, granaries and silos are part of the yard's inventory as well. The distribution of waste indicates in certain cases the existence of special areas used for threshing, butchering and forging iron in the open air (Fig. 7.29).

Class 1 is the least common. It is characterised by a house, set apart by its own enclosure from the living and working areas of the farmhands.

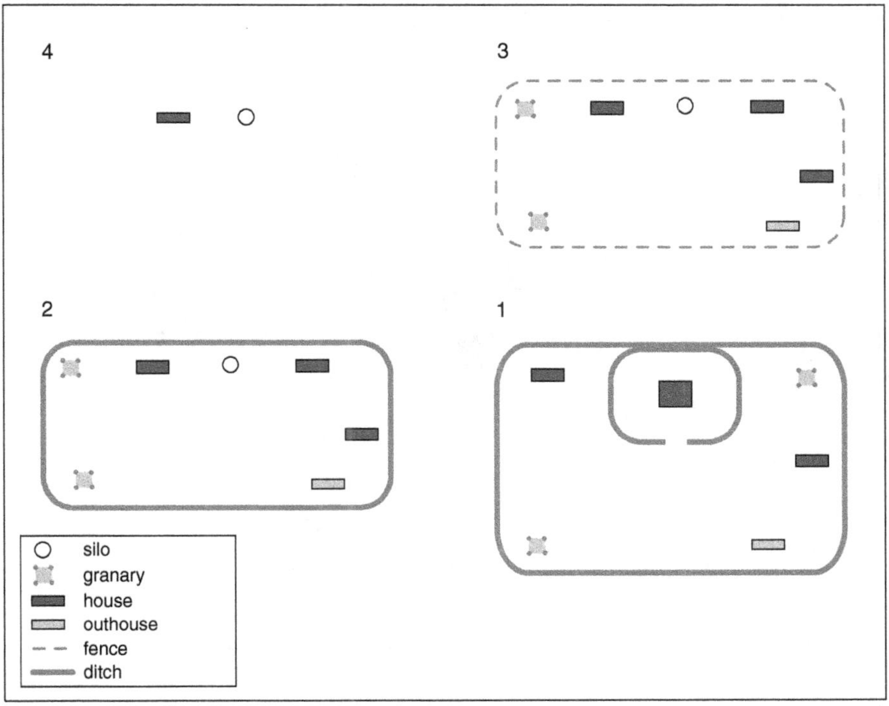

Fig. 7.28 Classes of farms

The exact architecture of the buildings varies according to the subperiod and/or subregion. Many different ground-plans are reported. Only the circular structures, pointed out in previous times in the west, have disappeared, at least in the loess region. Structures were made of wood and wattle and daub, and presumably with thatched roofs. It is rather surprising that even in this advanced period of the metal-using age, iron nails were not used in building.

Strikingly enough, most descriptions of farms and farmyards, again, do not mention a well. Wells were known at the time and a few have been found, but they are rare. Most establishments seem still to have obtained their water from surface water, most often a river or other stream nearby.

Ditches were not limited to the surroundings of yards. They are also found much further afield, in the shape of long lines, bending at right angles, and organising the landscape into neat parcels (Fig. 7.30). The smallest parcels are found near the yards. Their size increases with distance. Several French studies even suggest a standardisation of sizes, at least in some areas. It is this parcelling that is referred to in Section 7.3. In that section differences in size were connected with possible differences in cultivation regimes. But it is also possible that in this late stage of the period under review, the livestock was kept under more controlled circumstances than before. Certain parcels, enclosed by banks and hedges, may not have been used

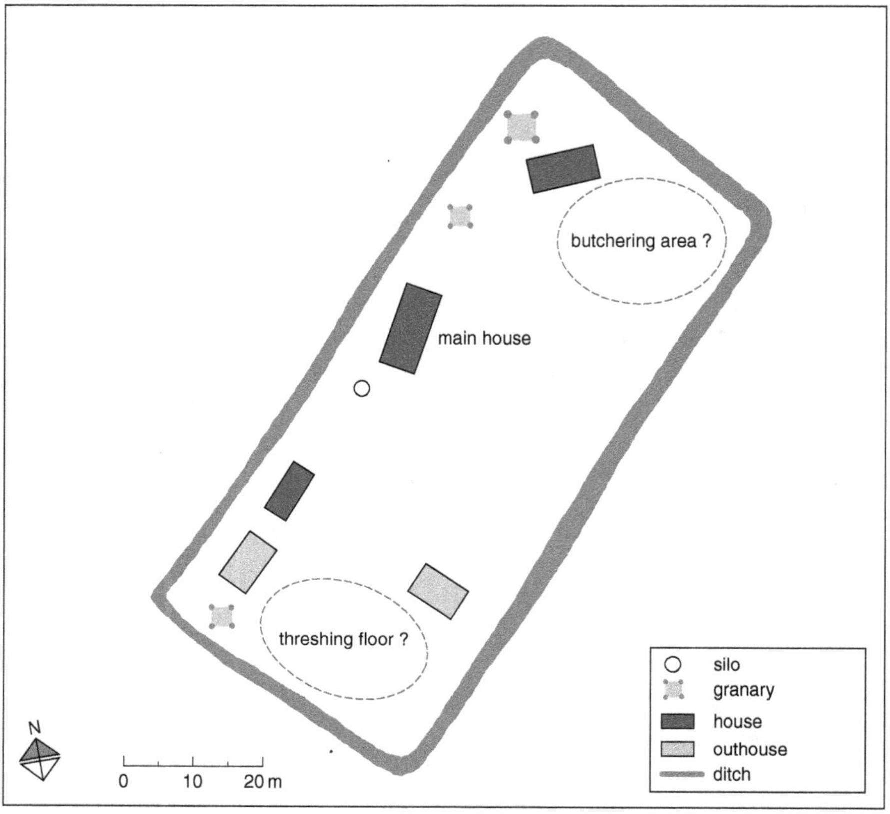

Fig. 7.29 The yard of a class 2 farm. Example from Verberie 'La Plaine Herneuse'(Dept. Oise), France

to grow things on, but to keep under control the most valuable animals, such as certain horses or the new, large animals introduced after 100 BC. It is possibly a period in which not every animal was roaming more or less freely in the surroundings.

The differentiation in size and importance of the single farm was not the only phenomenon of the last period. As mentioned above, in the north farms tended to cluster into small hamlets by now. But clustering of dwellings went further. Central places, called oppida, came into existence. But as these were not directly connected with the producer side of agricultural life, they will be dealt with in Section 7.6.

To sum up, there is not yet much to tell about development and details of the farms and farmyards. It is clear that they existed during the long stretch of time discussed here. But some outlines can be sketched. In the beginning the loess region was presumably divided into an eastern and a western subregion. The eastern is characterised by long main buildings with semicircular short sides. The western subregion is hardly known, but farms may have looked 'English', with round structures. Later on, this 'Atlantic Tradition' subregion seems to have become smaller

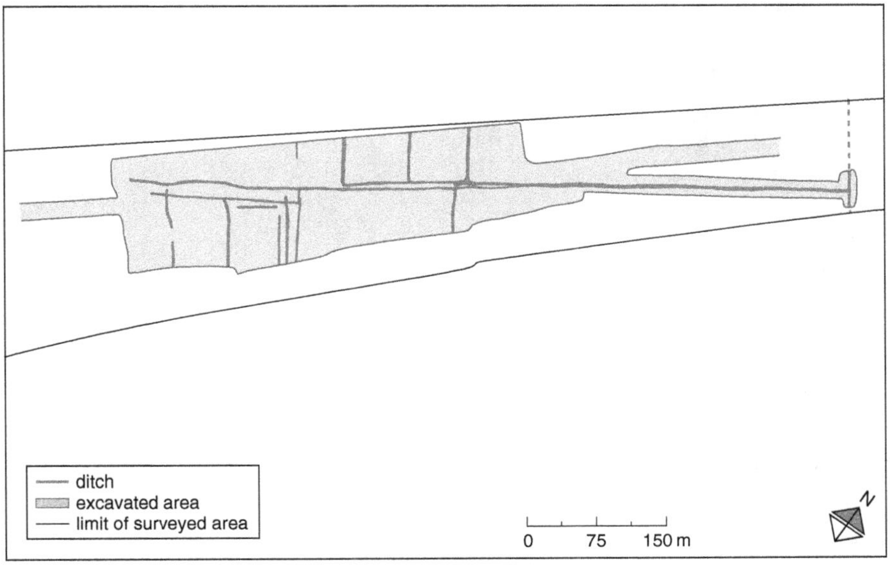

Fig. 7.30 Parcelling excavated at Fresnes-lès-Montauban (Dept. Pas-de-Calais), France

and smaller, to disappear as a separate entity in the end. In the course of time the main house shrank in size, mainly due to a reduction of its length. Analogous to the North-western European longhouses, half of the early buildings may have been used for the stabling of livestock and the other half for living in, although this has not been proven, but this style of building knew no continuation. A reason may have been that the two parts were split up into separate structures. One building, the main house, was intended for people, and the second one, an outhouse, for livestock. A comparison of the surface, covered by roofs, of houses and outhouses (granaries excluded) within one yard, for the period 1800 BC–450 BC, that is before the separation into farm classes, tells us that the total of roofed-over square metres has remained the same or shows a weak tendency to fall. This is at least the outcome of an analysis of the few yards that have been described for the two areas best known, Lorraine and the German Rhineland (combined with Sittard in the Netherlands) (Table 7.5). The amount of data is low, however, and this kind of analysis should be repeated when more yards have become available for study. After 450 BC the single farm is divided into classes, ranging from very simple and poor, to complex and rich.

7.6 The Farm in Its Setting

As a rule the farm was in many respects a self-supporting unit, dealing with both crop cultivation and tending livestock. This remained so during the whole period. Farmsteads were dispersed over the countryside. They were to be found everywhere,

Table 7.5 Floor-space covered by roofs in yards of different Metal Age periods

Space roofed-over in m^2	Main house	Outhouse	Total
Lorraine			
1800 BC –1100 BC	94	0	94
	88	12	100
	66.5	0	66.5
	50.3	22	72.3
1100 BC –800 BC	45	0	45
800 BC –450 BC	84	0	84
Rhineland			
1800 BC –1100 BC	94.5	0	94.5
1100 BC –800 BC	36.3	0	36.3
800 BC –450 BC	49.5	18.1	67.6
	90	16.8	106.6
	66	0	66

with the exception of vast plateaus far from open water. This does not mean that rural life went on without any changes. The attitude of the small farming communities towards their immediate surroundings did not remain the same. And the farm was always part of wide-reaching social systems. Developments elsewhere in Europe did not leave the loess region west of the Rhine untouched.

Most information has become available from the north-eastern and eastern subregions. Until 1100 BC the farmhouse and yard as a fixed material entity was a rather short-lived affair. Estimations of occupation vary between one and two generations of farmers. Buildings had a limited durability because of the kind of construction material and the way this was used. Wooden posts, set directly into the ground, do not last forever. People buried their dead next to the farm. Once the main building became too decrepit, the inhabitants did not rebuild their house in the same yard, but transferred the complete farmyard, buildings and all, to another location within their territory. In this way complete farms wandered over the landscape (Fig. 7.31). The distance between the old and new place was not that great, at most some hundreds of metres. The system could be detected, because people who died were now buried next to the new yard. And, as mentioned before, buried people are easier to find than traces of their farms. The distance of displacement is not important enough to suggest an economic reason connected with, for instance, the deterioration of arable soil as the sole reason. There must have been a cultural background to this shifting of settlement.

The number of people living on a single farm can only be guessed at. Unfortunately, it is not possible to calculate the living from the dead. Not everybody seems to have been buried in the graves belonging to an establishment. Children,

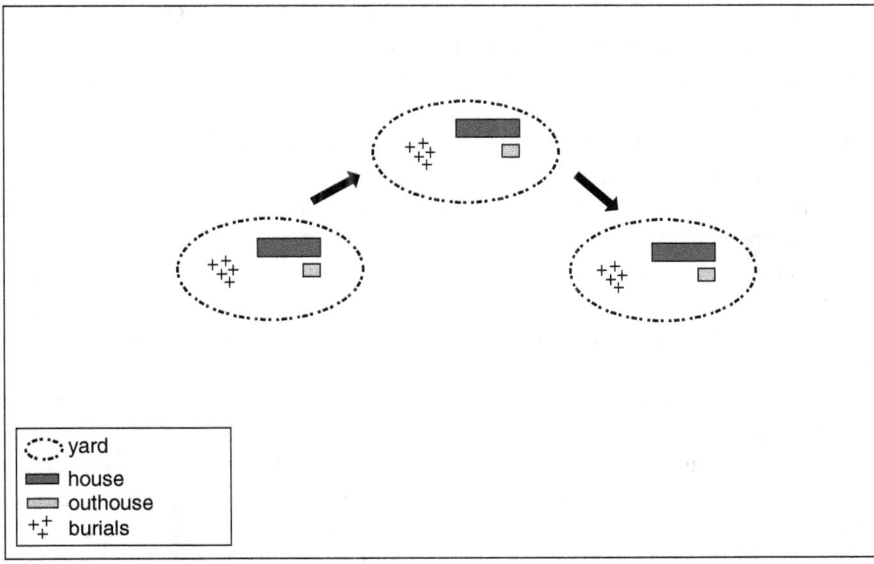

Fig. 7.31 Farms wandering over the landscape

for instance, are underrepresented. The large farms of the earlier part of the period could have been inhabited by eight to ten persons. This is, at least, the estimate for the longhouses of North-Western Europe beyond the loess region. The somewhat smaller buildings of the later part could have housed six to eight persons.

It is difficult to establish how many yards were present per square kilometre. Because of their loose dispersal they are hard to catch in archaeological research. Counting the small groups of graves provides no solution either, because, even if covered by barrows, a substantial number were erased by the action of farmers in later periods.

Although the single farms produced both crops and livestock, they cannot have survived independently of each other. They must have been united by a social network and even local networks cannot have been sufficient to provide for all needs. The most obvious need is for metal, in this subperiod bronze. The main constituents of this alloy, copper and tin, were not available within the loess region. The nearest areas where copper was to be found were the eastern and south-western Alps, and southern France. Copper was mined, for instance, in the neighbourhood of Montpellier. Other sources were England and Ireland. Tin could be obtained from the Erzgebirge in Central Europe, from south-western England, and, nearer, from Brittanny. The two metals were not alloyed in the region. Bronze came ready-made and was obtained via extended social networks. How these exactly functioned is not yet clear, nor who cast the first implements and where. When sufficient bronze came into circulation, the remelting of worn implements and scrap made the local casting of bronze possible and more widespread. But whether this was done regularly on farms in a domestic sphere, or was done by specialists, is not well understood. Other crafts, like spinning, weaving and the making of household pottery, were certainly local affairs. But the production of fine ceramic wares was possibly the domain of specialists, too.

It is clear that the farmers of the subperiod lived in a world of supraregional networks, which extended far into Europe. These contacts brought them not only metal and luxury wares, but also new crop plants, such as foxtail millet, and, of course, the horse and the spoked wheel.

The study of mortuary practices, and the gifts going with them, revealed that society was not completely egalitarian. Indeed, some people, or some families, seem to have acquired a higher status than the rest. Still, differences were only relative and society seems not to have been split up into separate levels.

Between 1100 BC and 800 BC this world of yards shifting around in a small territory became affected by a new cultural attitude towards territories and the marking of them, which came from regions east of the Rhine. Where before the immediate ancestors were buried near the actual yard, giving it thereby its identity, the dead were now buried in a more permanent central cemetery. The yards were shifted around as before, but shifted now around the central place of burial (Fig. 7.32).

At the same time a different kind of settlement appeared: the nucleated settlement with a different style of building, mostly located in a strategic place. They were to be found in the southern part only, and were scarce. They served as special nodal points in the social networks connected with the exchange of products and connected crafts. But most of the population lived in the country and on the single farms. Rural society was as mildly differentiated as it was before. Some cultural dif-

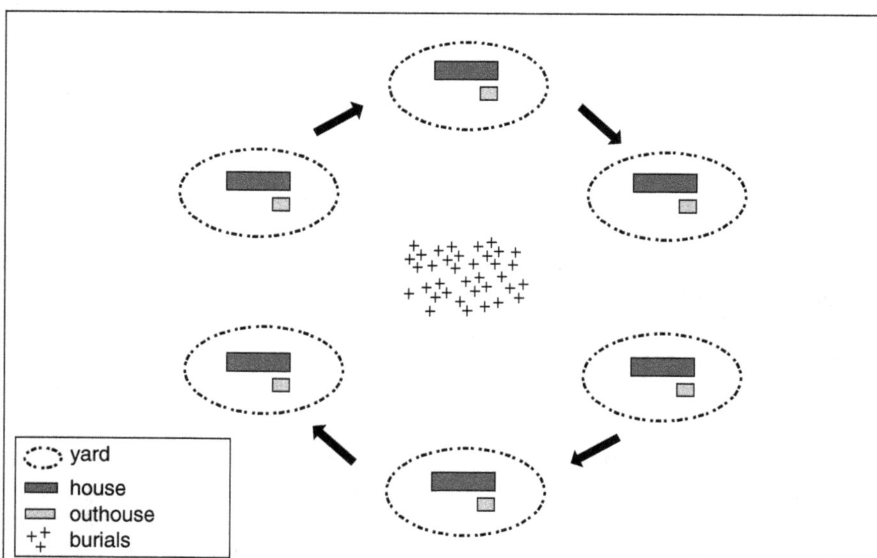

Fig. 7.32 Farms wandering around a central burial place

ferences developed between the north-eastern and eastern parts of the loess region, dividing it into two slightly different spheres, the north-western Urnfield Culture and the Rhin-Suisse-France orientale group (RSFO), but the basics of farm life remained comparable.

Of the western subregion, also called the Atlantic Tradition, much less is known. As described in Section 7.5, the main farm building seems to have been built following 'English' fashions. Following the English model, dispersed farms with round houses, rectangular outhouses and four-poster granaries in their yards lay scattered over the landscape. Also this subregion knew a different kind of site in addition to the single farm. The Etaples farm, mentioned in Section 7.5, was abandoned for a circular enclosure with a diameter of 160 m and provided with a double set of ditches, which resembles a British so-called ringwork. Ringworks were special status sites, partly farmstead, partly the place for specialised crafts. They commonly have a strategic location, for instance along important roads. The Etaples settlement may have been such a place along a route towards England.

The changes that affected the eastern subregion affected the Atlantic sphere as well. The RSFO tradition spread to the west, leading to a disappearance of the distinction between east and west in later times.

Around 800 BC Europe suffered a climatic deterioration. Summer temperatures are estimated to have been some 1.5°C lower, whilst winter temperatures remained more or less the same. The general climate became wetter, too. In several parts of Europe this dip affected rural life, even leading to the abandonment of certain areas and to a general decline in population density. However, in the loess region west of the Rhine such an influence has not been detected. Communities lived on, embedded in their local and supraregional networks. During this period iron was appearing as material for making all kinds of utensils. Iron ore is widely available and the dependence on ores from distant sources must have decreased. But this was a gradual process and bronze remained the metal for some implements and luxury products. Fine ceramic wares from specialised centres remained in demand. And in addition, another product with a definite non-local source became an important item for trade through extensive exchange networks: salt. It is not the salt itself that is found, because it disappears, but the special earthenware containers in which it was distributed. Salt is an important commodity for the preservation of food and hides, but is not present everywhere. It occurs as rock salt deep underground, in natural springs leaching salt from deeper occurrences, and in the sea. All three sources have been exploited in this period. People in coastal areas produced salt from the sea through evaporation techniques including heating brines in pots, and the same techniques were used inland in the case of salty springs. In certain areas in Europe rock salt was mined underground in vast systems of shafts and galleries, but the inhabitants of the loess region must mainly have had access to coastal and spring salt. Areas where salt was produced were, for instance, the Somme Bay along the Channel and the inland springs of Haute Seille (Dept. Moselle). Salt enabled the preservation of surpluses derived from animals.

Another indication of the production of surpluses is found in the centuries before 450 BC, when extra storage capacity turns up in the form of clusters of granaries and silos, at least in the southern part of the region (see Section 7.5). Strangely enough, the products stored in them represent only part of the crops grown at that time. Stored were mainly hulled cereals, at least as far as can be concluded from the carbonised remains (see Section 7.3). It may be that other products had a smaller chance to be preserved carbonised; but, and this is more probable, only hulled cereals were stocked as surpluses in the clusters. The question is, why? One reason may be that hulled cereals were easiest to grow and provide a surplus. Another may be that they were best to store and trade. Or, yet another reason, the surplus was created to serve special purposes, beer-making for instance, or to provide superior animal feed. Some archaeologists think that the surpluses were destined for communal celebrations and therefore stored in a communal place, not on individual yards. Cereals as animal feed may have been destined for the horses of the local leaders. At that time horses were not yet used for working on farms, but were ridden or used to pull ceremonial wagons. Be that as it may, the special places for storage disappeared around 450 BC. They served for only a century or two.

For the first time in the (pre-)history of the region, the people living there are not only called by a name given to them by archaeologists after some material characteristics, but bear a name of their own: the Celts. It is a name given by the Greeks to a part of a vast European complex of related tribes, and became the name of the complex as a whole. The name is applied to the inhabitants of the loess region from about 600 BC onwards, but most archaeologists agree that the Celtic heyday fell between 450 BC and 200 BC. The use of a new name suggests a population change by 600 BC, but this is not the case. Supraregional networks brought new elements all the time, perhaps not evenly distributed over time but in waves, but this does not necessarily imply invasions of new people as well. Local people absorbed new customs and ideas.

During the fifth, fourth and third centuries BC the region seems to have become an outer zone of the main 'world-economy' of Europe of that time, namely the Mediterranean world (Fig. 7.33). This brought wealth, as shown in certain subregions by the grave gifts of the time. It also brought the first kitchen herbs (see Section 7.2). Some people had rich burials, sometimes with complete chariots with spoked wheels, and such dead must have been important persons. In an area, where the pattern of burials is well studied, the Aisne valley in France, every cemetery counted one such grave. Small cemeteries were the rule, and they occurred every 15 km. On the one hand they show a hierarchisation of society. On the other hand, the territories controlled by these individuals cannot have been large and a picture emerges of a mosaic of small territories, covering the landscape, which picture may well apply to the loess region as a whole. The increase in remains of horses and the appearance of chickens, mentioned in Section 7.4, is clearly linked to this kind of society.

The Mediterranean world knew money. Introduced by the Greeks, the use of money was incorporated into the Italian economy during the third century BC. The Celtic world followed. The first coins were made of gold and showed the heads of Mediterranean rulers, often Philippus II of Macedonia, the father of Alexander the Great. Initially they may not have functioned as true money, but as something of

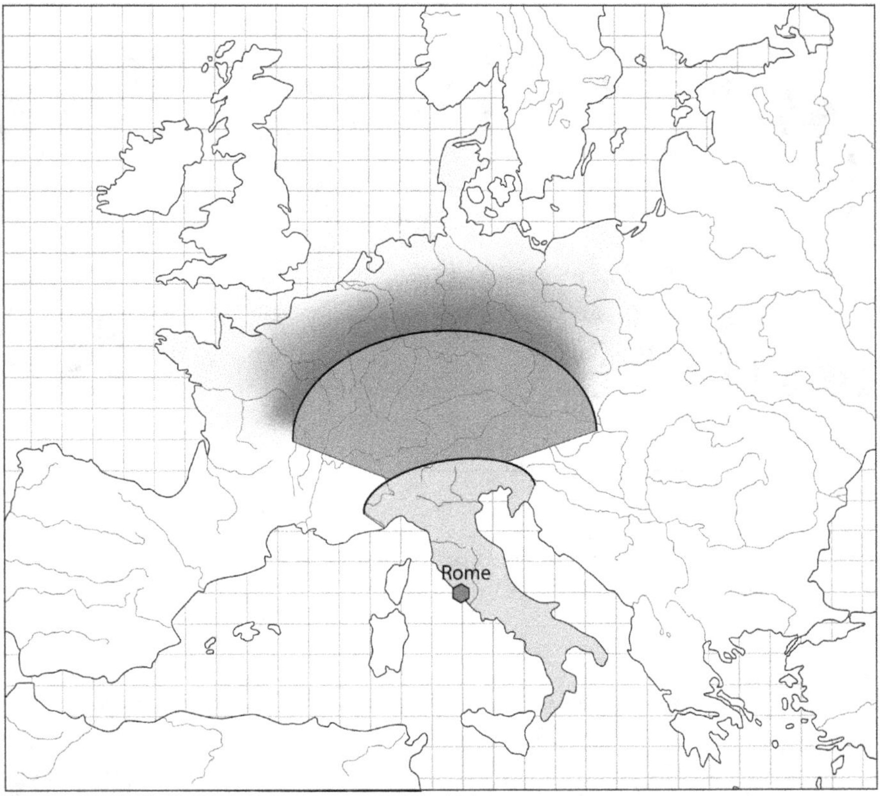

Fig. 7.33 The advance of the front of Mediterranean influence; the two relays in the wave of advance are dated first half of the fifth century and second half of the fifth century

value in the traditional exchange. Later on they came into circulation in something more like a money economy. Their design became adapted to Celtic fashion. In the second century BC every tribe struck its own coins (Fig. 7.34).

The Celts were very active and seem to have tried to gain control over the core of the Mediterranean economic world. Rome reacted by conquering, for instance, southern France, that became a Roman province. France was by this time known as Gallia, a name which includes large parts of Luxembourg and Belgium as well. In northern Gallia a socio-economic restructuring took place, which included the loess region. In its extreme north-eastern part this took the form of a clustering of farms into small hamlets with mutual distances of about 2 km. Farms were rebuilt within their own yards. In the remainder of the region a differentiation took place in the status of individual farms (see Section 7.5). A second development was that a larger, more nucleated type of settlement came into existence. Such settlements had already appeared in previous periods, see for instance Choisy-au-Bac mentioned in Section 7.5, but they now became truly common. They were surrounded by natural barriers, supplemented by artificial fortifications and covered areas of 4–15 ha, or even more.

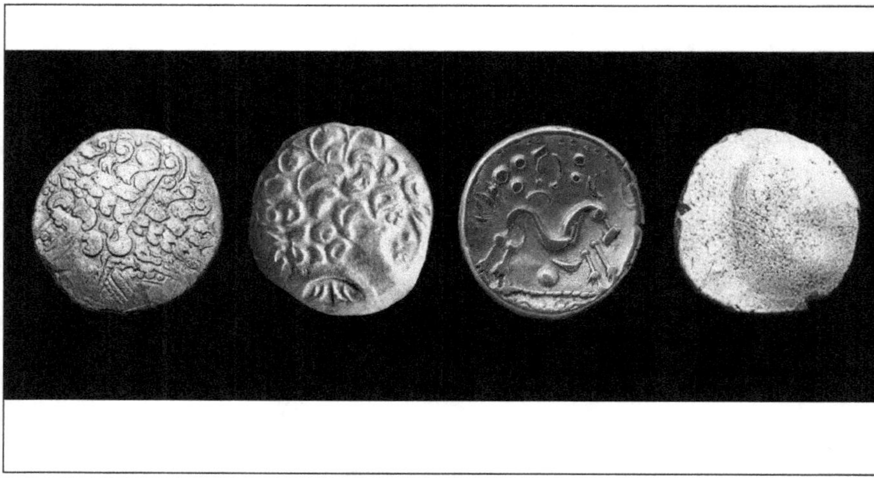

Fig. 7.34 Celtic coins from the loess region; the stylised heads on the two coins on the left are based on examples of Greek coins like those of Philippus II of Macedonia or the coins of the Greek colonies in southern Italy; the reverse has often the image of a horse; blank obverses occur too as is shown by the coin from Amiens on the right

Fig. 7.35 The *oppidum* of Chaussée-Tirancourt (Dept. Somme), France, from the air. The more or less triangular terrain (indicated by a *broken line*) was bordered by two natural barriers (the Somme valley at the back of the picture and a tributary valley on the *right*) and a man-made rampart and deep ditch on the side of the plateau

Another characteristic is a well-organised layout with streets and quarters. In some respects this kind of settlement was merely a structured agglomeration of farms as found in the country. But in other respects it has become clear that the inhabitants were not only concerned with agriculture and the crafts commonly executed on farms, but also with specialised crafts and trade. They are proto-towns. Such centres are known as *oppida* (singular *oppidum*), a name derived from the works written by Julius Caesar. There are lowland and hilltop *oppida*. The latter are the most common and stand out as strongly fortified settlements occupying strategic higher points in the landscape (Fig. 7.35). *Oppida* served as local capitals for population units known as tribes. They issued the money mentioned above and were the centre of territories of about 3500–12000 km^2. A system of roads linked them and linked the region to the outer world. It was the system that was later ameliorated and added to by the Romans. Witnesses of a supraregional orientation are all kinds of wares, imported from Mediterranean sources and found during excavations in the *oppida*. Nevertheless, the main population still lived on farms scattered over the landscape. It was their products that constituted the basis of the trade. One of the specialised jobs carried out in *oppida* was the butchering of cattle, delivered by the farms in the countryside. Cattle, and their products, may have been one of the pillars of the economy. Other animal products, and cereals, may have been important, too.

The Roman conquests did not stop in southern France. Between 58 BC and 51 BC Julius Caesar had conquered the main part of the loess region, and by 12 BC the complete area came under Roman rule. The book 'De Bello Gallico', attributed to Caesar, gives an account of this war.

Chapter 8
Part of the Roman Empire: 50 BC–AD 407

8.1 Roman Rule

Between 50 BC and 19 BC Roman troops invaded the region, first under Julius Caesar and later under several other commanders. The frontier of the Roman Empire shifted northwards until it reached the Rhine. After a far from successful attempt to push the frontier still further, beyond the Rhine and well into Germania, it was fixed along the Rhine (and eastwards along the Danube) in the form of a well-guarded border, called in Latin: the *limes*. The *limes* consisted of a series of smaller and larger forts, connected by a road, on the left bank of the river. The hinterland was kept under control with only limited military means. Initially a diplomatic approach sufficed. The earlier economic contacts had certainly facilitated this kind of rule. Local elites offered friendship and obedience to Rome, expressed in separate treaties between Caesar and individual chiefs and tribes. The further to the North, and thus the greater the distance to the core of the Mediterranean economy, the more difficult this mode of rule became. Caesar described this as follows (*De Bello Gallico* I-1): 'All Gaul is divided into three parts, one of which the Belgae inhabit, the Aquitani another, those who in their own language are called Celts, . . . the third. The river Garonne separates the Gauls (Celts) from the Aquitani, the Marne and the Seine separate them from the Belgae. Of all these the Belgae are the bravest, because they are furthest from the civilisation . . . and merchants least frequently resort to them, and import those things which tend to effeminate the mind'. In this text Belgae is the name by which the conglomerate of tribes was known that lived in the greater part of the loess region at the time. Only the people along the Moselle and the upper Rhine were not included and still designated as Celts (Fig. 8.1).

The system of separate treaties proved not to be very satisfactory, at least in the eyes of Rome. After local upheavals it was soon replaced by an efficient civil administration, as a result of which the region became truly part of the Roman Empire.

The first period of expansion of the empire was followed by a period of consolidation. This started halfway into the first century AD. Around AD 100 the true Pax Romana (Roman Peace) began, providing an economic climate in which the loess

C.C. Bakels, *The Western European Loess Belt*, DOI 10.1007/978-1-4020-9840-6_8,

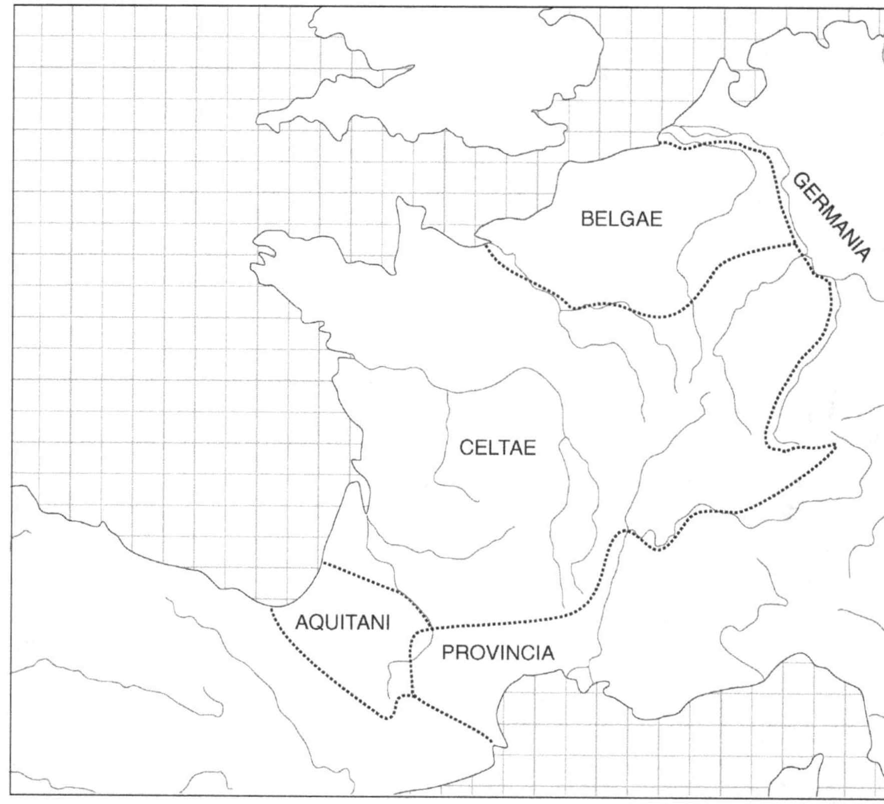

Fig. 8.1 Gaul during the time of Julius Caesar. According to him the land was divided into three parts, inhabited by, respectively, the Aquitani, the Celtae and the Belgae

region flourished. This heyday of the Roman Period lasted one and a half centuries. By AD 250 the long period of tranquility was over. Minor raids by Germanic warriors from areas east of the Rhine had regularly taken place, but the Roman army had always been able to restore order. But now a better organised tribe, the Franks, succeeded in destabilising the *limes*. It was the start of the Late Roman Period. Some authors think that a deterioration of the climate was one of the triggers of the disturbances, but this view is not generally supported. The army succeeded in restoring the frontier, but Germanic warlords crossed the Rhine again and again. The Roman government allowed some of them to stay, hoping that they could, and would, keep others out. The *limes* was given up as the sole line of defence. A new military strategy, defence in depth, was adopted to counter invaders when they had broken through the *limes*. Intervention troops in the hinterland, including a swift cavalry, should stop them. Military posts were constructed along the important roads. It did not always help, but when the empire was headed by strong emperors, such as Constantine the Great, life was tranquil again. Such periods saw a strong revival of the economy. Nevertheless, the decline had set in. In AD 402 most of the Roman troops were recalled to Italy, to defend the motherland against invaders. In AD 407 the army was withdrawn from the part of the *limes* bordering the river Rhine and in the first half of the fifth century the authority of Rome collapsed.

The four and a half centuries, known as the Roman Period, is the first period for which not only archaeological sources, but also written sources are available, although these are still very scarce. The longer texts were not written locally, but are by Roman authors. According to the generally accepted definition of history, that is that written sources are present, the loess region had entered history. The centuries prior to this belong to protohistory. The change in denomination, protohistory–history, suggests discontinuity in population, but there was none. The original population was conquered, but not driven away. The Roman way of life did not immediately replace the old ways, but was gradually absorbed, and then only to a certain extent.

8.2 Crops

Agriculture in the loess region started with seven crop plants (see Section 3.2). In the course of time species were added to the initial list. The widening of possibilities seems to have been a more or less gradual process. The introduction of agriculture, of the production of food, may be (and often is) described as an economic revolution after millennia of hunting, fishing and gathering. What followed was an evolution of the new way of life, but the Romans brought about a second revolution. They introduced a long list of new products. Moreover, the revolution did not express itself only as a wave of new species. Before the Roman occupation, agriculture was based on annual, herbaceous, plants. The Romans introduced trees. It is not that people before that time did not eat fruits and nuts, but they picked them from wild trees and shrubs. The gathering of wild apples and pears, for instance, or hazelnuts, is attested for in all protohistoric periods. Some archaeologists even maintain that wild stands of such useful trees and shrubs were tended in some way or another. Nevertheless, according to the archaeological records, fruits seem to have played a minor part in the farmers' lives. There are indications that wild apples and pears were dried in ovens to ensure preservation. This, at least, is the explanation for the presence of carbonised fruits, sometimes even cut into neat halves. Processing errors may have caused carbonisation. The only wild product that seems to have had a major role through the millennia is the acorn. Substantial numbers of carbonised acorns point towards a role as a staple for humans, and not as food for pigs as was common in, for instance, medieval times. Although they may seem unattractive as human food today, their nutritional value is considerable and they have good storability properties. The only problem is the tannin content, which give acorns a bitter taste and makes them poisonous for humans. But tannin can be removed by pounding acorns into flour and leaching this flour in water, or by roasting. Faulty roasting can lead to carbonisation. The collecting and preparing of acorns is known from Northern America, but from Europe as well.

The general idea is that acorns were gathered and stored as a buffer in the case of crop failure. It is noticeable that finds are more common in the sandy regions of Europe, while in the loess region they are rather scarce. Perhaps crop failure did occur there less often.

But, as remarked above, before the arrival of the Romans, farmers did not plant trees, at least not as cultivars in their yards or in orchards. What they also did not do was growing cultivar varieties of vegetables and condiments. This remark does not imply that people did not eat leaves, roots and the like, but that such additions to the diet were gathered from the wild. The traces of this class of plants are from indigenous species. If they were cultivated at all, those plants were not different from the wild species.

Table 8.1 presents the list of plants that arrived with the Romans. It is obvious that not every plant on the list can have been grown locally. It is, for instance, highly improbable that pistachio trees, date palms and black pepper could thrive in the region. Some plants, like the fennel and celery mentioned in Section 7.2, were known earlier, but it is not certain whether the seeds had been grown in the vicinity of the sites where they have been found. They probably belong to the set of Mediterranean products that preceded the actual arrival of the Romans.

A detailed study of plant remains found in continental Europe roughly north of the Alps, including the loess region discussed here, revealed that the entire list of plants was already present during the first phase of Roman occupation, which lasted until c. AD 50. The study also made clear that during this first phase the 'new' plants were mainly to be found in military contexts. Obviously the troops, and perhaps especially their Roman officers, would not forgo their Mediterranean habits. Some of these had quite a luxury aspect. Complete casks filled with fresh pomegranates, presumably intended for high-ranking officers, have been detected in a storehouse in Switzerland (Fig. 8.2). Rice was a luxury, even in the capital Rome. Such standards of living were not maintained for long, presumably because the people accustomed to Roman luxuries did not stay in the region.

Table 8.1 Plants introduced by the Romans and found in excavations

Group 1 Rarely found	Group 2 Uncommon	Group 3 Common fruits/nuts	Group 4 Common herbs
Rice	Olive	Grape	Beet
Chickpea	Date	Peach	Savory
Pistachio	Sweet chestnut	Walnut	Celery
Pomegranate	Medlar	Apple	Parsley
Black pepper	Mulberry	Pear	Coriander
Almond	Quince	Plum	Fennel
Gourd		Cherry/sour cherry	Marjoram
Pine nut		Fig	Anise
Melon			Dill
Garlic			

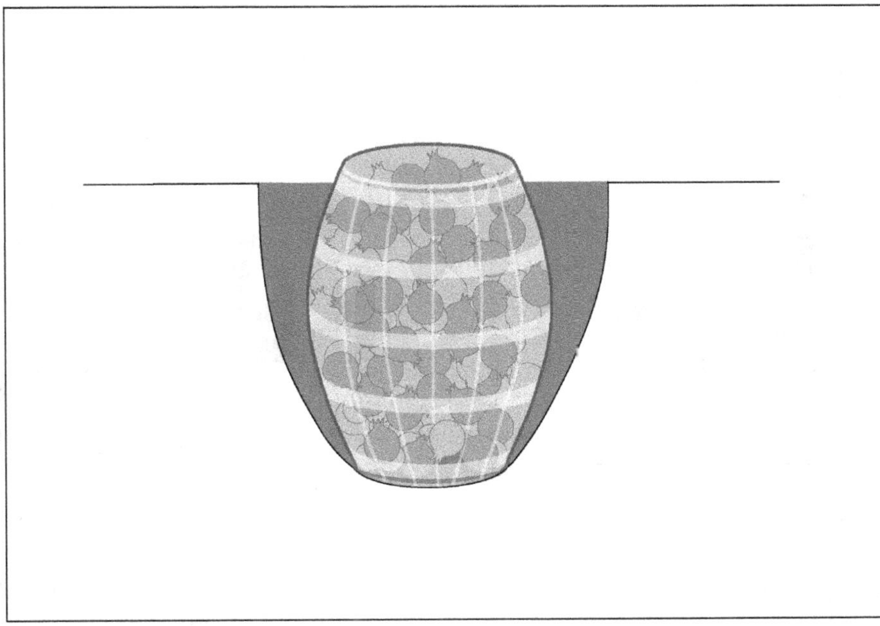

Fig. 8.2 Pomegranates kept fresh in a cask let into the ground. Found in Vindonissa, now Windisch, Switzerland

Four ‘exotic’ products, rice, chickpea, pomegranate and pistachio are indeed restricted to the first phase (Fig. 8.3). Two other species, which could not thrive either in the region, disappear before the heyday of the empire: olive and almond. The olive was mainly consumed in sites with a military character, and only during the first two phases of occupation. Only its oil remained common enough. Almond was also restricted to the first two phases.

Other exotics remained, for instance dates, pine nuts and pinecones, but these probably had a different place in society. They turn up in ritual contexts, such as temples and graves, and may have been essential components of ceremonies. As an element of the food chain they are rare, but as an element of ceremonies they are less so.

The fig was also a component of offerings, but otherwise fig lacks the special status of date and pine. As a matter of fact, fig seeds are very common finds. In sheltered places fig trees may have produced ripe fruit in the southern part of the loess region, but the bulk must have arrived, dried, from Mediterranean areas. The Roman author Columella ranks figs among the main foods of ordinary people in winter. Pliny classes figs under the staple foods. It is, therefore, quite possible, that figs arrived in large quantities along with the army of occupation. Most finds are restricted to military and civilian-urban contexts. Obviously the indigenous people did not include fig readily in their menu.

Surprisingly, black pepper, though like rice a luxury in the capital, did not disappear from the records. Pepper was grown in India. One of the established routes

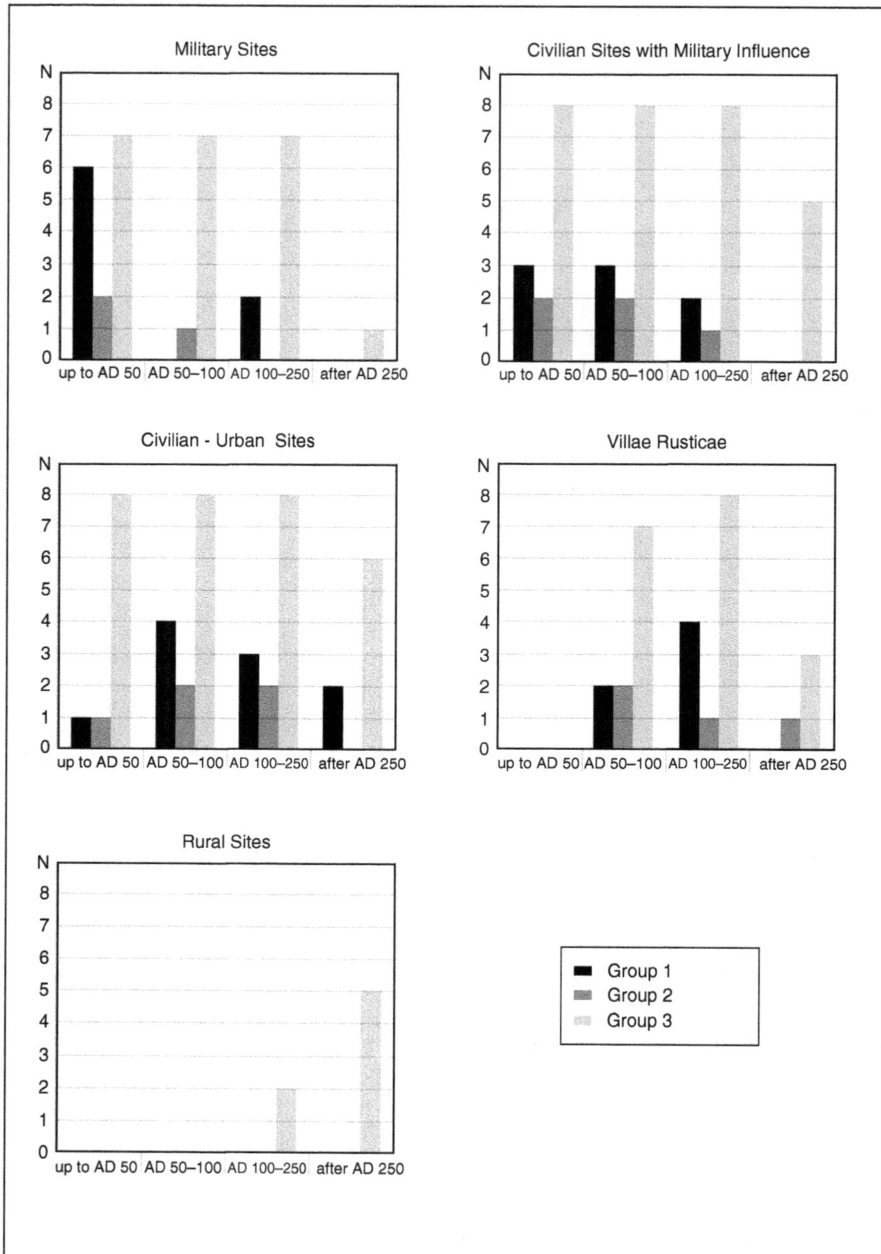

Fig. 8.3 Three of the groups of species mentioned in table 8.1 and their presence in four sub-periods and five settlement types. Sweet chestnut, medlar, mulberry and quince were omitted because there are uncertainties about their occurrence in the records. Four species belonging to group 1, i.e. rice, chickpea, pistache and pomegranate are not found after AD 50

leading to Rome, was to ship the products to ports on the Red Sea in Egypt, to send them from there over land to the river Nile, ship them to Alexandria in the Nile delta and there put them on another kind of ship bound for a port in Italy (Fig. 8.4). No wonder that black pepper was expensive. In his book *Historia Naturalis XII* the Roman author Pliny called pepper an unnecessary luxury. Nevertheless, black pepper was redistributed from Rome to the far ends of the empire. Such a luxury product is, of course, not readily wasted and therefore scarce in archaeological records. The few finds north of the Alps were made in military contexts and in ports on large rivers. A label, made of lead, and mentioning a certain amount of pepper, has been found in the river Moselle at Trier. It was presumably attached to a small shipment of this condiment (Fig. 8.5). Black pepper seems not to have been available, or not in quantities to be scattered about, in common civilian households. It was perhaps too expensive.

The species mentioned so far were imported and had to be imported, because they could not be grown in the region. At first, all the others mentioned in the table had to be imported as well, as local farmers did not produce them. But this situation changed fast. Already during the consolidation phase a new type of farm developed, which endeavoured to supply all wanted products (Fig. 8.3). These farms are known by their Latin name of *villa rustica* (plural *villae rusticae*) and were mostly run by Romans or Romanised people (see Section 8.4). The owners started to plant fruit and nut trees. It is supposed that they obtained the first seedlings

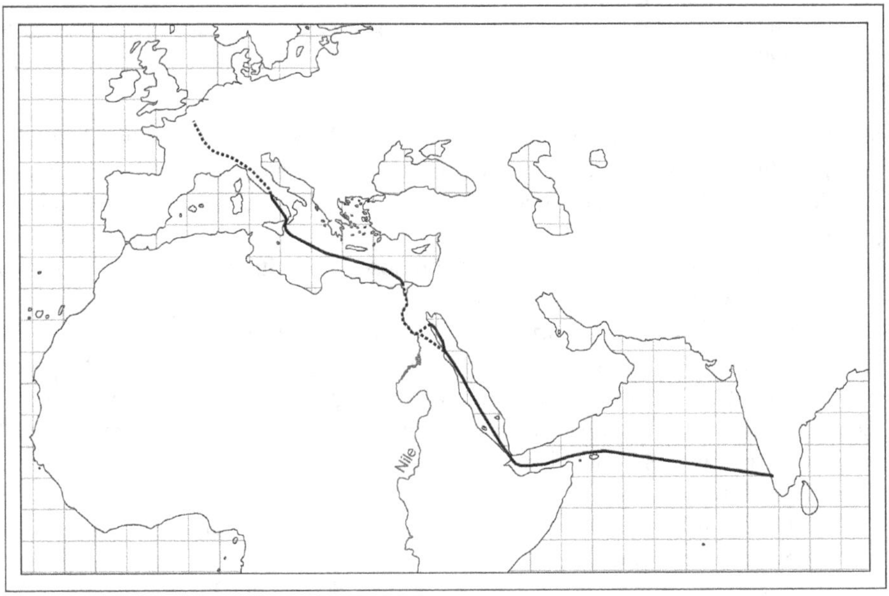

Fig. 8.4 Transport route of black pepper: by sea from India to ports on the Red Sea, from there over land to ports on the river Nile, shipped to Alexandria, transferred to sea-going ships bound for Rome, and finally over land or via the sea and rivers to the North-West

Fig. 8.5 Lead label found in the river Moselle near Trier, Germany. According to the inscription the label had accompanied a lot of pepper. One side bears the inscription NOVIILV, PIPIIR. . ., to be read as novel(l)u(m) piper. . . and the other PVIII, NSEXX, understood to mean p(ondo) VIII, n(?ummum) se(?estertium) XX. . . Or: new pepper weighing eight Roman pounds and its price in *sestertii* (Roman money)

and grafts from the Roman world. Apple, pear and sweet cherry occurred as wild trees, but their fruits cannot have been up to the quality standards of the customers. It is known that the Romans knew at least 22 kinds of apples and 39 kinds of pears.

The Roman fruit basket was mainly composed of fruits with origins in an area roughly between the Black Sea and the Caspian Sea. Plums are considered by some to be the result of a natural hybridisation of sloe and cherryplum somewhere in the Caucasus. The hybrid was taken into culture. Sour cherry is the result of a cross between sweet cherry, a wild tree with a wide distribution, and another wild cherry, the ground cherry (*Prunus fructicosa*), which grows in Central and Eastern Europe, and north-east Turkey. Populations in the latter area are held to be the source of the domesticated trees.

Sweet chestnut was also domesticated in northern Turkey and the Caucasus, as far as is known at present. The extensive stands of the tree in Italy, southern France and Spain are held to be feral, composed of trees that escaped from orchards. The same is said of walnut. Although the walnut tree is native to southern Europe as has been proven by finds of pollen and charcoal, the tree seems to have its origin as a cultivar in north-eastern Turkey, the Caucasus and northern Iran. In Iran, too, lies the origin of mulberry and medlar, whilst quince is native to the south-western parts of the Caucasus region.

Apple, pear and sweet cherry occurred as wild species in the loess region, but at least apple is said to have been domesticated on the basis of wild stocks growing in the area where most of the other fruits and nuts came from, and it is quite possible that this is true for the others as well. All trees belonging to this Black Sea – Caspian Sea complex were already cultivated before the Romans adopted them. Their exact history is imperfectly known.

One of the exceptions concerning origin is peach. The peach has its origin in western China, but became later widely appreciated in Iran, from where the tree spread westwards well into the Roman Empire. Some managers of *villae rusticae* in the southern part of the loess region seem to have tried to cultivate peach, though this is far from certain. One of the arguments in favour of local cultivation is that ripe peaches are difficult to transport, but the peach stones found could have derived from dried fruit or fruit conserved in honey or brine. The origin of grape presents another case. Although wild grapes are widely distributed, ranging from the western Himalayas to the Atlantic coast, most of the wild populations are composed of equal proportions of male and female individuals. The wild grape is so-called dioecious. However, in the area between the Black Sea and the Caspian Sea populations are found in which the flowers contain both pistils and anthers. The cultivated grape has the same property and therefore this area has for a long time been regarded as the sole place of origin. However, DNA analysis has shown that there must have been at least one other place of origin, situated somewhere in the western Mediterranean region. Therefore, the Roman races of grape, brought to the North, may have had a multifocal origin.

The farmers of the loess region even experimented with more difficult plants. Melon was one of those and bottle gourd another. Bottle gourd was eaten unripe as a vegetable. Both had a luxury status. Apicius, a high-society cook living in Rome, used them in his recipes. The prevailing climate in the region was not propitious to their culture, and if their culture succeeded, they were truly luxury products. Melons came originally from south-western Asia or Egypt, and the wild ancestor of bottle gourd grows in Africa.

The Roman way of life also asked for vegetables and condiments, although the eating of great quantities of vegetables was sometimes laughed at. There is a passage in the comic play *Pseudolus* by the Roman author Plautus in which a cook introduces himself as a cook who does not serve his guests 'complete meadows'. He does not want to treat his guests as oxen. Moreover, he is not addicted to seasoning all this herbage with other herbs like coriander, fennel and mustard. Nevertheless, traces of such plants are found everywhere where Romans have dwelt.

As remarked above, the much valued black pepper could not have been produced locally; and from at least one other condiment, aniseed, local production is not yet established. But the remainder was cultivated in the region from the times of the existence of *villae rusticae* onwards. It is certain that the list was much longer than the one presented here, because this category of plants can only be found when seeds or pollen have been preserved. When the plant was harvested for its leaves, stems or bulbs, it is normally absent from the records (see Chapter 2). The finds of garlic, in the form of carbonised cloves, are exceptional. Garlic must have been very common, as otherwise the chance of finding an incidentally burnt clove would have been close to zero.

The domestication history of vegetables and condiments has hardly been subjected to any serious study. Many species have ancestors in the eastern Mediterranean, Egypt or northern Africa. Although two plants, beet and celery, occur in the wild on the Atlantic coast, these populations are not the wild ancestors of the cultivated plants. The original stocks are Mediterranean.

Quite another category of plants are the dyeplants. Weld or dyer's rocket is one of them. There are some indications that the plant was already used in the centuries before the arrival of the Romans, but finds are regular from the Roman Period onwards. As its name indicates, it is a source of dye, yellow dye. It was mainly used to colour textiles. Dyer's rocket is one of the best natural sources of yellow. Another dyeplant is safflower, a source of yellow and red colours. Other dyeplants must have been present, but they have not yet been found.

The entire range of plants, introduced and grown by the first Romanised farmers, remains part of the archaeobotanical records during the heyday of Roman rule. The new species turn up in military, urban and *villa rustica* contexts, thus in every place where the Roman life style prevailed. The question is how far these plants made their way into the daily life of the original inhabitants, especially when not belonging to the elite. Unfortunately not much is known on the subject, because the main interest of investigations has been focused on military camps, towns and the main living quarters of large farms. But some indication can be derived from the finds dating from the period of the decline of the Roman influence, the period after AD 250. It is obvious that the imported wares disappeared, except for figs. What remained was what is seen today as common fruits and nuts: apples, pears, plums, sweet and sour cherries, grapes, and walnuts. They must have obtained a place in the common, daily life of the inhabitants of the loess region, as otherwise they would have disappeared as well. The fate of sweet chestnuts is less well known, but that is because chestnuts do not leave behind sturdy pips, stones or shells. But, as the tree became feral in the local forests, it did at least not disappear.

Locally grown vegetables and condiments were presumably retained as well, as they are present in medieval contexts following upon the Roman period, but their status as part of common diet is not very well known. Nevertheless it may be safely stated, that the Romans triggered the addition of orchards, vineyards and vegetable gardens to the common farm.

With all these new products to mention, it should not be forgotten that the main crops were still cereals, followed by pulses, oil seeds and fibre plants. This category also counts two newcomers, chickling pea (also known as grass pea) and rye. Chickling pea has a Mediterranean origin, but its history is not yet well established. It is nowadays grown mostly as animal food, but is known to have been consumed in the past by humans as well. Its status in the loess region is unclear.

Rye is a special case. It is occasionally present in cereal finds dating from well before the Roman Period, but is then regarded as a field weed. During the Roman

Period, grains of rye turn up, but still not very regularly and not in significant amounts (more about rye in Section 9.2).

Two crop plants were abandoned, einkorn and gold of pleasure. Traces of them are still there, but the number of seeds and the frequency of finds suggest that they had dwindled into weeds. But the loss of those two was more than compensated for by the introduction of a wealth of other species. The conclusion at the end of this section must be that the farmers of the loess region west of the Rhine had many more products to consider than their predecessors living in protohistoric periods.

8.3 Crop Cultivation

During the first phase of occupation, people continued farming as they were accustomed to. However, the Roman occupation triggered a reorganisation of the management of farms. As put forward in Section 8.2, a demand for new products arose. Moreover, the occupying army had to be fed. Many farms transformed into establishments that produced on a large scale those crops which did best on the market, and new farms were founded. The Roman Period knew a true market economy. On the loess, this resulted in a general orientation towards cereal cultivation, the main cereals being emmer wheat, spelt wheat, bread wheat and hulled barley. Specialisation went even further. Archaeobotanical research, carried out in northern France, revealed that farmers on the best soils, i.e. those with an optimal moisture regime, grew bread wheat. On less deep soils, those on top of a calcareous subsoil, the hulled wheats spelt and emmer were grown (Fig. 8.6). Spelt wheat was a speciality of farmers around Amiens. Barley was grown everywhere.

The large *villae rusticae* in the northern part of the loess region were specialised as well. A villa at Voerendaal (the Netherlands) produced mainly wheats and some barley. At first its wheat crops consisted of emmer, spelt and bread wheat, but in the end almost exclusively spelt wheat was grown. Reports on a villa in nearby Hambach (Germany, Hambach 59) mention spelt wheat as the main product, followed by barley. Emmer and bread wheat were present, but unimportant.

Growing one or two cereals as monocrops was the core business of the large farms. Their products turn up as single species lots in forts and towns. The spelt wheat grown around Amiens was for instance stocked in a large storehouse in the centre of the town. The building had accidentally burnt down at the end of the second century. A third century bakery in the same town, also destroyed by fire, made its bread with spelt flour. The wheats and barley produced by the northern farms turn up in large military granaries in Maastricht, which date to the end of the fourth and beginning of the fifth century. The cereals were stored separately. And, lastly, a ship, sunken with a cargo of emmer wheat on board, provides another example that cereals were monocrops. This ship was found in the Rhine near Woerden (the Netherlands), a place far downstream from the loess region, and is dated to the last quarter of the second century. The presence of the weed *Orlaya grandiflora* shows that the emmer cannot have been produced in the surroundings of Woerden, because

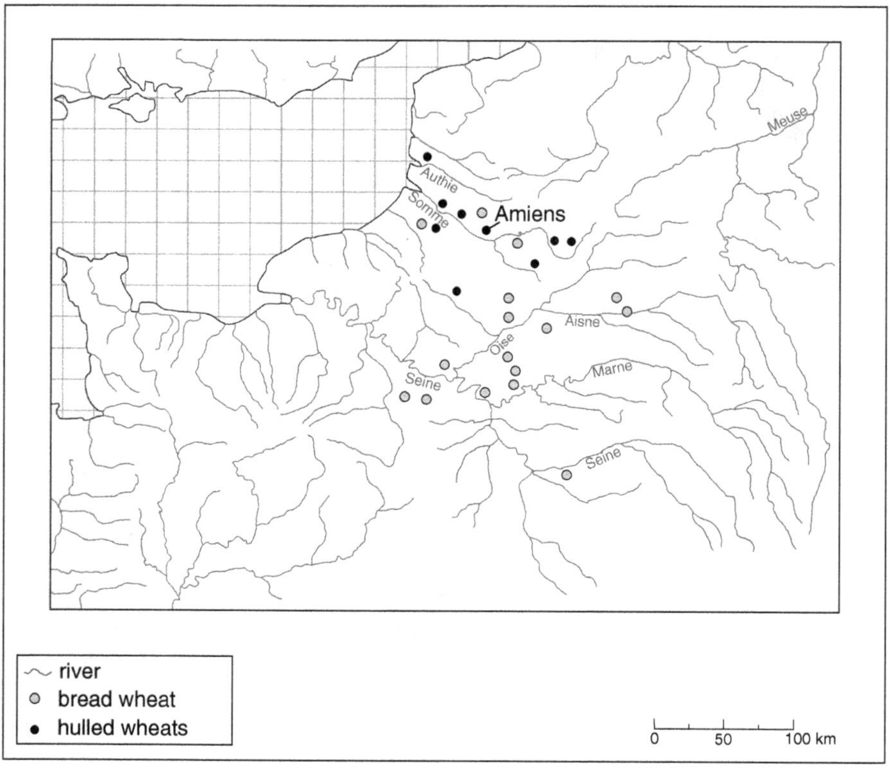

Fig. 8.6 The principal kinds of wheat retrieved from French sites: bread wheat is dominant except for the surroundings of Amiens where the hulled wheats spelt and emmer prevail

the local climate did not suit this thermophilous weed of crops. The nearest possible place was the loess region and the ship must have loaded its cargo somewhere there.

The grain was grown on a large scale (see Section 8.5). How the soil was prepared before sowing is not well known. Because the large farms were laid out following Roman models, the obvious way to learn about farming methods is to look at ancient Latin sources. But even then, the information on ploughs is not easy to interpret. Descriptions are not very technical and images are problematic. The ard, drawn by oxen, was still a very common implement. The ancient writers provide no information on different designs, only on the different types of shares. A wooden tip was used on loose and friable soils; but elsewhere the tip was fitted with an iron share, which was either fixed or detachable. The shares came in different types to suit different conditions. It is to be expected that the ploughs in the loess region were fitted with iron shares. An ard does not turn the sod. Something resembling turning can be achieved by holding the tip at an angle.

Another type of plough is the turning or mouldboard plough, which cuts the earth and inverts the clod, exposing the sods to sun and air, and destroying weeds. It achieves this operation in a single line (see Section 9.3). There is no indication that the mouldboard plough was already used on the loess soils during the Roman Period. The only finds concern remnants of ards.

The first ploughing was followed by a second ploughing at right angles to the first. This cross-ploughing may have been followed by a third, or by harrowing. Ploughing was heavy work. According to the Roman author Columella the oxen often had to be rested, also to recover from the continuous friction of the yoke.

Whether the fields were manured is another question. Animal droppings were the chief source of fertiliser. On farms oriented principally towards cereal production, manure may have been scarce. Just as in Section 7.3, an analysis of weeds, as part of carbonised cereal lots which, according to their archaeological context, must indeed represent one single lot, may provide insight into the condition of the fields. Table 8.2 provides a list. The first conclusion is that the list of weeds is very short and consists of species with seeds that cannot be discarded by sieving or winnowing, because their size and weight matches more or less that of the grain. Cereals were already as far as possible cleaned before transportation and storage. Except for the *Sambucus* species, they are well-known weeds in cereal fields. Some authors think that the *Sambucus* seeds were deliberately added, perhaps to hold some pests under control, but how this works does not become clear. But at least *Sambucus ebulus* can occur abundantly in cereal fields and may be considered a weed as well (Fig. 8.7). The table provides the acidity and nitrogen (R and N according to the Ellenberg system) values of the soils required by the weeds. R and N values are given in ranges from 1 to 9, with low values for very acid/very low N content, and high values for alkaline soils and soils high in nitrogen respectively (see the glossary). Only one weed points towards soils with a low pH, *Raphanus raphanistrum*. Nitrogen factors range from 4 to 8. Many species belong today in the phytosociological class of the *Stellarietea mediae* (Tüxen, Lohmeyer and Preising in Tüxen 1950) and it is perhaps permissible to apply this classification to weeds from the Roman Period. The farming techniques of earlier times are considered to have been too different to be able to provide the conditions under which the phytosociological class could arise and most people apply the classification only from the Middle Ages onwards. But if it is applied here, some species, first and foremost *Agrostemma githago*, but also *Buglossoides arvensis*, *Sherardia arvensis* and *Orlaya grandiflora* point towards the class' sub-unit of Caucalidion platicarpi Tüxen 1950. This unit is characteristic of cereal fields on rather alkaline soils, which are moderately rich in nutrients. The four species with either a low R, or a low N, or both grow on the poorer soils characterised by the order *Sperguletalia arvensis* Hüppe et Hofmeister 1990. The other plants can be found in both groups. The weeds show that poorer conditions existed, but that overall soils were not very depleted of nutrients. How fertility was maintained is not clear. In addition to a moderate application of organic manure, marling may have been used to correct for acidity. The Roman author Pliny enumerates several kinds of lime and marl, and some bear Celtic names. It looks, therefore, as though liming and marling was already practised before the Roman occupation.

Roman cereal sowing was broadcasting by hand. The seed was covered by subsequent light ploughing or raking or harrowing. Some idea of the implements used is provided by a set of miniature implements found in a grave near a *villa rustica* in

Table 8.2 Weeds found in concentrations of cereals. The sites of Valkenburg, Xanten and Woerden lie outside the loess region but on the bank of, or in the case of Woerden even in, the river Rhine and the grain may have been produced on loess. See for R and N values Table 3.1 or the glossary

	R	N
Valkenburg, Netherlands, AD 69		
Crop: Triticum aestivum		
Agrostemma githago	x	x
Avena fatua	7	x
Bromus secalinus type	x	x
Galium aparine	6	8
Secale cereale	–	–
Vicia sativa ssp. obovata	x	x
Valkenburg, Netherlands, AD 69		
Crop: Hordeum vulgare		
Avena fatua	7	x
Bromus secalinus type	x	x
Valkenburg, Netherlands, large granary, AD 69/70		
Crop: Triticum aestivum		
Agrostemma githago	x	x
Bromus secalinus type	x	x
Raphanus raphanistrum	4	6
Vicia sp.	–	–
Tongeren Kielenstraat, Belgium, AD 69/70		
Crop: Hordeum vulgare		
Plantago lanceolata	x	x
Vicia hirsuta	x	4
Xanten, Germany, second half 1st century		
Crop: Triticum spelta		
Agrostemma githago	x	x
Avena fatua	7	x
Lithospermum arvense	7	5
Vicia sativa ssp. angustifolia	x	x
Valkenburg, Netherlands, 2nd century		
Crop: Triticum spelta		
Agrostemma githago	x	x
Orlaya grandiflora/Caucalis platycarpa	9	4
Amiens, France, large granary, end 2nd century		
Crop: Triticum spelta		
Agrostemma githago	x	x
Avena fatua	7	x
Bromus secalinus type	x	x
Orlaya grandiflora	9	4
Sambucus ebulus	8	7
Vicia hirsuta/tetrasperma	x–5	4–5

Table 8.2 (continued)

	R	N
Woerden, Netherlands, ship, end 2nd century		
Crop: Triticum dicoccum		
Agrostemma githago	x	x
Cirsium arvense	x	7
Orlaya grandiflora	9	4
Sambucus ebulus	8	7
Maastricht, Netherlands, bakery, AD 100–300		
Crop: Triticum aestivum		
Agrostemma githago	x	x
Bromus secalinus type	x	x
Secale cereale		
Houdan, France, villa rustica, 3rd century		
Crop: Triticum aestivum		
Agrostemma githago	x	x
Knautia arvensis	x	4
Knautia dipsacifolia	(6)	(6)
Lithospermum arvense	7	5
Sambucus ebulus	8	7
Silene cf alba	x	7
Trier, Germany, large granary, Late Roman		
Crops: Triticum spelta and Triticum dicoccum		
Agrostemma githago	x	x
Avena fatua	7	x
Bromus secalinus	x	x
cf Vicia hirsuta	x	4
Fallopia convolvulus	x	(6)
Galium aparine	6	8
Orlaya grandiflora	9	4
Raphanus raphanistrum	4	6
Sambucus ebulus	8	7
Sambucus nigra	x	9
Sambucus racemosa	5	8
Secale cereale	–	–
Vicia tetrasperma	5	5

Köln-Rodenkirchen (Germany) (Fig. 8.8). Fields were certainly weeded, but how is not known.

Harvesting was done with hand-held sickles or by the reaping machine described in Section 7.3. Excavations have also revealed scythes, which according to the Roman authors, were used for mowing grass. The weeds, found in stocked cereals do not tell us much about reaping heights. As mentioned above, the cereals seem to have been stored partially cleaned and, as a result, show a shorter list of unwanted herbs. Nevertheless, the remaining species mostly belong to tall plants or climbers. If most of the reaping on the large farms was done by machine, no other kinds of

Fig. 8.7 Cereal field with danewort (*Sambucus ebulus*) in Northern France

plants could be expected, because the machine strips the ears off, leaving the stalks behind.

Large ovens within the boundary of the yard have been explained in the past as corn-dryers. They would have met the need for drying the harvest before further processing could take place. Experiments have shown, however, that the amount of unthreshed cereal, which can be treated at one and the same time, is rather restricted. As corn-drying kilns, the ovens are very uneconomical. Nowadays such ovens are interpreted as installations for the malting of cereals, a stage in the process of beer-making.

A first threshing took place within the boundaries of the yard. Excavations on several large *villae rusticae* have shown that the yards had outdoor threshing floors. The threshing methods are not known. Flailing, trampling by animals or the threshing-sledge, all were technically possible. However, the traditional Mediterranean sledge is not very probable, because the flint teeth, fitted to the underside of the board, have never been found. As they tend to come loose during action, they should have been present.

As mentioned already several times, the grain was as far as possible cleaned before it was put into storage. Hulled wheats were sometimes even dehusked, as was the case in the Woerden ship. But this is an exception. Most stocks of hulled cereals revealed their chaff still in place. Storage in the husks is better for preservation. The dehusked emmer in the ship was infested by all kinds of insects, mites and moulds. But also stocks with husks could be infested, as is, for instance, shown

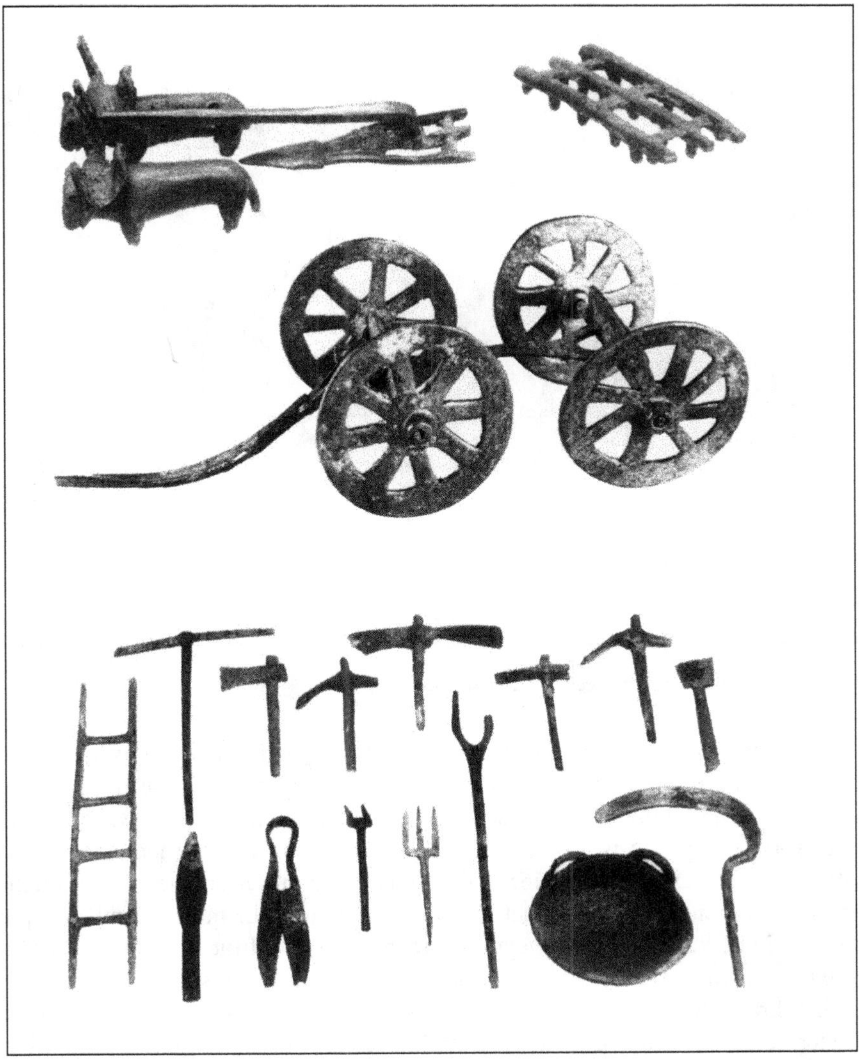

Fig. 8.8 Miniature implements found in a grave in Köln-Rodenkirchen, Germany: ard, harrow, wagon, ladder, diverse types of hoes and pitch-forks, spade, shears, winnow and sickle

by the contents of the town granary in Amiens. Very common were the weevils *Sitophilus granarius* and *Oryzaephilus surinamensis* (Fig. 8.9).

Cereals were, as far as is known, never stocked as flour. Milling on the farm was only done for its own inhabitants. Soldiers received their ration in grain, unmilled. They carried their own small rotary querns. Bakeries in towns milled their own flour. Towns and forts had their own large granaries, called *horrea* (single *horreum*), where large quantities of grain, but also other products, were kept in stock.

Large cereal-growing *villae rusticae* were not the only type of farms, and even these were not completely devoted to monoculture. All produced some pulses, oil

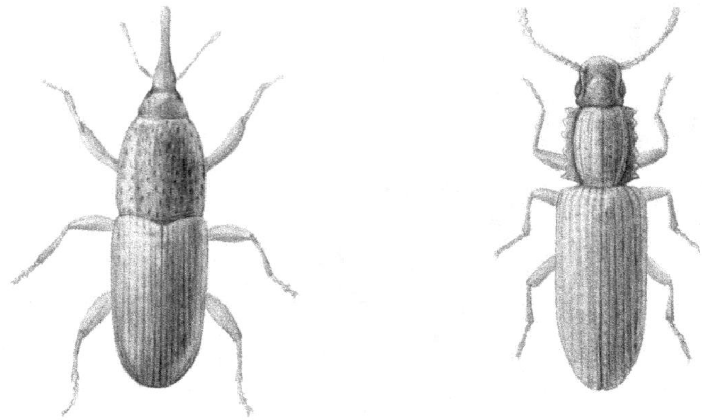

Fig. 8.9 Two very common weevils (modern specimens)

seeds, vegetables, fruit and condiments as well. It is noticeable that even the very large *villae* become more diversified in their main crops from the third century onwards. But in addition to the large establishments, smaller ones existed with a wider spread of products. Some in the north-eastern part of the loess region, for instance, revealed too few remains of threshing waste to warrant the conclusion that cereal cultivation was the main activity. Continuation of the mixed farming of previous periods, with a small vegetable garden and some fruit trees in the yard as an extra, was the course followed. But in some environments specialisation took another direction. Several specialised in vegetable growing. An example is provided by a terrain, excavated at Longueil-Sainte-Marie 'Le Bois Harlé' (Dept. Oise) which bears the characteristics of a large, commercial, vegetable garden. Situated not far from the river Oise, and provided with a system of small ditches and waterwells, the place corresponds to a large extent to the lines written by the Roman author Columella in his tenth *Book to Horticulture*: 'it should be situated near a running stream, for this kind of gardening is a continuous process, and irrigation is essential to quench the garden's ceaseless thirst'. This was written with Mediterranean circumstances in mind, but is also applicable to gardens elsewhere. In such gardens dozens of different kinds of vegetables and herbs for flavouring were grown. The Longueil-Sainte-Marie garden is much too large to have served only one single family. Its traces extend over 12 ha. Plant remains, retrieved from the bottom of ditches and wells, show a list of species, including bottle gourd (Table 8.3). Bottle gourd was difficult to grow and belongs to the products which farmers tried to grow

Table 8.3 Cultivated plants found, waterlogged, in the ditches of a large garden at Longueil-St. Marie, (Oise), France

Longueil-Sainte-Marie 'Le Bois Harlé' Cultivated plants	
Amaranthus lividus	Amaranth
Apium graveolens	Celery
Beta vulgaris	Beet
Brassica nigra	Black mustard
Brassica rapa	Turnip
Coriandrum sativum	Coriander
Cucumis sativus	Cucumber
Daucus carota	Carrot
Lactuca sp.	Lettuce
Lagenaria siceraria	Bottle gourd
Lens culinaris	Lentil
Origanum vulgare	Marjoram
Pastinaca sativa	Parsnip
Pisum sativum	Pea
Papver somniferum	Opium poppy
Reseda luteola	Weld
Sisymbrium officinale	Hedge mustard
Verbena officinalis	Vervain
Vicia faba var. minor	Horse bean

to meet the demand of Roman customers. All in all, this garden must be regarded as part of a rural establishment specialised in vegetables and condiments. Such gardens were heavily fertilised, and were tended with single or two-pronged hoes.

In certain cases, the orchard, too, has grown beyond an affair with just some trees. Straight rows of square pits, covering areas of 6000 m^2 are remnants of true, probably commercial, orchards. The pits are much larger and spaced wider apart than those found in the vineyards discussed below and represent the planting-holes for trees (Fig. 8.10). In Roman arboriculture trees could be produced from seedlings, from cuttings or by grafting onto another tree (Fig. 8.11). Grafting was considered to give the most reliable result, as seedlings did not always result in trees with fruit of the intended quality. Three grafting techniques were known: cleft-grafting, bark-grafting and patch-budding. In the first method the parent trunk was partly cleft, so that a scion matching the cleft in size could be inserted. In the second method the scion was grafted between the wood and the bark. This method was more difficult but ensured better results. The third was held to be the trickiest. A small part of the bark of the receiving tree was removed to be replaced by a piece of bark of exactly the same size and bearing a bud. The grafted parts were held together and protected by pieces of bark, cloth or string.

Pruning trees was standard procedure. Iron saws and bill-hooks are regularly found where conditions have allowed their preservation (Fig. 8.12). The products of the orchard were either sold fresh, or dried, or preserved in liquids like wine, honey or brine.

A special variety of 'orchard' is the vineyard. Wine was an indispensable part of Roman life and it is, therefore, not unexpected that local wine production was

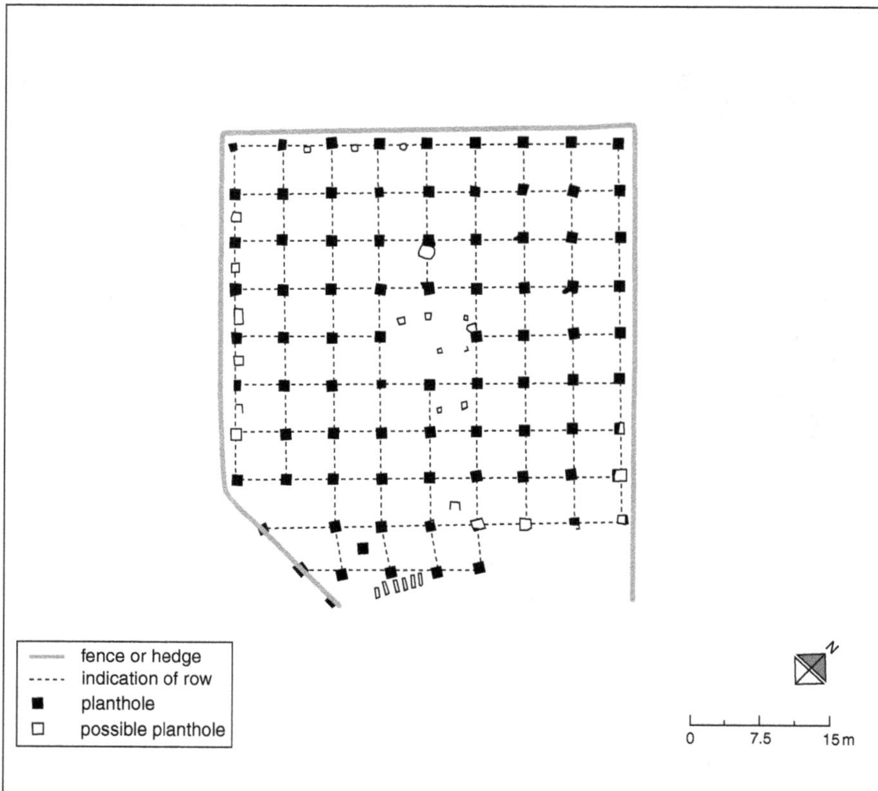

Fig. 8.10 Traces of a Roman orchard at Caurel near Reims (Dept. Marne), France

attempted, and with success. Long rows of small, equidistant pits found on a terrace of the river Oise (Dept. Oise) provide one of the proofs of viticulture. The rows of pits run in parallel lines for at least several hundreds of metres, indicating that this vineyard was not a small affair. It functioned between the second and the fourth century (Fig. 8.13).

Best known, however, are the wine-making establishments along the river Moselle near Trier, an area still renowned for its wine. The buildings housed crushing basins, a wine press, basins to receive the juice, and fermentation vats. The grapes were first crushed by trampling, during which the first juice was released. The mush was transferred to the press to extract the remaining juice. The press was provided with a wooden beam and a heavy counterweight of stone. If only one crushing basin was present, work had to be temporarily halted, because the basin had to be emptied before a second batch of grapes could be mushed. Several establishments had therefore more than one crushing basin to assure continuity of work (Fig. 8.14). Table 8.4 presents the size of the basins, found in several of such wine-making establishments. On the basis of the litres of juice the juice basins could hold

Fig. 8.11 Grafting of fruit trees; mosaic from Saint-Romain-en-Gal near Vienne (Dept. Rhône), France

and the length of the season for harvesting, an estimation of the size of the vineyards could be made. The size ranges from 5 to 6 ha for one of the smaller presses, to 60 ha for a large one. The smaller belonged to private *villae rusticae*. Most of them started in the middle of the third century, but the large establishments belong to the fourth century and were presumably owned by the state. Their presence is accounted for by the fact that, from the end of the third century onwards, Trier was one of the principal residencies of the emperors of Rome.

The wine was stored and transported in wooden casks of considerable size. Although some traces of casks have been found in connection with the presses, most of them are known from their ultimate use, as a lining of wells. In that role they are found far beyond the loess region, showing that the wine was transported over long distances. Casks also figure on stone reliefs. The cask is a latecomer in the world of packing. Although the technique of making vessels out of boards is older, the cask with its curved staves and substantial volume is only known from the

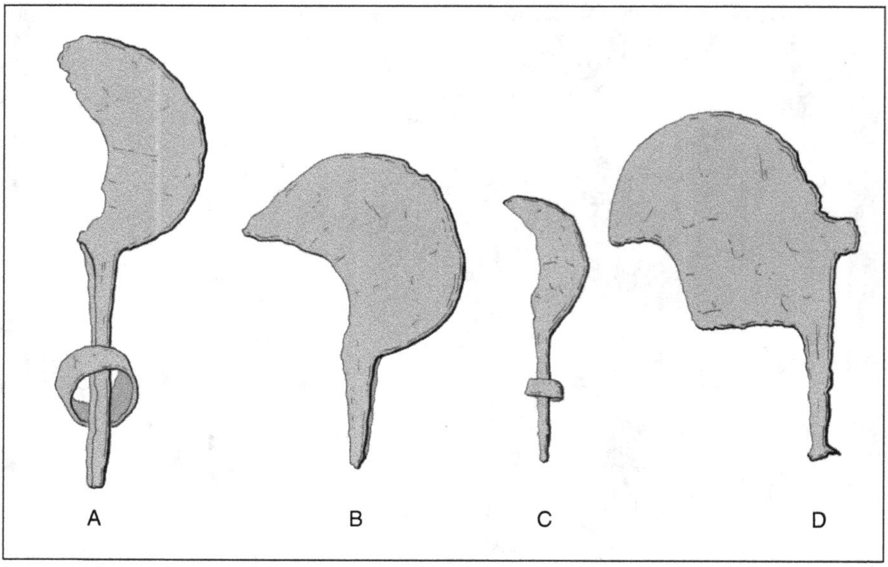

Fig. 8.12 Various types of bill-hooks. Sizes vary between 16 and 21 cm

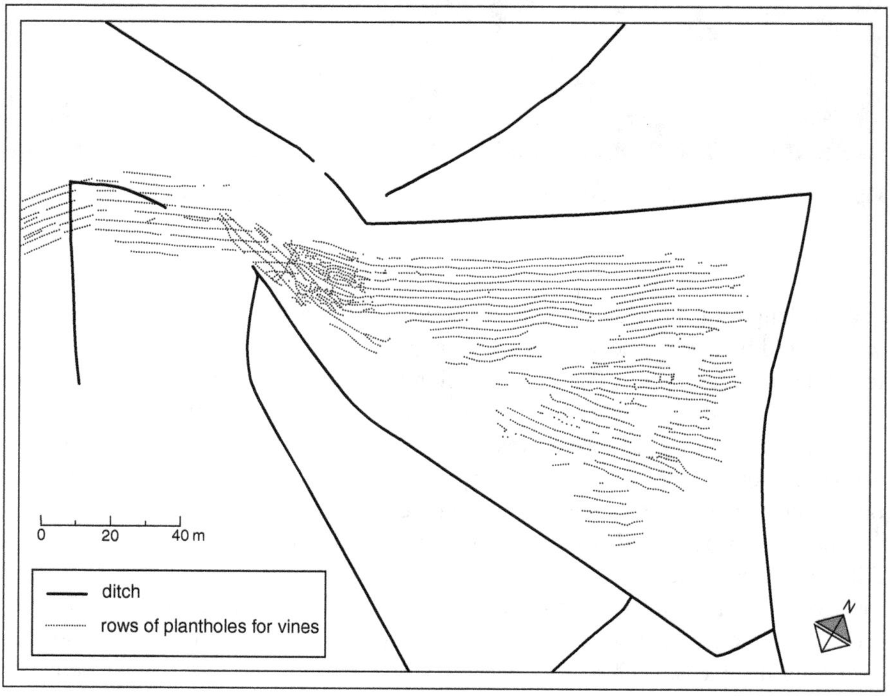

Fig. 8.13 Traces of vineyards at Bruyères-sur-Oise (Dept. Oise), France

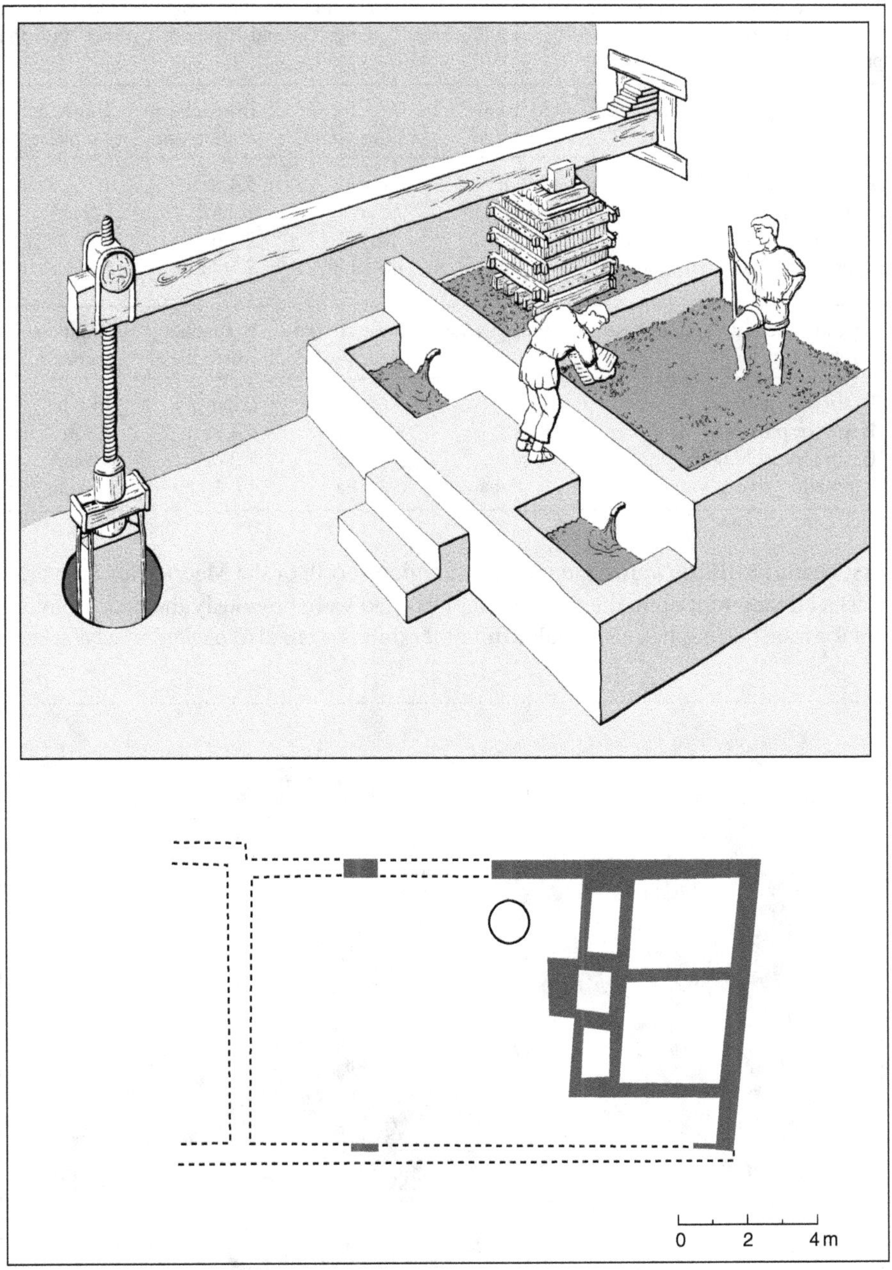

Fig. 8.14 Ground-plan of a wine press excavated at Brauneberg near Trier, Germany, (below) and its reconstruction (above)

Table 8.4 The size of the basins in the wine-making establishments of Maring Noviand, Lösnich, Brauneberg (an eastern and a western establishment), Ungstein and Piesport together with the presumed size of their vineyards

Establishment with 1 basin for mashing	Maring-Noviand	Lösnich	Brauneberg eastern est.	Ungstein 1st phase
Basin for mashing	5.3 m^3	9 m^3	5.4 m^3	?
Basin for pressing	3.9 m^3	7.5 m^3	4.6 m^3	3.5 m^3
Basin for catching the liquid	1.0 m^3	2 m^3	2.4 m^3	1 m^3
Estimation size of vineyard	7–8 ha	11–13 ha	8–9 ha	5–6 ha

Establishment with more basins for mashing	Piesport 1st phase	Piesport 2nd phase	Brauneberg western est.	Brauneberg together
Basin for mashing	23 m^3	30.5 m^3	8.2 m^3	13.6 m^3
Basin for pressing	8 m^3	8 + ? m^3	4.4 m^3	9.0 m^3
Basin for catching the liquid	5 m^3	5 + ? m^3	4.0 m^3	6.4 m^3
Estimation size of vineyard	c. 45 ha	c. 60 ha	17–18 ha	c. 26 ha

first centuries BC onwards and only from regions north of the Mediterranean world. Casks already turn up in the first Roman forts and were obviously the main containers for transporting liquids in bulk. But other things were also packed in casks, such

Fig. 8.15 Wagon with a cask presumably filled with wine, drawn by a team of oxen. Provenance Augsburg, Germany

Fig. 8.16 A Roman ship with casks on board. Provenance Neumagen, Germany

as the pomegranates mentioned in Section 8.2. Their volume ranges from 300 to 1500 L. Sepulchral monuments in Trier show how they were transported, loaded on a wagon drawn by oxen, or aboard a ship (Figs. 8.15 and 8.16). Next to casks, earthenware vessels, such as amphorae, were still known, but producers of bulk liquids north of the Alps obviously never used them.

8.4 Livestock and Animal Husbandry

Cattle, pigs, sheep, usually in combination with only a few goats, and horses were, as before, the main large animals to be found on the farms. But poultry held a considerable place as well. Poultry-yards were stocked with several breeds of chicken, ranging from small to large. Domestic geese are attested by their bones, which are larger than those of the wild species occurring in the region. Ducks, too, were kept and many *villae rusticae* had ponds to accommodate them. Apart from supplying meat, poultry was kept for their eggs and feathers. Goose-down was highly prized as the finest material for stuffing pillows.

A bird not yet mentioned in the previous chapter, and newly added to the livestock, is the pigeon. Its wild ancestor is the rock dove, which occurs in rocky upland areas and coastal cliffs around the Mediterranean Sea. Roman people kept them in dovecots for their meat and to serve as messengers. In the latter role they were much appreciated by the army. All the same, records of pigeon bones are very scarce in such rural settings as are described here, and traces of separate dovecots are absent.

Of course, a kind of dovecot may have been part of the upper floors of other buildings, but this is far from certain.

A still smaller, but important animal was the honeybee. There are many wild species of bees in the world. Most of them have been exploited for honey and beeswax, but only two species have been truly domesticated and only one of these, our honeybee, has been moved to any extent beyond its native range, which is eastern tropical Africa. Beekeeping had perhaps already been part of rural life during the Metal Ages dealt with in Chapter 7, but as the practice leaves hardly any traces in archaeological records, it is difficult to establish whether the honey of those times came from wild bee populations or from the actual keeping of honeybees on farms. Because Roman authors have written at length on this subject, it is known that most Roman farms in Italy kept bees in hives and it is safe to assume that the farmers in the loess region west of the Rhine kept bees as well. Honey was a most important sweetening substance.

Returning to the large animals, it should be noted that every kind of animal occurred in several breeds. Sizes range from small to large. Whether the breeds corresponded to different uses, such as breeds of cattle kept primarily for milk or for meat, cannot be attested by the archaeological sources, but the ancient writers mention them.

There is more that commonly escapes notice. Two animals, certainly known at the time, are the mule and the donkey. Mules, hybrids with a donkey as father and a horse as mother, were used extensively in vehicular transport and as pack animals. The mule has the advantage over the horse in being more sure-footed on difficult roads and having more endurance. The Roman army used them on a large scale. Mules were, for instance, victims of the famous battle in the Teutoburger Wald, where an entire Roman legion under general Varro was annihilated by Germanic tribes. Mules are often depicted drawing four- and two-wheeled vehicles (Fig. 8.17). They are also the animals that push the reaping machine depicted in Fig. 7.7. One would, therefore, have thought that the breeding of mules was one of the activities of the owner of a *villa rustica*. However, their remains are almost absent from the bone records. One of the possibilities is that they are not recognised. Bones of mules are very similar to those of horses. Another possibility, put forward by archaeozoologists, is that mules were not bred in the region, but were regularly imported from Italy and Spain.

Sparse evidence suggests that donkeys were domesticated c. 5000 years ago. DNA analysis points towards an African progenitor. Wild asses from Asia seem to be out of the question. Where the animal was actually domesticated is still unclear. Although donkeys are regularly mentioned in ancient texts, they are almost absent from the bone records in the loess region. There was obviously no role for them either as sire to produce mules or to help on farms and to carry loads.

Cattle, pigs and sheep were the main providers of meat (Fig. 8.18). A significant number of oxen and cows reached ages of more than four years, implying that they were kept for other purposes as well. The working ox was indispensable on the farms for ploughing and hauling heavy or bulky loads. They were common draught animals in the trading centres called *vici* (singular *vicus*) and in towns. Oxen appear

Fig. 8.17 Mule drawing a cart. Provenance Arlon, Belgium

on reliefs pulling wagons loaded with casks (see Fig. 8.15). Finds of foot bones with traces of former inflammations show that the animals were often burdened too heavily. Cows were used for traction as well, as is demonstrated by the finds of cow horns with deformations caused by traction. Nevertheless, many adult cows were kept for their milk, which was used for cheese making. Cheese was a common part of daily food. According to some ancient authors, cheese was even one of the principal items of food eaten by the Roman troops. Cheese-making equipment, found in military forts, suggests that the soldiers made their own cheeses. Nevertheless cheese will also have been produced on the farms, not only for domestic use, but also to be sold to non-farming customers. Indeed, most of the cattle found slaughtered on farms, were not young animals. Young individuals, slaughtered for their meat, are mainly found in trading centres, cities and military forts.

The extent to which milking of sheep and making sheep cheese was practised on the farms is unknown. Sheep were important for their wool. As a provider of meat they were less appreciated. Most of the bones are from older animals and the share of sheep in the bone records falls during the period, an indication that they were eaten less and less. However, one area presents an exception. In the west, near the Channel coast, the number of sheep/goat bones exceeds that of pig and reaches values of up to 45% of the bone spectrum. These values resemble those found on the other side of the Channel. Maybe the shallow soils on top of a chalky subsoil bore a vegetation that was best suited to the breeding of sheep.

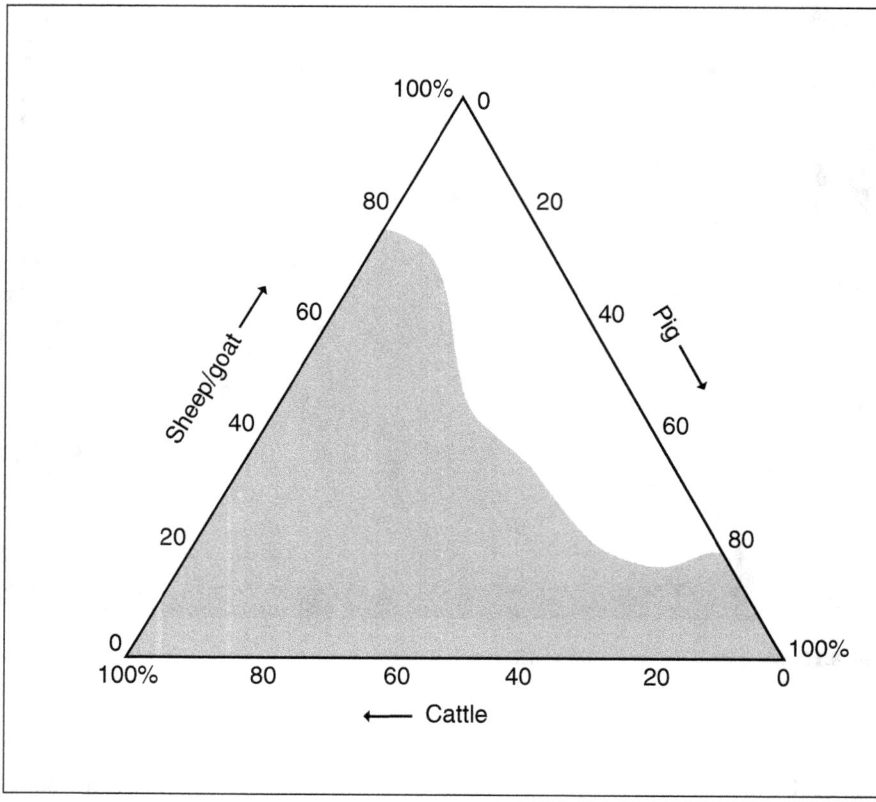

Fig. 8.18 Composition of the livestock

Pigs were destined for consumption. They were slaughtered young. Suckling pigs were very much appreciated and the survivors almost never reached ages beyond two years. Most of the remains are found in cities and military contexts. The army demanded bacon and lard as part of the daily rations for the troops. Lard was a substitute for oil, which is more difficult to transport. And, an aspect not to be forgotten, hams from Gallia were, even in Rome, much in demand.

In the later phases of the Roman occupation, pigbreeding seems to have fallen into decline. Their bones become less and less numerous in the finds, but some military posts are an exception. Pigs remained in demand there.

Horseflesh was not regularly consumed. Horses were chiefly kept as riding or carriage animals. The ancient writer Columella, living in the first century and a leading author on agricultural topics, divides horses into three classes: 1. horses for the circus and sports, 2. horses to breed mules and 3. the common stock. The third category will have been the common one in the loess region. Such horses were sturdy beasts with a height at the withers of 140–145 cm. An important customer for such horses was the army, though until late Roman times the cavalry arm was subordinate to the infantry, constituting ten percent of a legion as a maximum. Horses were ridden without stirrups. The saddle had four horns and the horseman could not fall off. Bits were of the bar or jointed snaffle type, but curb bits were also known and the military made much use of hackamores.

Some parts of Roman roads were truly paved and the hooves of horses are not well suited to such pavements. This led to the invention of the hipposandal, an iron plate with upturned sides and back, which was tied with leather straps onto the hoof. It was put on when necessary on rough stony roads or on diseased hooves. Normally horses went unshod (Fig. 8.19). Sometimes the hooves of oxen too were protected by a kind of sandal.

The hides of all animals were used for leather, dogs not excluded. Dogs were kept as watchdogs, to help in herding livestock, and as pets. Several breeds are known, from large to lapdog size.

It remains unknown how many animals were kept on a single farm. Judging from the relative abundance of bones from slaughtered animals, it is assumed that farms in the early phases of Roman occupation raised more animals than a true *villa rustica* in its heyday, at least where the establishments on loess soils are concerned. In the

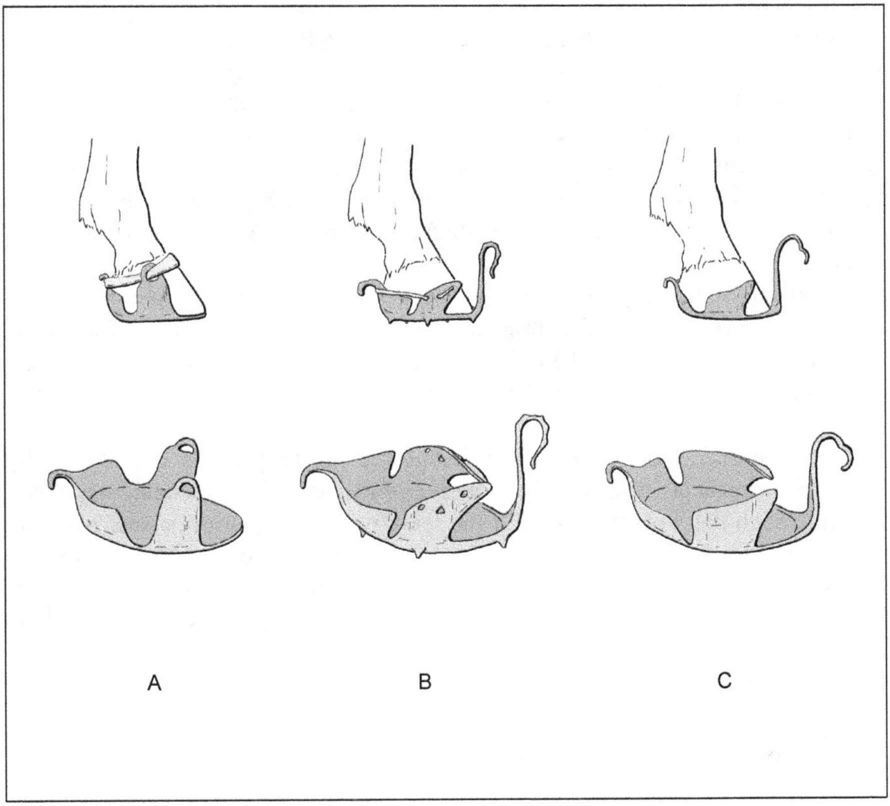

Fig. 8.19 Hipposandals. The central one has spines to provide a better grip on a slippery road

period of decline, animal husbandry once more gained importance. Nevertheless relative abundance of bones provides a weak argument. Animals may have been transported alive to be slaughtered elsewhere. Neither can the number of animals be estimated from the dimensions of stables and byres. Although such buildings are present on farmyards, they may have housed only working oxen, milking cows and riding or carriage horses. The classic authors are very much in favour of keeping animals year-round outside. As the farmers of the preceding ages kept their animals out of doors too, it is very probable that most of the livestock belonging to a *villa rustica* was never stabled.

It is unknown whether such herds were fed additional dry fodder in winter, but the stabled part of the livestock would have had to be provided with fodder year-round in any case. Fodder meant not only hay. The ancient writers mention fresh forage when available and this included tree leaves. According to Columella foliage of elm was best for oxen, followed by ash and poplar. Oak was the worst. In wine-producing areas grape-skins were valued as feed in autumn. Soaked acorns and, if all other kinds of feed were finished, chaff was given in winter. To keep oxen strong, vetches, including their straw, and good-quality hay were preferred, however. The author Cato remarks concerning oxen: 'you must not put them to grass, except when they are not ploughing; when they have once eaten green fodder, they are always expecting it, and they have to be muzzled when ploughing to keep them from going for the grass'.

It is quite feasible that the feeding strategies, recommended for farmers in Italy, were also followed on the *villae rusticae* in the loess region. On horse feed much less has been written. On Italian farms, horses were far less important than oxen and this may have been the same north of the Alps. Fodder, found in the stables of a cavalry camp in Dormagen, near Cologne in Germany, gives however a hint of horses' feed. It consisted of oat and hay. The hay was obviously cut in several types of grassland. Some of it had a near-wetland origin as could be found on terrains close to the river Rhine. Other origins were drier stretches of pasture. Hay seems to have been cut in May and June, after a period during which the grassland was kept free from grazing animals. There are no indications of the existence of true meadows.

To return once more to poultry, this was kept in the yard. In Italy especially, chickens were kept in special henhouses, which were kept meticulously clean. Whether such hen-houses existed in the loess region is open to debate, but some parts of the sheds may have served the purpose. As mentioned above, a pond for ducks and geese was normally present. The ancient authors mention millet and coarsely ground barley as the best feed, but chaff with weed seeds served the purpose as well.

8.5 Farmbuildings and Yards

The type of farm most characteristic of the period looked rather unlike the establishments known from the period before the Roman occupation. Some farms were new creations, founded on land not inhabited earlier. Nevertheless, many evolved from existing farms. During the first phase of occupation such farms were still of the type described in Section 7.5 and consisted of relatively small buildings set in an enclosure. During the consolidation phase, the old structures were pulled down. New buildings appeared, laid out according to a ground-plan in which traditional elements were combined with Roman ideas (Fig. 8.20).

A characteristic element of the main building was a rectangular plan with one of the long sides as front. Both corners of this frontal part consisted of square rooms slightly jutting out. A veranda was constructed in between, with the entrance to the building in the middle. If the roof of the veranda was supported by columns, as it often was, it is called a *porticus*. The corner rooms had an upper storey. On approach, the building had the appearance of a solid home, with a colonnade in front, flanked by two square towers. It was the most common basic type (Fig. 8.21).

The interior was divided into separate rooms. Original plans are easy to trace because walls were founded on a stone footing set in a foundation trench. The remainder of the walls consisted of a wooden frame filled with wattle-and-daub work. A Roman touch in such constructions is the use of iron nails. An indication of the height of the walls is provided by a villa, found at Ahrweiler (Germany), which was destroyed and subsequently buried by a landslide. A toppled-down wall had an original height of 2.88 m. Walls had windows and fragments of glass suggest that these windows were fitted with windowglass.

Most buildings were modernised, improved and added-to during their lifetime. Principal rooms were rebuilt in stone and provided with floor heating. Cellars were dug, Roman-style baths built and second storeys added. Walls were plastered and decorated following the prevailing fashion in Italy. Floors were covered with mosaics. The roof was tiled. In some cases buildings remained modest, but others radiated sheer luxury in the end. Some buildings had facades with a length of 190 m.

Most rooms in the main building were for the owner, his family and his servants to live in. This part of the farm is called *pars domestica* or *pars urbana*. It was erected in a yard with other buildings, a threshing floor, a vegetable garden, an orchard, a well and very often a pond. This part of the farm is called the *pars rustica*. The more or less rectangular yard was enclosed by a hedge or a wall. The main entrance was commonly opposite the main building. The size of the yard ranged between 0.75 and 4 ha. *Pars domestica* and *pars rustica* together form the *villa rustica*.

The main building still functioned, besides providing living quarters, as part of the farm. In establishments of a modest size, part of the livestock seems to have been stabled there. Storage of threshed or semi-threshed grain also took place under its roof. Nevertheless most of the agricultural functions were fulfilled in the secondary buildings. This was not new, but already quite common in the previous period. As in the past, excavations centred on the main building, while the exact nature of the secondary buildings, erected with timber, wattle and daub and provided with a thatched or shingle roof, is not always precisely known. Some must have been barns, byres, stables or sheds. They will have provided housing for slaves and other

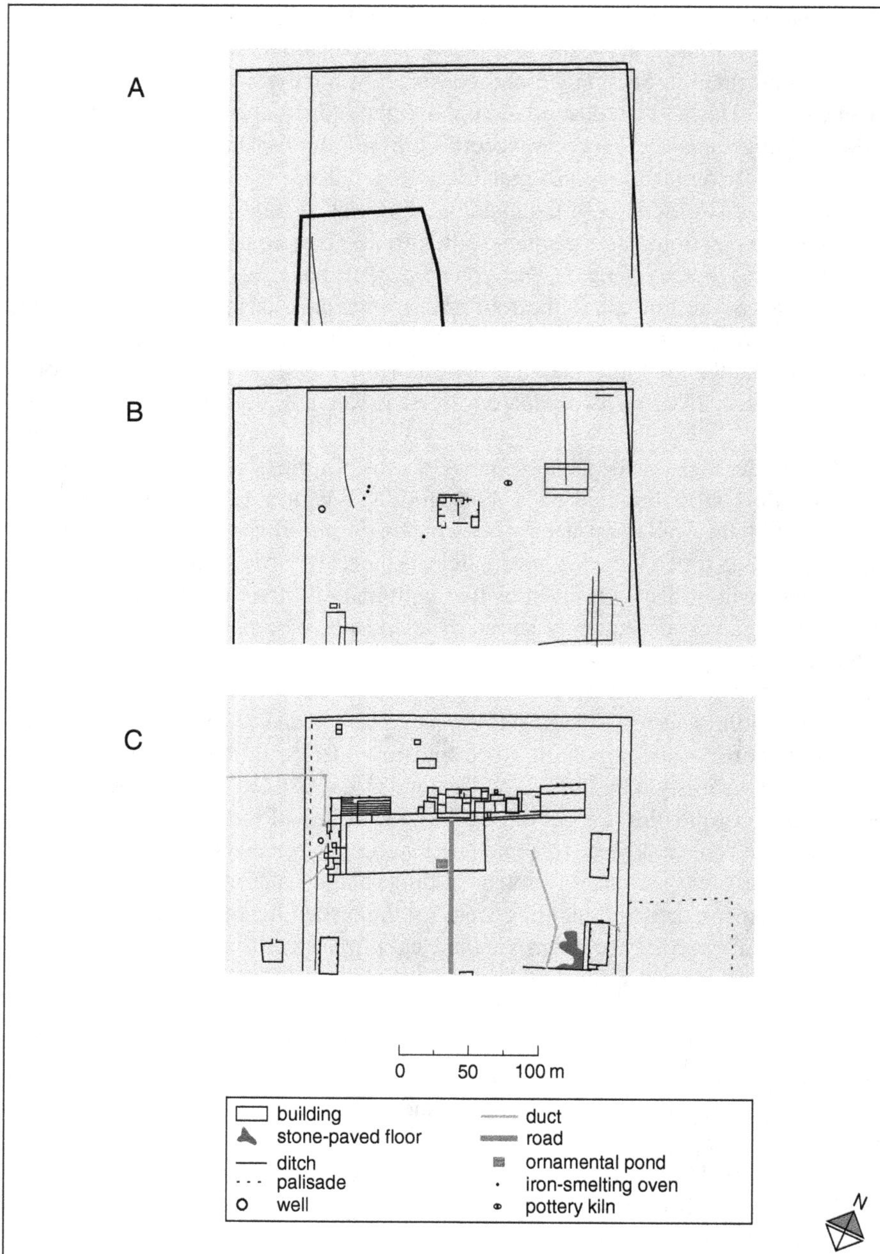

Fig. 8.20 The transformation of an Iron Age farm into a large *villa rustica*: the case of Voerendaal, the Netherlands. **A**: period 50 BC–AD 50, **B**: second half of the 1st century AD, **C**: 2nd–3rd century AD. The farm buildings of the first period must have been located outside the excavated area

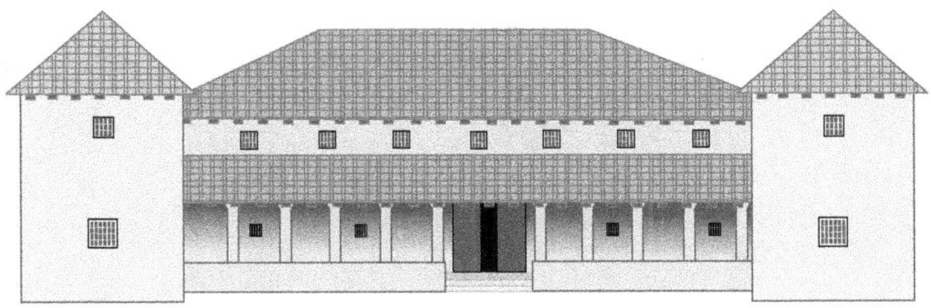

Fig. 8.21 Facade of an average *villa rustica*

personnel as well. Other constructions housed the crafts, which were also part of the activities on a *villa rustica*. Next to farming most of the villas were engaged in other types of business, such as the production of pottery, bricks and tiles, glass or iron and iron implements. Some exploited quarries. Nevertheless, these activities were always secondary to farming.

The presence of an open-air threshing floor is attested by concentrations of chaff, which is found embedded in the soil at specific places within the yard. What is missing in the yards is the underground silo, so common in pre-Roman times (Fig. 8.22).

Gardens and orchards reveal their existence by the planting holes discussed in Section 8.3. The terrain in front of the *pars domestica* of truly luxurious villas had lost its farmyard aspect and showed formal ornamental gardens instead.

The pond, for ducks and geese, and possibly also for watering livestock, is a common feature. Some villas had moreover a true horse-pond, a rectangular structure with steep walls at three sides and a downwards sloping entrance on the fourth.

The well provided cleaner water, but some villas also drew water from a duct. These ducts, leading from a natural spring to the main house, could be several kilometres long. They were underground constructions made of wood, brick or stone. At regular distances settling tanks, preventing blocking by mud and accessible for maintenance, ensured their continuous functioning. In certain areas, for instance in the Moselle area, they even had the form of true *qanats*, a kind of underground duct invented in Persia. The *qanat* has spread from its country of origin to other cultures, as far east as China and as far west as the western Mediterranean world. It consists of a series of well-like vertical shafts, connecting sections of a subterranean tunnel hewn out of the bedrock. The tunnel taps water from an underground aquifer, not

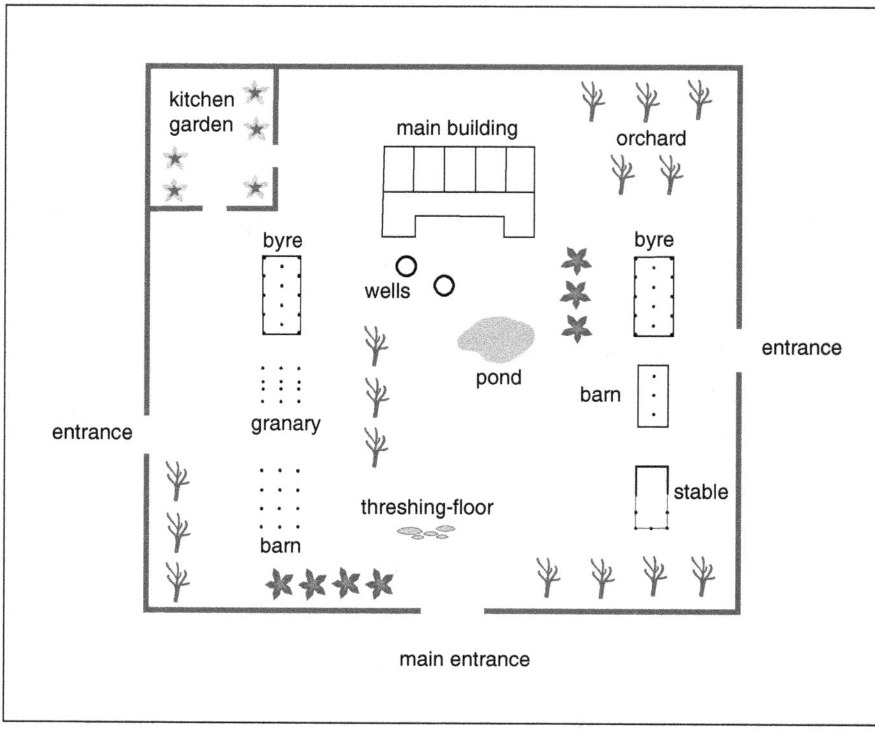

Fig. 8.22 Plan of a *villa rustica*: main building, outhouses and yard

from a surface spring. The *qanat* was developed for regions where natural springs are absent and groundwater levels too deep for ordinary wells. Though the construction of ducts or *qanats* looks simple, it requires quite a degree of understanding of engineering. Too shallow a gradient yields no flow, too steep a gradient a too strong, erosive, current (Fig. 8.23). Roman *qanats* in the area reached lengths of up to 1660 m with shafts of a maximum depth of 35 m.

The owners of the villas were often local families that became integrated into the Roman socio-economic structure. Formerly it was thought that all the land was taken by the Roman authorities, redivided and allotted to their own people, or to local people who had worked for them, as a kind of pension after service. But, although this practice certainly existed, it was not the chief trigger of villa rustica development, at least not in the areas intensively farmed before. In areas, where such was not the case, the land was indeed newly developed and given out to new people.

Before the Roman conquest, farmers were already part of a society where surplus production was required to meet the demands of a tribal organisation. This was based on a system of 'patronage', a complex of relationships between persons with a higher status, the patrons, and those with a lower status, the clients. It was a system of reciprocal services in which people with access to certain commodities were paid for these commodities by services in return. These could be of a politi-

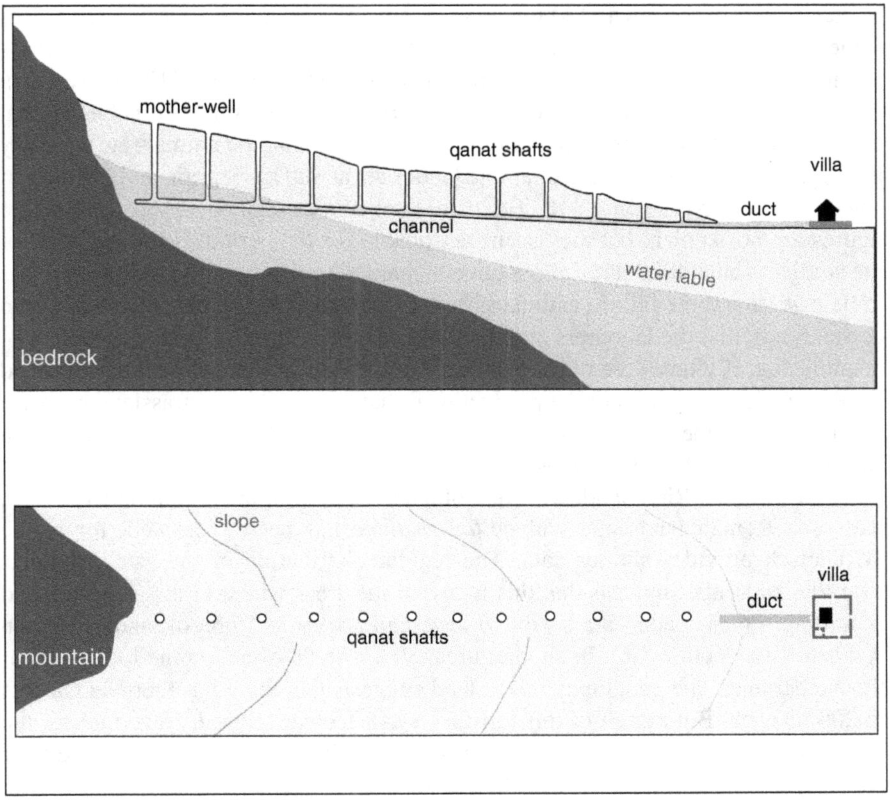

Fig. 8.23 *Qanat*, in section and from above

cal, military or material nature. The system led to a hierarchically organised society, in which people with the best access to some much-valued goods reached the top. Access to Mediterranean objects, breeds of cattle and food plants were such items. A substantial part of the goods playing a role in the system consisted of agricultural products. The better a farmer could do in the system, the higher he could rise and the richer his establishment. This is expressed in the categories of farms described in Section 7.5. The upper strata of this rural society are thought to have converted their establishments gradually into a *villa rustica*. And this developed into the provider of food and other products to the Roman army, Roman civil authorities and the remainder of a non-rural population.

Required tributes and taxes were paid in kind and in money, obtained by marketing part of the products. The Roman Period knew a market economy in which money played an important role.

The demand was greater near military sites and large administrative centres (see Section 8.6). What was needed to feed the army of occupation has been calculated for a comparable region just outside the loess region under review, namely the

Wetterau, east of the town of Mainz in Germany. The calculation by A. Kreuz gives some insight into the requirements to be met by the 250 *villae rusticae* in this area. In the year AD 165, food had to be provided for 8312 soldiers, 2121 horses, 976 pack animals (mules) and 450 head of cattle. From ancient sources it is known that a soldier required 1 kg of grain per day. It is assumed that the animals ate 8 kg hay per day. The yield of a hectare of arable land is set at 800 kg of grain and the ratio of sowing seed/crop is set at 1 : 10. The values are educated guesses, because the true figures are not known, but they seem reasonable for the period. It follows that for the soldiers some 4000 ha of productive arable land were required. Allowance has to be made for some fallow, estimated here as one year. And it has to be taken into account, too, that the labourers and the families on the farm had to eat as well. The result is that, if villas were of the same size, every one of them should have farmed 35–40 ha of arable land. For the production of hay some 20 ha of grassland per farm was needed. As the Wetterau area measured 750 km^2, such a large-scale production was indeed possible in that landscape.

In the German Rhineland, west of Cologne, and within the region under review, part of a Roman landscape with *villae rusticae* has been excavated, for which W. Gaitsch provides similar data. The regular distribution of the establishments over the landscape suggests that this is one of the areas where farms were founded according a plan. Each villa seems to have had access to 50 ha of land: a Roman *centuria* (see Section 8.6). In another area, 30 km to the west, on the loess soils of the Netherlands, the geography of the land suggests that the villas there had access to 200 ha each. But not all of this territory was necessarily used. Nevertheless, the main buildings of these farms are larger than those in the Rhineland case, and their size may correspond to the size of their land.

The number of farmhands required to till the land has been estimated as well. The Wetterau establishments could handle almost all necessary work with four workers and two draught oxen per farm. Only at the time of harvesting and haymaking were more people needed. The estimate mentions a minimum of six and a maximum of 15 persons. Calculations made for the 200 ha farms in the Netherlands leads to 14–18 workers and seven teams of draught animals. At harvest time some 40–60 persons would have been needed. In both estimates it is the number of ploughmen, required for the actual ploughing of the land, which determines the number of resident workers. The extra labourers, needed to bring in the harvest, are supposed to have been seasonal workers. Where these extra hands came from remains unspecified.

It should be noted that the models deal only with crops and not with livestock. As mentioned in Section 8.4, the size of the livestock is unknown, but crop cultivation does seem to have been the principal activity of the farms in the Wetterau and the Netherlands. Nevertheless, the large villa at Voerendaal in the Netherlands had, notwithstanding its specialisation in cereal growing, space for 50 animals in its byres. And for the Rhineland farms, a more diversified agriculture with both crops and animals has been proposed. Their byres are supposed to have held about 25 head of cattle. All those animals required personnel as well and if the villas also had animals to be herded year-round in the open air, additional hands would have been required. Moreover, there was poultry to be looked after. It has been suggested that such activities may have been carried out by the families of the farmhands, but this is far from certain.

In addition to the people connected with farm work in a strict sense, there were artisans living on the farm. They were occupied with the crafts mentioned earlier. All in all, a *villa rustica*, especially a larger one, was an enterprise with quite a number of resident people. Who were they? According to the ancient writers, a large part of the personnel consisted of slaves. If the owner was rich and non-resident, the farm was run by a steward, called *vilicus*, who could be a free man or a slave. The question is, however, whether this was also the case in the loess region west of the Rhine. As far as can be deduced from graveyards connected with the farms, the owner was living on the farm. At least, he and his nearest family were buried there. He may have had slaves, but many of the people working for him are considered to have been free. It is assumed that they were the clients of the social system mentioned above, while the owner was the original patron. They may have possessed smaller farms in the neighbourhood, which never made it into *villae rusticae*. In the course of time their dependence dwindled to the status of labourers. Indeed, small farms of a non-Romanised, indigenous type, are no longer reported during the heyday of the villa system, at least not on the loess soils, though they persisted on its fringes. However, they may have escaped archaeological notice.

Archaeological traces of true slaves, human beings regarded as movable possessions, are absent or near-absent from the records. Rows of small structures, aligned along the wall surrounding the yard of very large and wealthy villas, have been interpreted as their dwellings; and of some cellars it has been suggested that they served as a prison for unwilling slaves, but such instances are very rare. Of course, slaves will leave almost no traces behind, but it is, nevertheless, assumed that the slave on the villa was not as common as in Italy, for instance.

8.6 The Farm in Its Setting

Section 7.6 has explained how society gradually entered the Roman sphere of influence, which preceded the actual conquest. Farms were at that time not the only inhabited sites. Well-organised nucleated settlements occupied a central place in the economic and social life of the people, who were organised into groups, more or less loosely knit together, called tribes. The tribe was characterised by a mild hierarchical structure, built on patron-client relationships. In such relationships one party has more to give and the other more to be thankful for. Tribes worked together, or made war on each other. The Roman conquest was based on the famous principle of ‘divide et impera’, divide and rule, alliances with the one, war against the other tribe. It was aided by the pre-conquest contacts in which Romans were at the patron side of the relationships.

During the first phase of Roman occupation the old economic and social systems remained intact. They were, however, not suited to meet the demands of the new rulers who created a new system in which the local tribal elites were incorporated. The old tribal territories were converted into administrative districts with newly-founded central places which developed into real towns. A well-kept system of roads, partly based on existing roads and partly new, provided the lines of swift communication necessary to keep the country together.

Small towns, called *vici* (singular *vicus*) sprung up near important crossings. Such *vici* arose also in close connection with the military forts, near springs with medicinal water, or rural holy places, thus everywhere where there was something to do. Waterways were important as well, and inland ports emerged on the navigable rivers.

Towns, *vici* and *villae rusticae* were the pillars on which society of the Roman Period was founded. Towns were focal points in the administrative, economic, social and religious life. They had regular ground-plans with a square, main streets and secondary streets, all laid out in a chess-board pattern. Some were open, others were surrounded by a town wall with towers and gates. One of the main amenities was the public waterworks. These represented true specimens of the high standard of Roman engineering. Cologne, for instance, received its water from a series of springs in the Eifel mountains. The duct was 95.4 km long and had a transport capacity of 20000 m^3 drinking water per day. It was constructed around AD 80, functioned until around AD 260 and served public taps, public bathhouses and private houses. It is the largest Roman construction north of the Alps. The several functions of the town had their own buildings: a building for the court, temples for religious duties, bathhouses for hygiene and social life, and a theatre and/or amphitheatre for cultural life. Redistribution of the products, collected from the farms, was another function to be met and required large-scale storage. This was provided by large town granaries, within or just outside the town wall. The public system was upheld by taxes. People had to pay them partly in coin and partly in kind, especially foodstuffs. The latter was called *annona* and served to feed the town-dwellers as well as the army.

The *vici* were centres of trade, crafts and industrial enterprises. They seem to lack earlier, pre-conquest roots. They provided market facilities in addition to that of the towns. Most *vici* consisted of a single row of buildings on one or both sides of a road. The rectangular, oblong houses were built of timber and wattle and daub, and faced the street with their short sides (Fig. 8.24).

The basis of economic life was, however, provided by the rural establishments, where the greater part of the population lived. According to an educated guess c. 3 million people lived in the country, ten times more than the population of towns and *vici* together. Farms were not quite equally dispersed over the landscape. Though Roman technology made it possible to settle in areas not settled before because of problems with access to water, plateaus with a very deep water table were still avoided. But, more importantly, villas tended to be located in the vicinity of the main axes of traffic. Also villas occurred in higher density around towns and forts. Transport facilities, markets and concentrations of consumers did influence the distribution of villas over the countryside.

Even if they had roots in older establishments (see Section 8.5), not every villa was built on the exact locality of the former place. Moreover, some of the land had

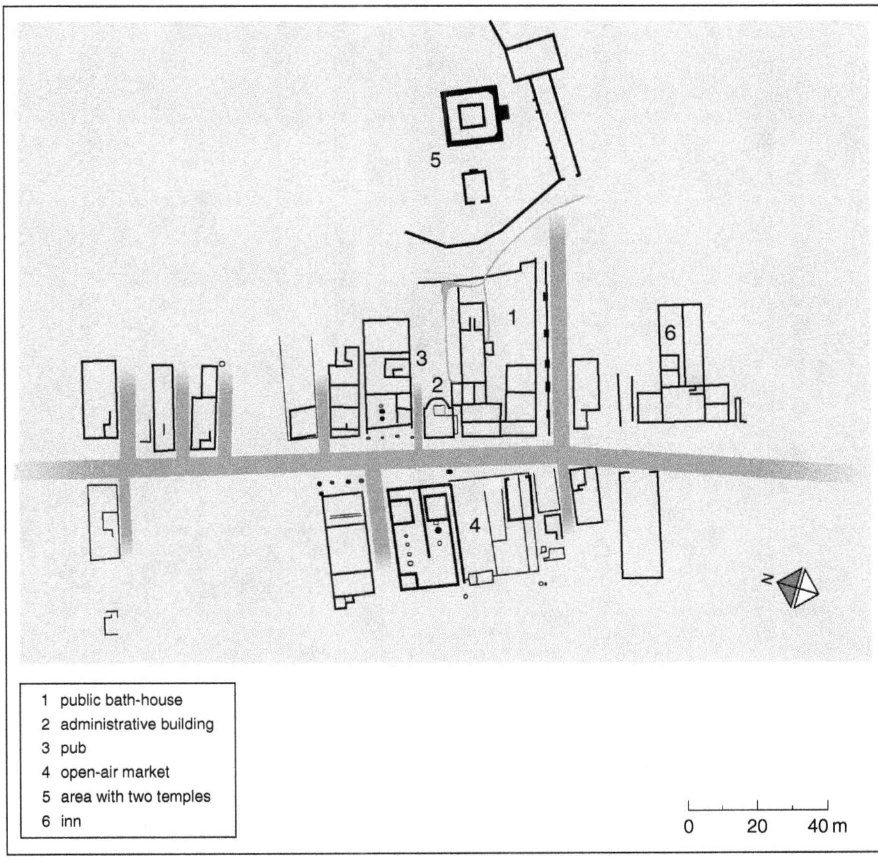

Fig. 8.24 The *vicus* of Clavier-Vervoz, Belgium

undergone new parcelling. This is especially evident near newly-founded central places and along new highways. In certain areas, for instance around the town of Tongeren in Belgium, Roman parcelling is still visible in the recent landscape, as revealed by cadastral maps and aerial photography (Fig. 8.25). At least, this was thought to be the case until quite recently. Land was divided into lots of 710 by 710 m, which were subsequently divided into smaller square or rectangular lots. The Roman surface unit in question is the *centuria* (504576 m^2 or slightly more than 50 ha). The application of the *centuria* to the land is called *centuriatio*. It is quite possible, though not proven, that the centuriatio created those estates, which were given out as a kind of pension to persons who had served in the army or civil administration. Nevertheless, the location of the villas is not strictly related to the lay-outs still visible. And nowadays it is called into question whether the traces in the landscape are indeed Roman. They are perhaps of much younger date.

The serious disturbances which set in during the second half of the third century had not only their effect on the military forts along the river Rhine and on the large

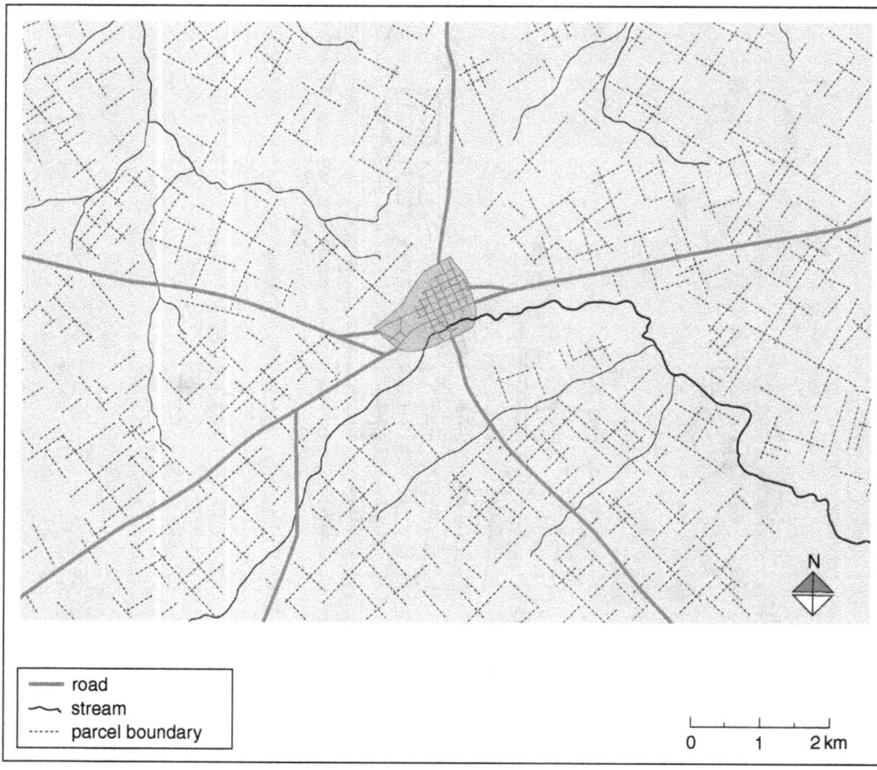

Fig. 8.25 Traces of a possible Roman *centuriatio* around Tongeren, Belgium. The traces consist of field boundaries in the modern landscape; the town is depicted as it was in Roman times

towns (Fig. 8.26), but also on the countryside. The new military strategy, adopted after AD 250 (see Section 8.1), led, after forts, towns, *vici* and villas, to the foundation of a fifth kind of settlement: the small inland fort. These settlements had as their main function, or as one of their principal functions, to serve as depots of food, destined for the swift armies of intervention. Large granaries occupied an important part of their surface. These held fodder in addition to grain and other foodstuffs, as proven by granaries that had caught fire and left their carbonised contents behind (Fig. 8.27).

Concerning rural life: aerial photography by R. Agache in the Picardie (north-western France) revealed a change in settlement pattern. The great majority of the villas on the plateaus seem to have been abandoned. Rural settlement concentrated around the cores of human activity, along roads and rivers for instance. This has led to the general opinion that in the early decades of the fifth century the countryside had been almost deserted. This view of the matter is contested at present. An analysis of an area north-west of Trier showed that the survival of rural settlement lies around 25% if all sites are taken into account; but that, if only well-dated sites are considered, survival rises to 83%. However, Trier itself was at that time one of

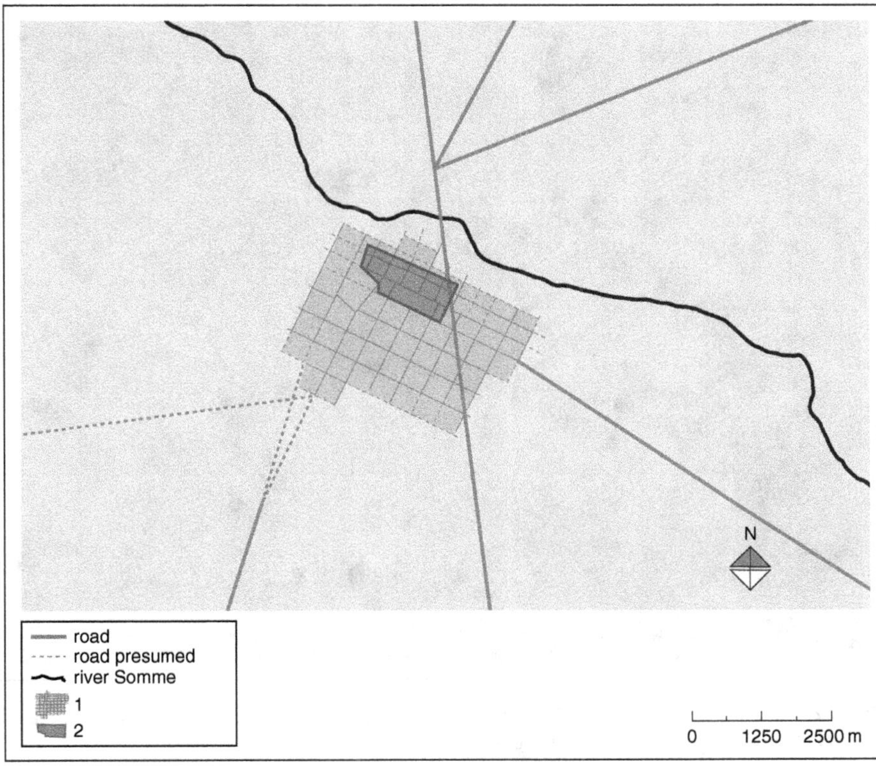

Fig. 8.26 Plan of Amiens, France, in its heyday (1) and after its decline (2). The grid pattern shown in the picture represents the old street plan

the towns that had survived a range of troubles rather well. In the fourth century Trier even became the capital of the Roman province of *Gallia Belgica prima*. The emperor Constantine the Great had his residence there from AD 306 to AD 337. Therefore, the picture may be biased. Comparable studies in other areas show, however, that in the loess region west of the Rhine c. 75% of the farms survived. Near large towns the decline was even less, but resettlement around important centres, roads and waterways seems not to have taken place. In general rural life remained dispersed.

Nevertheless, rural life did not continue in the manner it had during the heyday of Roman rule. The Romanised aspect of the main farmbuildings disappeared more or less gradually. New constructions were more and more built in timber and wattle and daub. Also yards are seen to shrink. Structures of brick and stone were not well maintained. The first to suffer were the heating systems and the baths. Former principal buildings were gradually abandoned, but rural establishments kept the spatial organisation of the villa for a long time.

Some areas did not follow the general pattern. Some very large luxury villas in the southern part of the loess region kept their monumental character and the

Fig. 8.27 Reconstruction of Late Roman Maastricht, the Netherlands. The two largest, parallel orientated buildings are storehouses for grain and other products

establishments in the vicinity of Trier appear to have escaped the decline very well, at least temporarily. The reason is undoubtedly the presence of the Roman imperial court. The vineyards and large wine-making installations mentioned in Section 8.3 belong to this period. Nevertheless, repeated turbulences asked for some adaptations even there. This resulted in the appearance of towers with thick walls, integrated into existing buildings or set apart and interpreted as tower-granaries.

There remains the question why rural life was declining as it was. Wars and raids were not unique in the (proto)history of the region. It is possible that such declines escaped general notice, because the rural settings of the earlier periods are less easy to trace than those of the Roman Period with its conspicuous buildings, made partly of brick and stone. Another possibility, often put forward, is that the economic structure, developed during the heyday of Roman rule, was one of the principal causes of the decline. According to current views, the development of large estates, which absorbed smaller establishments, caused the disappearance of independent farmers. The estates were managed by the owner, autochtonous or allochtonous, but mainly resident, and run by the local population, as tenants or as labourers. This ended in a system that could not but collapse when the general pattern of a well-run country, with a market economy, was gradually destroyed. It is quite possible that the large units broke up into smaller units and that the tenants went on, on a smaller scale.

The *pars domestica* of the villa was abandoned, but this does not imply automatically that agricultural production went into the same decline. The countryside still produced surpluses, which is made obvious by the finds in towns. The towns, still present though shrunken, were still provided with cereals, pulses, and animals to be slaughtered. Certainly, with the final departure of the Roman troops part of the demand fell away. However, the old, traditional view of a deserted countryside and total collapse of farming activities has to be abandoned.

A second factor affecting the country is the arrival of new people with a different background. Germanic people came to settle. Some were invited to do so, in the hope that they would keep out countrymen with less peaceful intentions. Others came without invitations. The transformations of the rural society of the last century of Roman rule are still not well understood. There is still a gap in our knowledge. What came after the decline is dealt with in the next chapter.

Chapter 9
The Early Middle Ages: AD 407–AD 1000

9.1 The End of Roman Rule and Thereafter

On the first of January AD 407 the defence of the Roman border was given up once and for all. What followed is generally known as the 'Dark Ages', dark, because both written and archaeological sources concerning the loess region (and other regions as well) are scarce.

The end of the Roman military presence led to a collapse of the internal structure; and the central government, that was the steering agent behind the flow of goods, lost its power. The town was one of the principal sufferers. Those towns that survived did so on a much smaller scale. Some Roman and Romanised civilians seem to have left the region. Others tried, more or less successfully, to rule over subregions, such as the region around Trier. But the people who were truly in control were the Germanic tribes, which the Romans in the end had failed to keep outside. In the loess region the most important group was a confederation of tribes, called the Franks.

At first, Frankish authority was a rather local affair. Small kingdoms developed around old centres such as Cologne, Trier, Soissons, Cambrai, Tournai, and Paris. The socio-political structure of these kingdoms seems to have been Germanic with Roman traits. The new leaders made use of the services of the old officials. At the end of the fifth – beginning of the sixth century, however, one of the kings, Chlodovech, also known as Clovis, succeeded in uniting many of the small kingdoms. He founded the Merovingian kingdom. From this time onward the number of records increases. After his death his kingdom fell apart into smaller units ruled by his descendants, but the territorial fragmentation of the previous century did not return. During the following period the land was split up into but a few Merovingian kingdoms. It was, however, not a quiet period, because disputes about successions were not settled without violence.

In AD 750 the Merovingian dynasty gave way to the Carolingian dynasty, a family of high Frankish officials, which were already de facto in power during the rule of the last, weak, Merovingian kings. The most renowned Carolingian king was Charlemagne, born according to most sources in Herstal, in the loess area near Liège (Belgium). He extended the kingdom until it covered large parts of central and western continental Europe (Fig. 9.1).

C.C. Bakels, *The Western European Loess Belt*, DOI 10.1007/978-1-4020-9840-6_9,

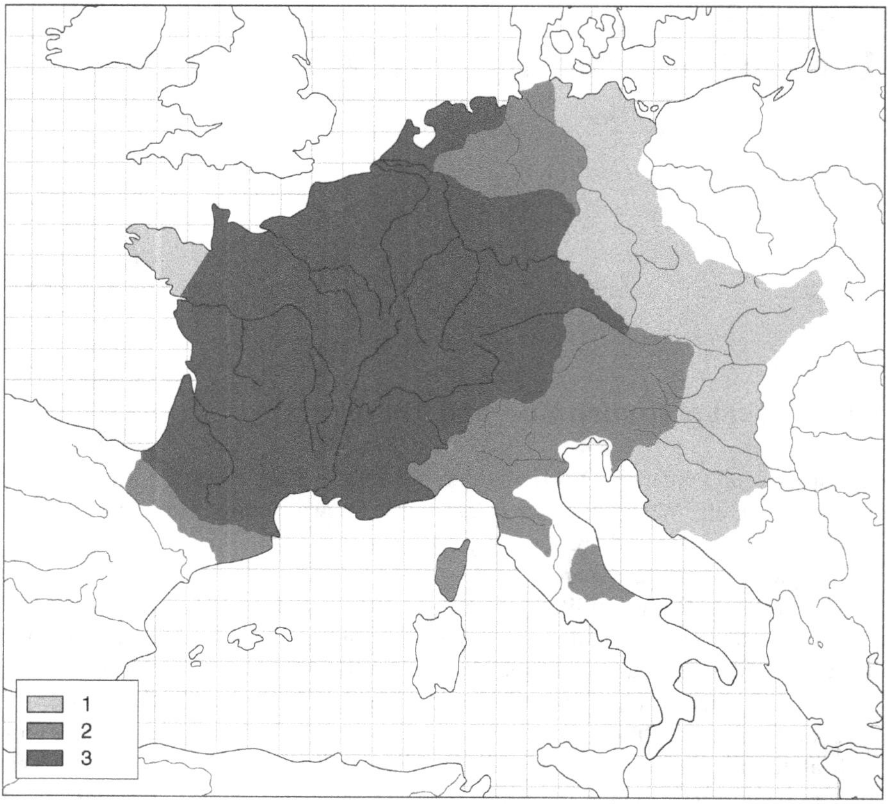

Fig. 9.1 The realm of the Franks: situation in AD 768 (3), territories added (2) and paying tribute (1) during the reign of Charlemagne

As mentioned above, Frankish rule incorporated Roman traits. Rulers referred also to Rome when their power was at stake. In AD 800 Charlemagne was crowned by the pope in Rome as emperor of the Roman Empire. A new coin, minted on this occasion, bore the text 'Renovatio imperii romanorum' (renovation of the Roman Empire). The Roman way of life was still an example of how things should look.

The Carolingian empire did not last. In the ninth and tenth centuries it fell apart. Problems regarding the succession were part of the loss of cohesion. Local rulers started to behave as independent men. It was also the main period of the raids of the Vikings. The Frankish kings were not able to protect rich towns and religious centres against them, as their armies were not specialised in coastal defence and swift reactions. Local rulers succeeded better and this accelerated the process of decentralisation of power.

The Frankish kingdoms, Frankish Empire and the smaller civil and military units that followed after the Empire, were not the only segment of society to wield power. The Christian Church was the other. In the beginning, Christian religion was only one of the numerous cults practised within the boundaries of the Roman

Empire. This low status ended with the mysterious conversion of emperor Constantine the Great in AD 312. The religious disposition of his successors varied, but at the end of the fourth century Christianity was raised to the status of official religion of the State. The bishop of Rome, the pope, became a leader. More than the Romanised elite of the original population, it was the Church which seems to have provided the continuity of elements of Roman culture. Towns with a bishop's seat survived the Dark Ages. The art of writing was preserved, together with the use of Latin and the custom of keeping records of affairs through written administration. The Church attended to the education of the new Frankish elite, whose first generations had been illiterate. The Church also provided officials for the civil service. From Carolingian times onwards the number of written records rose steadily.

As usual, common rural life remains in the shadows. Nevertheless, most of the population did live in the country, even more so than before, because the towns had shrunk to smaller settlements and the large military forts were abandoned. Conditions prevailing during the Dark Ages and the Merovingian period are not very well known. What emerges from the Carolingian written documents is that many farming households were part of larger establishments owned by the new elite, that is by either the military-civic leaders of the Franks, or the Church. The big landowners protected the rural population against robbers and raids. The farmers paid mostly in produce and labour. The Roman market economy was lost and a central system for levying taxes was lost as well. The only way to extract a surplus from the population was to extract it directly. The so-called domanial system was the result (see Section 9.6).

When ownership of the land stabilised, the phenomenon of the town reappeared. Near the new centres of power, be it local headquarters or religious centres, urban settlement emerged, or when based on Roman towns, saw a revival. This started in general towards the end of the seventh century. At first, towns were mainly administrative centres, but later on they gained economic importance. They represent the first phase of the medieval town with its economic, social and political characteristics. In the tenth century their role was such that artisans, who in the preceding centuries worked in a rural setting and were employed as part of the domanial system, emigrated to the towns, provided of course that they could go as free men. Artisans with the status of slaves were not free to leave. This process was aided by the destabilisation of affairs as the Carolingian empire fell apart. Crafts became definite parts of the towns' activities, thereby initiating the division of labour and industrial production.

With the rise of the town came an end to the Period of Direct Consumption in the sense given it by the historian Slicher van Bath. It is a system in which the non-agrarian consumer part of the population obtains the produce of the land directly from the producer, without any middleman. The Roman Period was a small interruption of only a few centuries of a way of life that prevailed in the loess region west of the Rhine from c. 5300 BC to c. AD 1000, or for those who want to see fully developed towns as its end: c. AD 1150. As explained in the Introduction, this book ends there.

9.2 Crops

The first crop that should be mentioned is rye. As put forward in Section 8.2, Frankish tribes and rye as a main crop are seen to arrive simultaneously. The ancestor of rye is to be found in an aggregate of annual weedy ryes, distributed over south-west Asia. In contrast to most grain crops which are self-pollinating, rye is a cross-pollinating cereal. One of the factors determining yields is effective wind pollination. The aggregate contains at present a range of wild types with fragile rachises, shattering into spikelets when ripe, types with a semi-brittle rachis, weedy types with a non-brittle rachis, and fully cultivated plants with plumper grains. All are inter-fertile. The non-shattering weedy types are common weeds in the wheat fields of south-west Asia and the Balkans. Their grains resemble wheat in grain size and weight, and cannot therefore be separated by sieving or winnowing. Consequently, they always form part of the sowing seed for the next year. Farmers, particularly those on elevated plateaus such as Anatolia, also tolerate the rye. In bad years with extreme climatic conditions, rye survives when wheat does not.

Rye was a tolerated weed, a man-dependent weed and an early crop in the Near East from early times of crop cultivation onwards. Outside the Near East information is fragmentary. Grains of rye turn up in European sites from the first beginnings of European agriculture onwards, but whether these kernels represent true crops is uncertain. Most probably they were weeds. Rye grains become more numerous in Central Europe after 1800–1500 BC, suggesting that the plant became established as a tolerated weed. Rye as a crop in its own right appeared in Western and Central Europe just before or around the beginning of the Roman Period. It is assumed that the crop evolved in contexts with a relatively harsh climate on poor soils. Experiments on a poor soil near Berlin have shown that a mixture of initially half wheat half rye, harvested and sown again, ended up as almost exclusively rye after only three years. It is quite possible that the rise to a true crop took place in several regions, independently of each other. One of them was north-western continental Europe beyond the boundaries of the Roman Empire. But the Romans knew rye as a crop, too.

The cereal is described by the Roman author Pliny as growing in the area around Turin. The local name was asia, a non-Latin name which would suggest a non-Latin origin. He finds it poor food and only fit for averting starvation. According to him mixing with wheat made the cereal more palatable, but even then it was hard on the stomach. Its only positive attribute was, according to this author, that rye gave good yields on every kind of soil and that it enriched the soil as well. How he arrived at the second attribute remains unexplained.

The earliest Roman find of the actual crop in the loess region dates from the beginning of the second century and was excavated in an important Roman *vicus*, Jouars-Pontchartrain (Dept. Yvelines), south of the Seine, in France. The find is strange, because all other finds of rye in the loess region are from the last quarter of the fourth century or later. The rye is interpreted as horse-feed. Feeding horses may also have been the purpose of the stocks of rye present in the large military granaries at Maastricht (the Netherlands), which are dated to the end of the fourth and to the fifth century. Cavalry was then, much more than before, part of the swift armies of intervention. Nevertheless, the rye may also have been food for humans, for instance destined for Frankish military units serving in the Roman army. It is a pity that the food preference of the Franks at the time is not known, but the fact remains that early instances of rye in civilian contexts are precisely found in a Frankish type of dwellings, known as huts with sunken floors. Such a hut revealed the earliest rye found at Maastricht. Similar huts, constructed on the terrain of the large Roman *villa rustica* of Voerendaal (the Netherlands), contained the first rye at this archaeobotanically well-documented place. It is also a fact that only from the fifth century onwards rye did become a common crop plant in the loess region. It was a cereal to stay (Fig. 9.2).

A second crop plant to be mentioned is woad (*Isatis tinctoria*), a dyeplant. Woad provides a blue colour that was very much appreciated for dying textiles, until it was replaced by indigo in later times. Its history as a cultivated plant is not well known, but its ancestors are to be found in southern Europe, the Levant and northern Africa. The plant easily escapes archaeobotanical notice because it is not the seeds, but the leaves that after fermentation give the blue dye. And, as is explained in Chapter 2, plants used for their leaves are not likely to be detected in archaeological records. The earliest traces in the loess region are seeds, found in a ninth century context in Dury (Dept. Somme) near Amiens. But this does not mean that the plant was not

Fig. 9.2 Carbonised rye, found in a military granary at Maastricht, the Netherlands

cultivated earlier, as the plant was well known in the Germanic world. According to the Roman author Tacitus, Germans went to war with their bodies painted blue. If this story is true, this must have been done with woad. The Franks used woad to dye their traditional cloaks. A Frankish writer, known as Notker the Stammerer, noted at the end of the ninth century that the cloaks, worn daily by Charlemagne, were either white or blue 'as those of the Franks of old'.

The fact that traces of the plant are, so far, of a relatively late date has possibly to do with the scale on which it was cultivated. If cultivation takes place on a garden-plot scale, the chance that seeds are preserved are rather small, especially so when the plant is harvested before seed-setting.

Other dyeplants were used as well. Remains of dyers' rocket, provider of a valued type of yellow (see Section 8.2), turn up in archaeobotanical lists, and texts mention madder (*Rubia tinctoria*) as the source of red. A garment found in the grave of Arnegundis, buried in the seventh century in St. Denis near Paris, was dyed with madder, but archaeobotanical traces of this plant, of which the roots are used, have not yet been unearthed.

Otherwise, the spectrum of crop plants remained more or less the same as it was in the period before. Very few traces are left which date from the fifth or sixth centuries, but these few reveal that 'Roman' plants like coriander, beet, sweet cherry, chestnut, and walnut were still cultivated. The late Merovingian and the Carolingian periods are better known. They, too, show continuation of vegetables, kitchen herbs, and fruit trees. The written records of those times confirm this. Merovingian texts on agriculture are near-absent, but the importance of fruit trees is obvious from the old Frankish law, known as *Lex Salica*. This law was originally an oral law, but Clovis ordered it to be written down and copies of the manuscript have survived. The law states:

> Title 27,8 'He who takes away the grafted twigs from an apple or pear tree shall be liable to pay one hundred and twenty denarii (in addition to return of the grafts or their value plus a payment for the time their use was lost)'.
> Title 27,9 'If they were in a garden, he shall be liable to pay six hundred denarii'.
> Title 27,10 'He who strips the bark from an apple tree or pear tree shall be liable to pay one hundred and twenty denarii'.
> Title 27,11 'If it was in a kitchen garden, he shall be liable to pay six hundred denarii'.

From this text can be deduced that the craft of grafting fruit trees had survived the break-down of the Roman Empire. Fruit trees were valued; orchards and gardens still existed. To get an idea of the value of a denarius: a burglar who copied a house key, entered, but fled before he could steal anything, had to pay twelve hundred denarii.

More is known about the Carolingian period. Appropriate information is, for instance, provided by an ordinance concerning the management of royal domains. This 'Capitulare de Villis vel Curtis imperii' is commonly attributed to

Charlemagne, and was in any case written by somebody at his court. The ordinance enumerates the plants to be cultivated on the royal estates. It need not reflect reality, but gives at least an insight into the state of affairs to be striven for. Orchards, for instance should contain a long series of species: 'We wish to see several varieties of apple trees, several varieties of pear trees, several varieties of plum trees, service trees, medlars, chestnuts, several varieties of peach trees, quinces, hazel shrubs, almond trees, mulberry trees, laurels, pines, figs, walnuts, several varieties of cherries.'

It will be clear that almond, laurel, pine, fig and possibly peach as well, could not thrive on the estates in the northern parts of Charlemagne's realm. The list is inspired by what a Roman-type orchard should have. Nevertheless, other texts of the period, like the 'Brevia Exempla' (see Section 9.3), confirm the presence of all species that could be grown. And the archaeobotanical remains confirm their presence too.

The list of vegetables and herbs in the *Capitulare* is still longer (Table 9.1). It even enumerates some flowers. This list also reflects the Roman garden.

Unfortunately, the *Capitulare* does not specify the crops to be grown on fields. This segment of agriculture belonged clearly to a different sphere. But the text refers to separate vineyards.

The written texts show continuity of plant cultivation between the Roman Period and the more or less well-organised Early Medieval states. The art of maintaining an orchard and vegetable garden was not lost during the Dark Ages. Most probably it was the Church which preserved the knowledge, possibly together with Romanised autochtonous civilians who had stayed. Monasteries, which were founded in the

Table 9.1 List of plants (left column) which, according to the *Capitulare de Villis*, ought to be sown or planted in royal gardens. The translation of the names into scientific names (middle column) is difficult and sometimes only tentative

lilium	*Lilium candidum* L.	Madonna lily
rosas	*Rosa*	Rose
fenigrecum	*Trigonella foenum-graecum* L.	Fenugreek
costum	*Tanacetum balsamita* L.	Cost-mary
salviam	*Salvia officinalis* L.	Sage
rutam	*Ruta graveolens* L.	Common rue
abrotanum	*Artemisia abrotanum* L.	Southernwood
cucumeres	*Cucumis sativus* L.	Cucumber
pepones	*Cucumis melo* L.	Melon
cucurbitas	*Lagenaria siceraria* (Mol.) Standl.	Bottle gourd
fasiolum	? (not Phaseolus)	A kind of bean
ciminum	*Cuminum cyminum* L.	Cumin
ros marinum	*Rosmarinus officinalis* L.	Rosemary
careium	*Carum carvi* L.	Caraway
cicerum italicum	*Cicer arietinum* L.	Chick-pea
squillam	*Urginea maritima* (L.) Baker	Squill
gladiolum	*Iris germanica* L. or *I. pallida* Lam.	Common or pale iris
dragantea	*Artemisia dracunculus* L.	Tarragon
anesum	*Pimpinella anisum* L.	Anise

Table 9.1 (continued)

coloquentidas	*Cucurbita colocynthis* ?	Colocynth ?
solsequiam	*Calendula officinalis* L. ?	Marygold ?
ameum	*Ammi copticus* L. ?	Ajowan ?
silum	*Laserpitium siler* L.	Laserwort
lactucas	*Lactuca sativa* L.	Garden lettuce
git	*Nigella sativa* L.	Black cumin
erucam albam	*Eruca sativa* Mill. or *E. vesicaria* (L.) Cav.	Rucola or rocket
nasturtium	*Lepidium sativum*	Garden cress
parduna	*Arctium lappa* L.?	Greater burdock ?
puledium	*Mentha pulegium* L.	Penny-royal
olisatum	*Smyrnium olusatrum* L.	Alexanders
petresilinum	*Petroselinum crispum*	Parsley
apium	*Apium graveolens* L.	Celery
levisticum	*Levisticum officinale* W.D.J. Koch	Garden lovage
sabinam	*Juniperus sabina* L.	Savin
anetum	*Anethum graveolens* L.	Dill
fenicolum	*Foeniculum vulgare* Mill.	Fennel
intubas	*Cichorium endivia*	Endive
diptamnum	*Dictamnus albus* L.	Burning bush
sinape	*Sinapis alba* L.	White mustard
satureiam	*Satureja hortensis* L.	Summer savory
sisibrium	*Sisymbrium officinale*	Hedge mustard
mentam	*Mentha*	Mint
mentastrum	*Mentha longifolia* L.	Horsemint
tanazitam	*Tanacetum vulgare* L.	Tansy
neptam	*Nepeta cataria* L.	Catmint
febrefugiam	*Tanacetum parthenium* (L.) Schultz Bip.	Feverfew
papaver	*Papaver somniferum* L.	Opium poppy
betas	*Beta vulgaris* L.	Beet
vulgigina	*Asarum europaeum* L.	Asarabacca
mismalvas (altaea)	*Althaea officinalis* L.?	Marsh mallow
malvas	*Malva sylvestris* L.	Common mallow
carvitas	*Daucus carota* L.	Carrot
pastenacas	*Pastinaca sativa* L.	Parsnip
adripias	*Atriplex hortensis* L.	Garden orache
blidas	*Amaranthus blitum* L.	Amaranth
ravacaulos	*Brassica rapa* L.	Turnip
caulos	*Brassica oleracea* L.	Cabbage
uniones	*Allium*	? Onion family
britlas	*Allium schoenoprasum* L.	Chives
porros	*Allium porrum* L.	Leek
radices	*Raphanus sativus* L.	Radish
ascalonicas	*Allium cepa* L. var. *ascalonicum*	Shallot
cepas	*Allium cepa* L.	Onion
alia	*Allium sativum* L.	Garlic
warentiam	*Rubia tinctorum* L.	Dyer's madder
cardones	*Cynara cardunculus* L.	Cardoon or artichoke
fabas maiores	*Vicia faba* L.	Horse bean or broad bean

Table 9.1 (continued)

pisos mauriscos	*Pisum sativum* L.	Pea
coriandrum	*Coriandrum sativum* L.	Coriander
cerfolium	*Anthriscus cerefolium* L.	Chervil
lacteridas	*Euphorbia lathyris* L.	Caper spurge
sclareiam	*Salvia sclarea*	Clary
jovis barbam	*Sempervivum tectorum* L.	Houseleek

region from the sixth century onwards, had gardens. Such institutions represented continuity in times of turbulence.

9.3 Crop Cultivation

At the end of the Roman Period the scale on which crop cultivation was practised had shrunk, but crop cultivation did not come to an end. How much land was tilled in the fifth and sixth centuries is not known, nor how it was tilled. It is sometimes presumed that agriculture returned to the ways of the centuries before the Roman occupation. It cannot have been as simple as that. The rural population had known several centuries of Roman influence and it cannot have been replaced entirely by the newcomers. Moreover, reconstruction of the landscape has shown that the post-Roman Period did not look the same as the pre-Roman Period (see Chapter 10). This implies that the use of land must have been different.

Rural settlements where agriculture can be followed through the centuries are very few. One of them is Savy 'Le Bois de l'Abbaye' (Dept. Aisne) west of St. Quentin, although even there a time gap is present, spanning the fourth and most of the fifth century. The carbonised seeds present a picture of a community that cultivated a broad spectrum of crops during the first half of the first century, specialised in spelt wheat during the heyday of the Roman Period, to return to a broad spectrum cultivation at the end of the fifth century (Fig. 9.3). Whether there was continuous occupation by the same people may be questioned because of the time gap, but in favour of such a continuity is the fact that they cultivated spelt wheat all the time. As mentioned in Section 8.3, the Roman Period farms in northern France cultivated bread wheat instead of spelt wheat if the soil allowed this. The soil at Savy was good enough for growing bread wheat and, moreover, the Merovingian (and Carolingian) farmers of a site nearby, Athies 'Le Chemin de Croix' (Dept. Aisne) did indeed grow bread wheat as part of their broad spectrum crop cultivation (Fig. 9.4). The spelt wheat crop of Savy must have been a deliberate choice.

How the arable land was used during the fifth, sixth and seventh centuries is not very well known. The *Lex Salica* devotes a complete Title to damage to a cultivated field or other kinds of enclosure. Another Title mentions fines for setting fire to fences or hedges. Still another is concerned with the stealing of flax from flax fields and the stealing of turnips from turnip patches, beans from bean patches,

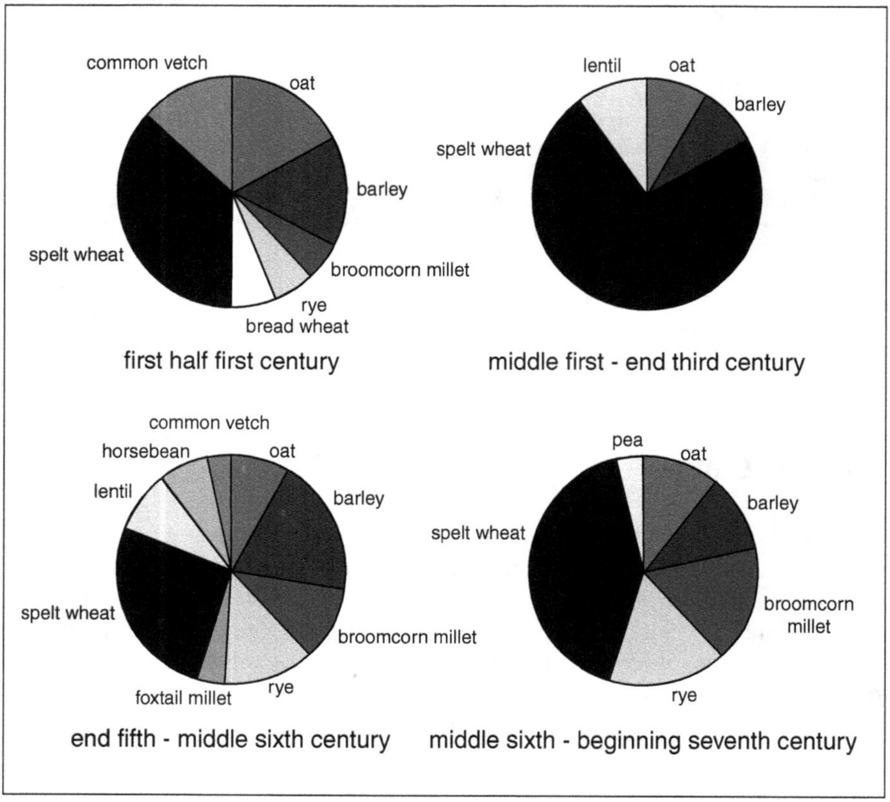

Fig. 9.3 Composition of the crops grown at Savy (Dept. Aisne), France, according to their carbonised remains

peas from pea patches and lentils from lentil patches. This implies the existence of fields and more garden-like plots, surrounded by a kind of fence. It implies as well that those crops were cultivated separately. Vineyards are mentioned, too. Another problem, dealt with in the Law, is preventing a plough from entering another man's field. The *Lex Salica* as a whole was obviously meant to function in an agrarian society.

The written texts of the Carolingian period offer more. There is a hint of a three-year system, in which the soil was tilled for one year and left fallow for two. But a three-course rotation seems to have been more important. In this system a winter cereal is sown in the autumn to be harvested in the following summer. The second crop, called summer crop, is sown in the next spring, to be harvested in the summer of the same year. After this came a fallow year. Three-course rotation was the predominant system in the Carolingian empire north of the river Loire, at least from the eighth century onwards, but it may be older. A third system was a two-course rotation, in which the soil was tilled and left fallow in alternate years.

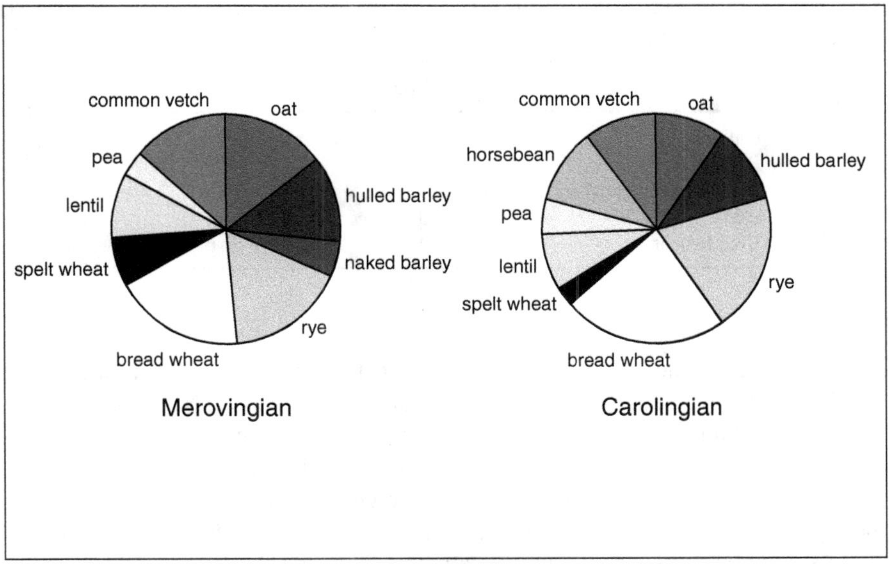

Fig. 9.4 Composition of the crops grown at Athies (Dept. Aisne), France, according to their carbonised remains

As autumn-sown crops, bread wheat, spelt wheat and rye may be considered. Oat, millets and pulses are typical summer crops, whilst barley was obviously a summer crop at the time as well. Flax may have been grown as summer or winter crop. The archaeobotanical records show that the farmers had enough choice. Rye was grown everywhere. If a choice had to be made for a second winter crop it was which wheat to sow, the free-threshing bread wheat or the hulled spelt wheat. The other hulled wheat, emmer wheat, had become a very minor crop and need not be considered here.

According to archaeobotanical investigations, the farmers inhabiting the German Rhineland cultivated both wheats, in addition to rye which was by far the most numerous in the finds. The share of spelt wheat decreased in the course of time, but even in the tenth century and later this wheat was still grown. Thirty kilometres to the west, the people who continued farming on the land of the former *villa rustica* of Voerendaal (the Netherlands) started with both spelt wheat and bread wheat, but shifted during the fifth century to the cultivation of rye. From then on traces of wheat become scarce. The growing of spelt wheat was, however, never completely abandoned in this subregion, because an underground silo in Maastricht, dated to the seventh century, was full of spelt. Both in the Rhineland and there, spelt wheat was never given up completely until late historical times. Next to rye it was the most winter-proof cereal. In severe winters bread wheat can be seriously damaged by frost. Spelt wheat is also less susceptible to diseases, but spelt wheat had less market value. This was already so in the Carolingian period. Prices of cereals at the Frankish market place of Nijmegen (the Netherlands) show this. Although the market town

Table 9.2 The price, in *denarii* per muid, of five cereals sold at the markets of Frankfurt (in AD 794) and Nijmegen (in AD 806)

	Frankfurt	Nijmegen
Oat	1	1
Barley	2	3
Spelt wheat, dehusked	–	3
Rye	3	4
Breadwheat	4	6

was not located in the region considered here, it still provides an appropriate insight (Table 9.2). Spelt wheat had the same value as barley, at least in Nijmegen.

In other subregions another choice was made. The table mentions also the prices set at the market in the German town of Frankfurt am Main. Spelt was not sold there, though this market served a loess area. The town is situated near, but not quite in the region under review in this book, but the absence of spelt demonstrates that not every subregion grew spelt. This was already so during the Roman Period, when spelt production in northern France had retreated to the surroundings of Amiens (see Section 8.3). The preference for bread wheat persisted in later times, at least as far as the botanical remains are showing. Merovingian finds reveal an absence of spelt in almost all excavated sites. If this wheat is present at all, it concerns only a few grains. The only exception is Savy, mentioned above. The same applies to the Carolingian period, when spelt is rarely found.

But, most surprisingly, the archaeobotanical results do not always agree with written texts. Relevant texts in this case are the inventories of crops, stocked on royal domains or sown on land belonging to abbeys. They date from the Carolingian period. A charter from Charles the Bald, dated AD 862, mentions the delivery of spelt wheat to a complex belonging to the monastery of St. Denis near Paris. The *Brevium Exempla ad describendas res ecclesiasticas et fiscales*, a Carolingian text from c. AD 810, contains, among others, an inventory of four royal domains in the region of present-day Lille (Dept. Nord). It lists the crops, expressed in units of volume, encountered during an inspection tour. The inventory mentions both the crops produced in that year and the crops still in stock. The difference is due to seed already sown. The seven products, presented as percentages of the total volume, are given in Table 9.3. Spelt was the main product, followed by barley. Bread wheat and rye are unimportant. The same table shows archaeobotanical data from four settlements in more or less the same subregion and of more or less the same age. It presents the frequencies of the products in the finds: that is, how often the product has been found in the features examined for plant remains. In this way the data of the manuscript and of the excavations are comparable. In the settlements spelt wheat is near-absent, bread wheat, rye and barley being the important cereals. Moreover, pulses are more important in the settlements than in the written text. The discrepancy is difficult to explain. Of course, the inventory gives the state of one year only, while the settlements present a mix of perhaps some decades. But

Table 9.3 Comparison of the relative importance of crops, as noted in a written source and as found in excavations. Inventories of four royal estates mentioned in the *Brevium Exempla* and data from four farms of a more or less comparable date and social status are compared. In three instances the text mentions grain produced and grain actually seen during the inspection. The difference is due to the fact that some of the seeds had already been used for sowing. To allow comparison the volumes of the crops mentioned in the text are converted to percentages of the total volume of all products in the respective establishments. The archaeological finds are expressed as frequency of occurrence in the samples of the respective sites. The more often they were found, the more important they were. Values of the main products are printed in bold

Written source(%)	Annappes Produced	Found	Vitry Produced=found	Cysoing Produced	Found	Somain Produced	Found
Spelt	**49**	**57**	**61**	**37**	**63**	**43**	**60**
Wheat	2	1	–	–	–	–	–
Rye	2	0	–	12	11	–	–
Barley	**39**	**27**	**39**	**35**	**26**	**57**	**41**
Oat	9	15	–	16	0	–	–
Beans	0.02	0.03	–	–	–	–	–
Peas	0.2	0.4	–	–	–	–	–
Excavations (frequencies) Century AD	Serris Vp 7e–8e		Serris Fd 7e–8e	Athies 8e–9e		Dury 9.5e–10e	
Spelt	–		–	22		–	
Wheat	**64**		**83**	**89**		**92**	
Rye	27		**50**	**67**		23	
Barley	**45**		33	33		15	
Oat	27		33	44		**31**	
Beans	6		17	22		–	
Peas	6		11	22		23	

it is difficult to see why the inventory should have been made up in an exceptional year. It cannot have been a difference in status of the farms either, because at least Serris-Ferme domaniale and Athies were important domains. It is possible that spelt wheat, produced at the excavated farms, did not get the chance to carbonise, because its use hardly ever included exposure to heat or fire. Indeed, the only instances of lots of carbonised spelt wheat concern germinated grains which suggest that they were allowed to sprout in order to produce malt for beer-production (Fig. 9.5). Malting involves roasting of germinated grain and roasting going too far ends with carbonised matter. Making beer may have been an activity that was not carried out everywhere. Another explanation may be that the written inventories concern only stocks intended for long-term storage, because the two main products are hulled cereals and especially those, if not threshed and dehusked, keep well over longer periods. Or, still another explanation may be that the inspected farms stored mainly animal feed. They may have been orientated towards a production based on livestock (see Section 9.4 and Table 9.6). However that may be, the case serves as a warning that neither the archaeobotanical material, nor the texts may represent the actual truth.

Fig. 9.5 Germinated spelt wheat, excavated at Savy (Dept. Aisne), France

Another text, written in the middle of the ninth century and concerning the Abbey of Saint-Remi-de-Reims at Reims mentions the amounts of cereals sown on the different establishments belonging to this abbey (Table 9.4). As many of these farms are in Champagne, where the soil is shallow and far from optimal for cereal cultivation, it is perhaps not surprising that spelt wheat, with rye as second, were the dominant cereals. Summer crops (barley and oat) are hardly mentioned. It must be assumed that these farms grew almost no summer crops. A crop rotation including summer crops seems not to have been practised (yet?) on the abbey's large fields, called *culturae* (see Section 9.6). The texts indicate that on the truly large fields, those with areas of several hundreds of hectares, exclusively spelt wheat was produced. Rye, some bread wheat, and very small amounts of barley and oat were sown on smaller fields, called *campi* (single *campus*) and enclosed fields, called *avergariae* (single *avergaria*).

Texts dealing with the same abbey, but now from the eleventh century, show that later on spelt wheat culture was totally abandoned. Its place was taken over by bread wheat. Rye remained. Oat had gained as much importance as rye. This development might imply a change towards a crop rotation practice, even on the large fields. It is quite feasible that crop rotation originated on the smaller fields and expanded from there to the extensive fields.

All in all, the general conclusion is that in the southern part of the loess region, spelt wheat was one of the winter crops, but only in certain areas, and that the cereal lost importance towards the end of the Carolingian period, until vanishing altogether. The abandonment of spelt wheat went rather fast. A farm in Isle-sur-Suippe, belonging to the abbey of Saint-Remi, contributed still one *modius* (63 L,

Table 9.4 Cereals sown on the mid-ninth century fields of establishments (villae) belonging to the abbey Saint-Rémi-de-Reims. Cereals expressed in muids

Villa	Denomination	Bread wheat	Rye	Spelt wheat	Barley	Oat
Aigny	46 campi	24	30.5	85	–	–
Muizon	11 campi	–	21	–	–	–
Petit-Fleury	17 campi	28	30	–	–	–
Baconnes	3 avergariae	–	5	–	–	–
	terra forastica	–	–	200	–	–
Louvercy	3 avergariae	–	20	–	–	–
	3 culturae, 2 campi	–	–	320	–	–
Ville-en-Selve	21 campi	–	70	350	–	–
Chézy	5 campi	–	31	64	–	–
Courtisols	2 avergariae	–	24	–	–	–
	9 culturae	–	–	2848	–	–
Beine	4 avergariae	–	20	–	–	–
	6 culturae	–	–	1204	–	–
Bouconville	32 campi	–	107	–	–	–
Viel-Saint-Remi	1 avergaria	–	–	–	6	–
	6 culturae	–	–	600	–	–
Villers-le-Tourneur	field	–	–	–	–	48
Gerson	17 campi	30	30	–	–	–
Sault-Saint-Remi	19 campi	–	140	1385	–	–
Vesle	17 campi	8	8	56	–	–

see below) of spelt wheat in the second half of the tenth century, whilst fifty or at most one hundred years later, it no longer contributed spelt, but oat and rye.

The main summer cereals were obviously barley and oat. Broomcorn millet is occasionally present and foxtail millet is rare. But next to cereals, pulses must have been important as well. Beans and peas are mentioned in texts concerning field crops, and horse beans, peas and common vetch are common finds in excavations. Although the *Capitulare de Villis* mentions beans and peas as garden plants, it is quite possible that this text refers to certain varieties, which were consumed as a vegetable and were not dried in quantities to serve as a staple. In the *Capitulare de Villis* the plants are referred to as 'fabas maiores' and 'pisos Mauriscos'. Especially the name *faba maior* implies that there was also a *faba minor*. The latter may have been the variety of the bean that was grown on a larger scale. Horse beans are the common beans in excavations. A bean of a type maior (broad bean) has been found at Dury (Dept. Somme) and is dated to the twelfth century. It is the earliest find of this larger bean, but it may already have been grown earlier in royal gardens. It is curious that common vetch is not mentioned in texts, because, according to the finds, it was regularly cultivated and not on a small scale either. Perhaps this vetch was regarded as a kind of pea.

As mentioned earlier, flax fields were already part of the early legislation written down in the *Lex Salica*. Although flax is not a crop turning up regularly in the

Carolingian texts, linen is mentioned regularly and the seeds (linseed) are found quite often, indicating that flax/linseed was one of the standard crops. In some subregions the plant may have been cultivated on a larger scale than in others, but details await further research.

The texts mention the existence of maslins (see Section 7.3), but sowing maslins seems not to have been common practice before the eleventh century and then only on poorer soils. Where maslins are mentioned they are mixtures of wheat and rye. They have not been found yet in the archaeobotanical material.

On large fields the soil was prepared by ploughing. Small plots were worked with spades made of wood and provided with an edge sheathed in iron, and/or with two-pronged hoes (Fig. 9.6). The ard was still in use, but by and by another type of plough, the mouldboard plough, took over. This plough cuts the soil with a vertical iron knife, called coulter. Directly after the coulter comes the horizontal iron share. In the period under review this share was broad, but still symmetrical. Asymmetrical shares occur only after AD 1000. The shovel-like share widens the cut, deepens it

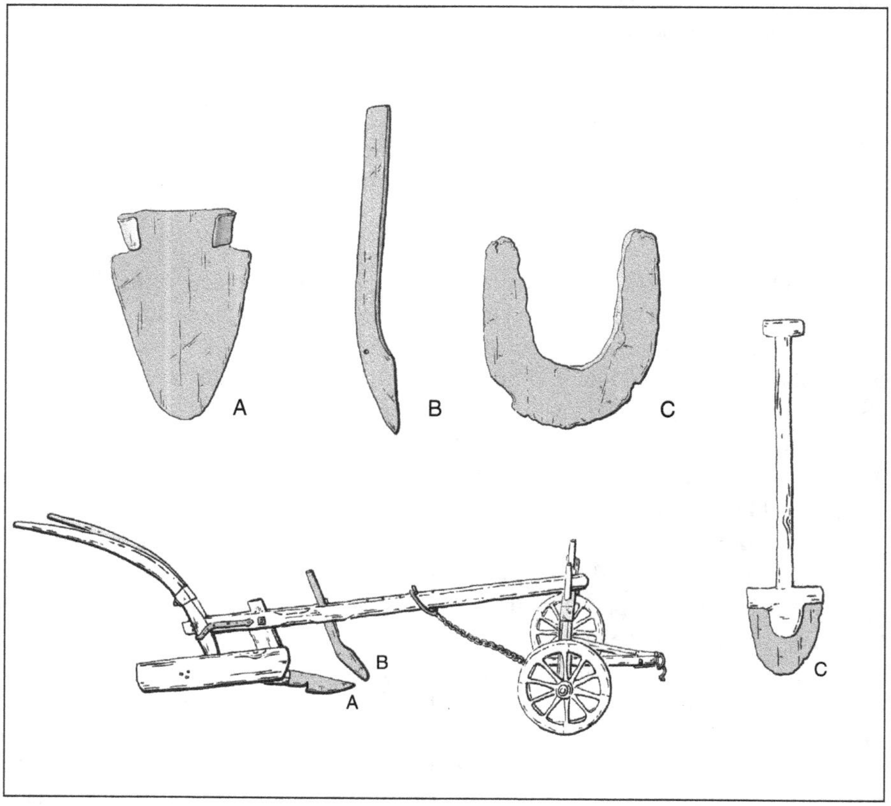

Fig. 9.6 Upper row: remains of tools intended for loosening the soil. **A**: iron sheath for the tip of a plough, **B**: iron coulter, **C**: iron edge of a spade. *Lower row*: a wheeled plough and a spade, both with the parts shown in the *upper row* indicated

slightly and creates clods. It is followed by the mouldboard, a wooden board set at an angle, which turns the clods to one side, and upside down. The whole is mounted on a wooden frame with handles to steer the tool and a beam to allow attachment to the yoke of the draught animals. The beam could rest on two wheels, and the plough was then called a wheeled plough. Most of this is known from pictures, but the share and coulter are found during excavations. The shape of the shares shows that they were meant to be fixed to the plough, but the coulters seem to have been separate elements which could be adjusted to the job at hand. This is supported by a Title in the *Lex Salica* in which the theft of coulters from somebody else's plough is subject to serious fines. Obviously they could be easily taken away. Although coulters could have been part of advanced types of ards, loose coulters are regarded as typical of mouldboard ploughs.

An early picture of a wheeled mouldboard plough is depicted on the Bayeux tapestry, a cloth embroidered on the occasion of the Battle of Hastings in AD 1066 (Fig. 9.7). On this cloth the plough is drawn by a mule, which is highly unusual. Normally the wheeled plough was drawn by a team of oxen, or, in the case of heavy soils or breaking up of fallow land, by two or even more teams. The oxen wore a yoke, fastened to the head and horns. Such yokes are unsuitable for equids, and indeed, the mule does not wear one. Equids draw best by means of a horse-collar, and it has been put forward in the past that the collar was already in use during the Carolingian period, making the swifter working and better-at-turning horse the draught animal for ploughs at the time. But this is now held to have occurred much later, in the thirteenth century at the earliest.

Cross-ploughing, necessary for loosening the soil when using the ard, was not necessary when using the plough. The new tool could also work heavier soils. Exactly when the mouldboard plough came into use is not yet known. Some authors

Fig. 9.7 Bayeux tapestry: ploughing, broadcast sowing and harrowing. The wheeled plough is, quite unusually, drawn by a mule. The harrow is drawn by a horse

state that it was already there during the Roman Period, but evidence is scarce and sometimes even questionable; and as mentioned in Section 8.3, the plough is not thought to have been used on loess and equivalent soils at that time. The earliest parts were found in the north-eastern and eastern part of the region, suggesting a technological influence from outside the reach of the traditional agriculture as practised in Roman Gallia.

After ploughing came harrowing. The harrow was a rectangular implement, made of wood and provided with wooden teeth. Trapezoidal and triangular harrows are of later date. On the Bayeux tapestry it is drawn by a horse. Horses work faster than oxen and it is possible that towards the end of the period considered here, horses were used for the lighter work. Nevertheless, oxen remained the principal draught animals throughout.

It is not known to what extent manuring was practised. Texts refer to the spreading of dung every three years. Animal dung must have been as scarce as it was in previous periods. Marling seems to have taken place on the fields of northern France, but otherwise it was mostly the year of fallow which restored fertility. Next to the cultivation of pulses, of course, as pulses improve the nitrogen content of the soil.

Field weeds may give hints about the soil condition. An analysis of single harvests, in the sense explained in Section 7.3, shows that the weeds found in the seventh century batch of spelt wheat from Maastricht (the Netherlands) give an indication that this spelt was grown on a rather poor, acid soil (Table 9.5). The sixth to seventh century spelt wheat from Savy (Dept. Aisne) was harvested on a field with a slightly better soil, although here too some weeds point towards acidification. Merovingian rye from Sissonne (Dept. Aisne) provides a similar picture. But

Table 9.5 Cereal concentrations and the Ellenberg values of associated weeds. The share of the R and N classes is expressed in percentages. Total provides the total number of species on which the percentages are based. See the glossary for Ellenberg values

R and N values of weeds in cereal concentrations, number of species in %

	R 1-4	R 5-6	R 7-9	R indiff	Total	N 1-4	N 5-6	N 7-9	N indiff	Total
Maastricht										
Spelt wheat	0	17	25	58	24	9	26	35	30	23
Athies										
Oat	0	5	20	75	20	5	25	20	50	20
Bread wheat	0	0	0	100	5	0	14	43	43	7
Savy										
Spelt wheat	6	6	18	70	17	18	18	18	46	17
Goudelancourt										
Bread wheat	0	17	0	83	6	33	17	17	33	6
Bread wheat/rye	7	7	33	53	15	19	25	25	31	16
Serris										
Bread wheat	0	12	12	76	8	13	50	13	25	8
Oat	0	14	14	72	7	0	28	28	43	7

Carolingian rye, grown around Dury near Amiens (Dept. Somme) shows no acidification at all. This may be explained by the presence of a calcareous subsoil underneath the relatively shallow topsoil of the locality. Stocks of bread wheat, excavated in Merovingian Athies (Dept. Aisne), and Carolingian Goudelancourt (Dept. Aisne) and Serris (Dept. Seine-et-Marne) contained only few weed species; but among those present there is hairy tare (*Vicia hirsuta*) that shows that the soils were perhaps not always in optimal condition. Merovingian oat from Athies came from a rather good soil. A lot of oat from Serris contained no weeds with clues as to soil condition. To conclude, the spelt wheat from Maastricht was grown under far from optimal conditions, but the cereals from France were not harvested in fields with truly bad soils.

Cereals were sown broadcast. This is inferred from early illustrations, such as the Bayeux tapestry. Whether other crops were also sown broadcast is unknown. It is possible that they were sown in rows, especially when sown on smaller plots. Written sources dealing with the estates of the abbeys of St.-Germain-des-Prés (near Paris) and St. Amand (near Valenciennes) mention that for the sowing of one 'bonnier' of land, four 'modii' of winter cereals or six 'modii' of summer cereals were needed. Ideas on how large a 'bonnier' was vary, but in these cases it is held to be 1.38 ha. The 'modius' is problematic, too. Sometimes the modius represents 52 L, and in other cases 63 L. In the case of the two abbeys it is generally set at 63 L. The outcome is that 197 L of winter cereal and 295 L of summer cereal were sown per hectare.

Weeding was practised and was part of the chores of the workers on the farms, but this kind of labour is only summarily dealt with in the manuscripts. Harvesting was far more interesting. It is depicted, for instance, on one of the pages of the Utrecht Psalter, a book of psalms, written and illuminated in Reims between AD 816 and 834. People cut the stalks halfway, with a sickle, stooping all the time (Fig. 9.8). The scythe for harvesting cereals came into use only centuries later.

Fig. 9.8 Psalter of Utrecht: cutting cereals with a sickle

The *Brevium Exempla* with its enumeration of seed harvested and seed already sown provided the historian Slicher van Bath with the opportunity to calculate the sowing seed/yield ratio. The proportion is about 1 : 3, which is very low. The outcome may be too pessimistic. In later medieval times this subregion had better results. The cereals found at the four estates mentioned in the *Brevium Exempla* may perhaps not represent the harvested product in its entirety. And the results of the extensive exploitation of the large fields belonging to large estates dominate the records and may provide a distorted picture of the true state of affairs elsewhere.

The short stalks were bound into sheaves and brought to the farm. Women and children gleaned the fields to gather the last stalks. In late fall or in winter the cereals were threshed, with threshing flails or sticks, and winnowed. Next, spelt wheat, and the hulled variety of barley, had to be dehusked, if intended for human consumption. Some settlements grew naked barley, but hulled barley was by far the more common cereal. Dehusking is only successful if the glumes are dry and brittle. If not, this condition has to be brought about by fire, for instance by singeing the ears. After this, the grain may be pounded, or milled in a special way. The pounding can be achieved by treading, as was still done until quite recently in Asturias (Spain), or by beating. Milling to knock off the husks is done with the aid of millstones, more widely spaced apart than when used in the making of flour. A final winnowing separated the fragments of the husks from the grain kernels.

When the hulled grain was intended for sowing, as animal feed or for making beer, dehusking was not necessary. Next to barley, spelt wheat was renown for its beer making qualities.

Part of the cereals were consumed in porridges; part was milled to flour for baking bread. The mill was of such economic importance, that it turns up in most texts concerned with the Carolingian abbeys. The records left by the abbey of St. Remi de Reims mention 13 mills. The ones concerning the large estates of the abbey St. Germain-des-Prés, dated c. AD 825, even mention 84 mills. Only large estates owned mills. The reason is that the mill of this period was a watermill, and the construction represented a large investment of capital. Although the Romans already knew the watermill, its use on a large scale occurred not before the Carolingian period. Why the watermill did not spread earlier is open to debate. Some maintain that the delay was due to the required size of the millstone, which was much larger than stones for handmills. The transport of the stones from the quarries would have been too difficult. But this cannot have been the reason. Some millstones, used by bakeries in Roman towns, also had a considerable size. Moreover, millstones were often transported over water. Others think that the delay was due to the presence of enough slaves to work the handmills. When slave labour became less available, the watermill was more readily adopted. It is not that slavery was unknown in Early Medieval times. Slaves were caught during raids, and poor people sold themselves or their children as slaves. The Church was not against slavery as such, but there were slaves and slaves. When the slaves were Christian, they were considered as people. In the course of time the distance between master and slave decreased and this may have triggered the introduction of labour-saving technologies, such as the watermill. The abandonment of handwork may also have caused the decline of spelt cultivation. The rise of the watermill and the decline of spelt wheat as a crop occur at more or less the same time.

Details on the cultivation of other crops are hardly available. The only type of other agrarian activities dealt with to any extent in the written texts is wine-growing. Vineyards are mentioned separately. They were mostly to be found in the area around Paris and along the rivers Moselle and Rhine, and larger vineyards belonged mainly to abbeys. Looking after vines and the vintage was done by specialised tenants. Half or one third of their produce had to be handed over to the owners of the land. The wine press was on the main farm of the wine-producing estate from where the wine was transported in casks. The wine production of many abbeys exceeded their own needs and was sold. The tenants received their own share which they traded for grain, meat and manure. One of the largests producers was the abbey of Prüm (Germany), which did not lie in the region under review, but had vineyards along the Moselle. Its annual production was c. 120,000 L. Its largest estate, comprising eight vineyards covering an area of 30 ha, was situated in Mehring. This estate alone was good for 72,000 L. The wine was transported by boat.

Fields, vineyards, gardens and orchards were not the only parts of the land to produce a crop. Meadows were important as well. The *Lex Salica* mentions a fine of eighteen hundred denarii to be paid by a man who had cut another man's meadow. The tool for cutting the hay was the scythe. Since its earliest appearance in the centuries after 800 BC (see Section 7.3), the blade of this implement had gradually become longer and thinner. The Early Medieval scythe was thus longer and thinner than the Roman one and resembled closely the modern type.

Hay was valuable. It was needed to feed oxen and horses, especially in winter. Meadows were not just pasture land. In later documents meadows and pastures are mentioned separately. Meadows served first and foremost as hayfields. They were private property and were fenced in. A good meadow was more expensive to rent or to buy than a similar stretch of arable land. Pasture land was every kind of land covered in grass and herbs, fallow land included. Until the eleventh or twelfth century everybody was free to use it.

9.4 Livestock and Animal Husbandry

The animals kept on Early Medieval farms were the same as in the previous period. Cattle, pigs, sheep, goats and horses were the main large ones. Their remains turn up in excavations and they are mentioned in the old manuscripts. The Frankish *Lex Salica* does not refer to donkeys, but bones of these animals are found regularly from the sixth century onwards and they are also mentioned in Carolingian texts. The *Brevium Exempla*, for instance, reports on the presence of donkeys in two of the four royal estates visited. Each of them possessed a pair. Donkeys held perhaps no place in the world of the Germanic tribes and are therefore not mentioned in the ancient legislation. The animal will have been part of the Romanised world where the Franks came to live. Mules must have been present as well, but their importance is difficult to assess.

Excavations of Merovingian sites have shown that in general both cattle and pigs were the main providers of meat. The decline in breeding pigs, noticed during the final stages of the Roman occupation (see Section 8.4), was obviously reversed. A mixed livestock composed of cattle, pigs and some sheep/goats is interpreted as being typical of the small-scale farming of the period, serving only local demand. In addition there were farms which were less occupied with pig breeding but concentrated on the raising of cattle, an activity maintained and expanded during the Carolingian period, when this kind of animal became the most important member of the livestock (Figs. 9.9 and 9.10). Half of the cattle were slaughtered between their first and fourth year, indicating that they were raised for their meat. The other half were kept as draught animals or as dairy cows. As mentioned in Section 9.3, oxen were the preferred animals for ploughing and haulage. Milk was used for making cheese and, at least in the Carolingian period, for making butter as well. Milking and churning in a large churn is depicted in the Psalter of Utrecht, but is also mentioned, as is cheese, in the texts of the period (Fig. 9.11). The history of butter

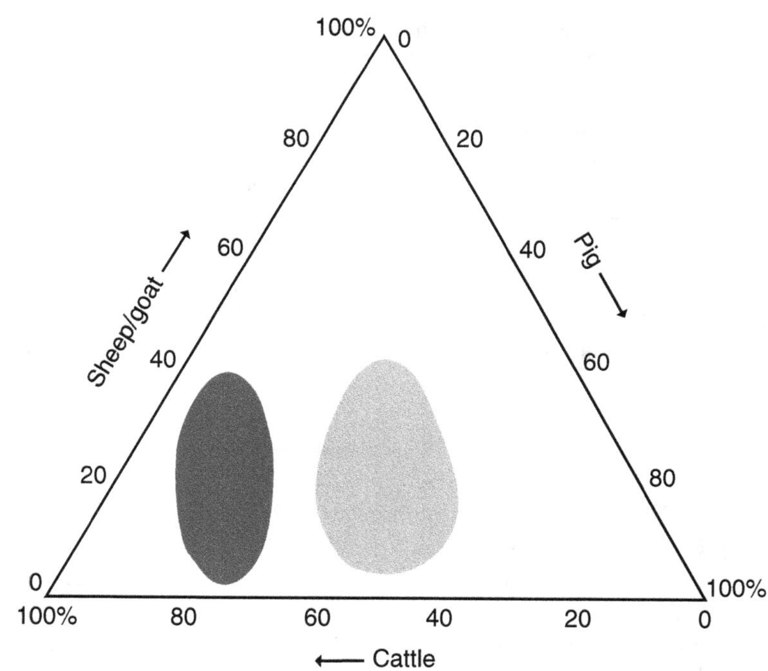

Fig. 9.9 Composition of the livestock during the Merovingian period in Northern France. Two traditions are observed, the one on the *left* more oriented towards cattle breeding than the one on the *right*

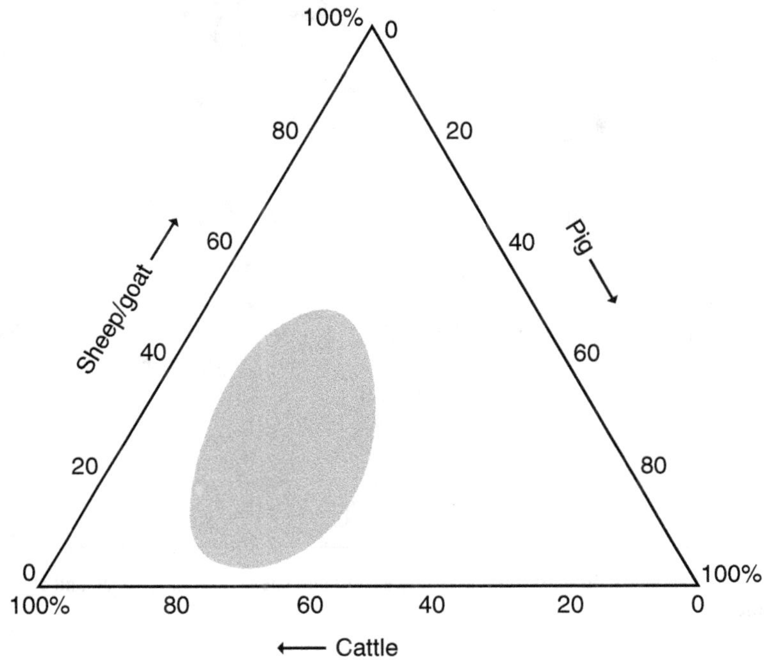

Fig. 9.10 Composition of the livestock during the Carolingian period in Northern France

making in this part of Europe has unfortunately still hardly been studied, but some authors state that churning milk is a technique that came with the Germanic tribes.

In one subregion, the Somme area, sheep/goats came second after cattle instead of pigs. As far as the information provided by excavated bones can tell, less than half of the animals were slaughtered before their fourth year, obviously for their meat. The remainder lived until beyond their sixth year. Such animals were kept for their milk and wool. The milking of a goat or sheep is depicted in the Psalter of Utrecht (Fig. 9.11). This relative importance of sheep/goats is clearly characteristic of the subregion Somme, since it is an aspect already noted with respect to the Roman Period (see Section 8.4). The local landscape, with its shallow soils on top of a calcareous bedrock, was presumably best suited for the raising of sheep and goats. Another explanation is sought in the development of a local wool industry, but the two are surely not independent and the landscape was presumably the main trigger. There were, however, exceptions.

A seventh to eighth century more aristocratic site in the Somme area showed a larger proportion of pigs, some slaughtered very young, but most kept until an age of

Fig. 9.11 Psalter of Utrecht: milking of cows (*left*) and sheep or goats (*right*), and churning in the middle

one-and-a-half to two years which is the age at which they are grown and fattened. The difference with the rural sites is interpreted as a social difference. The same is seen in several settlements dating to the end of the Carolingian period. In truly rural sites sheep/goat bones come second after cattle, but in the wealthier parts of towns such as Amiens it is pig.

Next to the information provided by bones there are the written records, which, as should be stressed, refer mainly to the situation on large estates. An example is provided by the inventory of animals kept on the royal farms written down in the *Brevium Exempla* (Table 9.6). The text mentions large numbers of pigs and sheep, but male pigs were limited to five animals in one farm only, probably because their meat was not highly valued. On the contrary, with sheep the proportion of rams is high, presumably because wool was the main product to be obtained and the wool of a ram is as good as the wool of an ewe. The high number of sheep and goat in the royal estates' list resembles the situation as reported from excavations in the *departement* Somme, though the estates in question lie north of this region. The number of cattle is lower than would have been expected on the basis of the bones found in contemporary settlements. The report refers to only eight to twelve teams of oxen. Eight teams seem a rather low number for a large estate like Annappes. It has been suggested that the extra teams needed for ploughing came with the tenants, who did this kind of work as part of their duties, and were therefore not kept in the main buildings of the estates.

Table 9.6 Composition of the livestock on the four royal farms mentioned in the *Brevium Exempla*

	Annappes	Vitry	Cysoing	Somain
Mares	51	79	44	c. 30
Foals	18	18	7	c. 20
Stallions	3	4	2	2
Donkeys	2	2	–	–
Oxen	16	20	24	24 ?
Cows without calves	50	30	6	6 ?
Heifers	20	–	–	8
Yearlings	38	–	–	3
Bullocks	3	3	–	1
Other animals, cattle ?	–	10	5	–
Pigs	360	250	160	250
Boars	5	–	–	–
Ewes without lambs	150	80	150	150
Once shorn sheep	200	58	200	200
Rams	120	82	8	100
Goats	63	27	41	130
Geese	30	40	10	20
Chickens	80	100	–	–
Ducks	–	6	–	4
Peacocks	20	8	–	–

From the list it becomes clear that an important part of the royal estates were devoted to horse breeding. This is understandable, because the upper class was mounted, and cavalry formed a most important part of the armies. The warrior on horseback physically dominated the land, not only in his own territory but everywhere (Fig. 9.12). A warhorse was already valued in an old Frankish law, the *Lex Riparia*, at four times a milk cow or an ox, and its value did not depreciate in the course of time.

In the beginning horses were still ridden without horseshoes and without stirrups. It is still not very clear when the first horseshoes appeared, nor where they were invented, but they were certainly used from the ninth century onwards. Early horseshoes are small, with lengths of three to six cm and a width of two cm. The stirrup was developed on the Asian steppes and was introduced sometime during the seventh century. Whereas the horseshoe was an efficient protection of the hoof, the stirrup was the start of the medieval mounted shock combat. Horses seem hardly to have been used for drawing carriages. Carts, drawn by oxen, even provided the transport of kings and their households. A cart, the *carpentum pompaticum*, was the sign of royal and religious dignity of the Merovingian dynasty, for instance.

Traces on horse bones found during excavations in rural settlements show that horses were butchered as well. Still, the consumption of horse flesh seems not to have been on equal footing with the consumption of beef, pork or mutton. Some archaeozoologists suggest that the meat was fed to dogs. Others think that the meat was consumed during rituals only. A few bans, issued by the pope in Rome, on the consumption of horsemeat, hints at the latter, but as the eating of horsemeat in the

Fig. 9.12 Charlemagne (or possibly his grandson Charles the Bald) on horseback. The statue is in the tradition of statues of Roman emperors and emphasizes the status of mounted people

classical Roman world was almost taboo, the bans may also indicate a general reaction of the now Christian Roman world to formerly non-Roman, barbaric, customs.

Meat, wool, milk, cheese and animal power were not the only products people asked from their livestock. The *Capitulare de Villis* makes clear that pig and sheep were also kept for their fat, for instance as a source of tallow. This text further mentions goat as a source of horn. Horn was used as raw material for several kinds of objects, such as knifehafts.

What is not much stressed in the texts but must have been important as well, is the production of hides, for leather, but surely also for making parchment. It is known that around AD 700 the export of papyrus to the land of the Franks broke down. This had to do with a shift in the infrastructure in Europe, bringing with it a shift in trade routes. Egyptian papyrus, which was the main type of writing material in the Merovingian period, was not supplied anymore. Parchment, made from goat, sheep and cattle hides, took over.

Sizes of animals have not been mentioned so far. They were variable, one of the reasons being that there were many breeds at this time. Moreover, the size seems to be linked to the general economy. In times of prosperity people seem to have possessed more animals and also larger individuals. In difficult times or in areas with a high population pressure the reverse was the case.

The second important category of farm animals was domestic fowl. Chickens, geese and ducks were kept everywhere. Pigeons, however, seem to have been hardly important at the time. In manuscripts they are listed together with peacocks, pheasants, swans and the like, birds which at least according to the *Capitulare de Villis* were kept as ornamental birds, for showing off. The *Capitulare de Villis* states: *pro dignitatis causa*. Occasionally fragments of eggshells have been found in addition to the more common finds of poultry bones. Chickens were the most important type of fowl. The *Capitulare de Villis* asks for henhouses with no fewer than 100 chickens and no fewer than 30 geese in the main complex of the estate, and for henhouses with at least 50 chickens and 12 geese on the farms dependent on the main house. The *Brevium Exempla* shows that such numbers were indeed present. Chickens and eggs were common products. In a society where money played only a secondary role, they were used as payment in all kinds of transactions. Feathers were valued too. The *Capitulare de Villis* mentions cushions stuffed with feathers.

A third category of farmyard animal is represented by the honeybee. Bees were kept in hives. The *Lex Salica* has a whole title devoted to the theft of hives and swarms. The large Carolingian estates knew specialised beekeepers. Bees provided honey and wax. Both products are mentioned in the texts and must have been of considerable value. Honey was the main source of sugar. Wax was not only important for making high-quality candles, waxing cloth and implements etc., but also provided the top layer of wax tablets, which were still in use for informal messages.

A fourth category is fish. It is known that fish was farmed in fishponds. But as fish is notoriously badly represented in archaeological remains due to preservation problems, the kinds of freshwater fish farmed there is largely unknown, but carp was one of them.

It is quite obvious from the written sources that yards included both henhouses and sheds or racks with hives, and that poultry and bees were kept close to the human dwellings. But the size of the outhouses found during excavations suggests that not many of the other animals could have been kept there as well. Presumably teams of working oxen, and saddle horses were stabled within the yard. The remaining part of the livestock must have been kept in the open. Texts refer indeed to pigs, driven into the woods to forage for themselves. They state that three hectares of woodland could carry one to three pigs. Texts referring explicitly to other animals are lacking.

Access to woodland and land lying waste was initially free. Only towards the end of the period did the use of such terrains become more and more regulated. The same applies to other uses, such as woodland as a source of wood and game. The demand for wood was not yet a problem. Game seems not to have been hunted so much for food but for skins and furs. Hunting provided also sport. It was problems connected with the disappearance of game, that led to regulations on livestock herding. Especially pigs seem to have brought about too much damage to the woods in the long run. Game was given a more or less protected status. After AD 818 hunting became a royal privilege. Only the king or emperor could assign the right to hunt.

9.5 Farmbuildings and Yards

The decline of the Roman Empire and the loss of control that went with it brought about a change in the use of the buildings and yards of the *villa rustica*. Buildings fell into disrepair. Nevertheless, parts of them continued to be inhabited or used by farming people. In addition, new structures were erected in their yards.

Most commonly encountered is a new kind of dwelling, called sunken floor huts. As their name indicates, they are characterised by having a floor below the surface. In excavations the huts turn up as rectangular pits with depths of 0.2–0.6 m, but floors may have been lower, because erosion may have removed some of the soil between the time of construction and the present time. Their general size varies between 2 × 1.5 m and 4.4 × 3.2 m. The most common type reveals traces of a sturdy post in the centre of both short sides. These two posts are explained as the supports of the roof. Others had a post at each corner. The largest type had six posts, one at each corner and two in the middle of each short side (Fig. 9.13).

Reconstructions of the buildings vary, but they are similar in providing them with steep, thatched roofs. In one version the eaves are resting on the surface just outside the pit. In a second the eaves are also resting on the surface, but with the eaves ending at some distance from the edge of the pit, leaving some kind of benches around a sunken floor. A third version allows for a low wall of wattle and daub, without foundation and thus not found, to provide more height. Ideas about the place of the door vary as well. Both short and long sides are considered. Ditches, dug close to the end of the eaves, must have protected the huts from the intrusion of

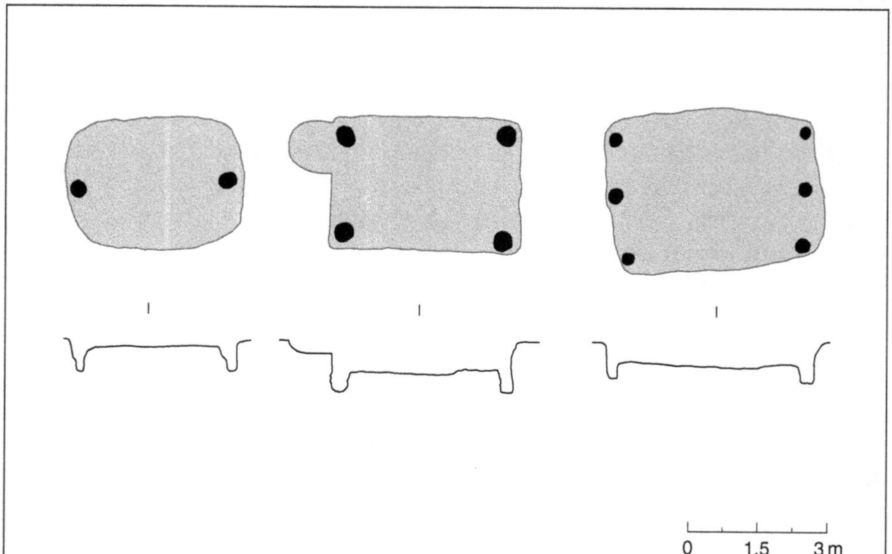

Fig. 9.13 Ground-plans and sections of sunken floor huts. The bulge in the plan depicted in the *middle* is an oven. Examples from Goudelancourt-lès-Pierrepont (Dept. Aisne), France

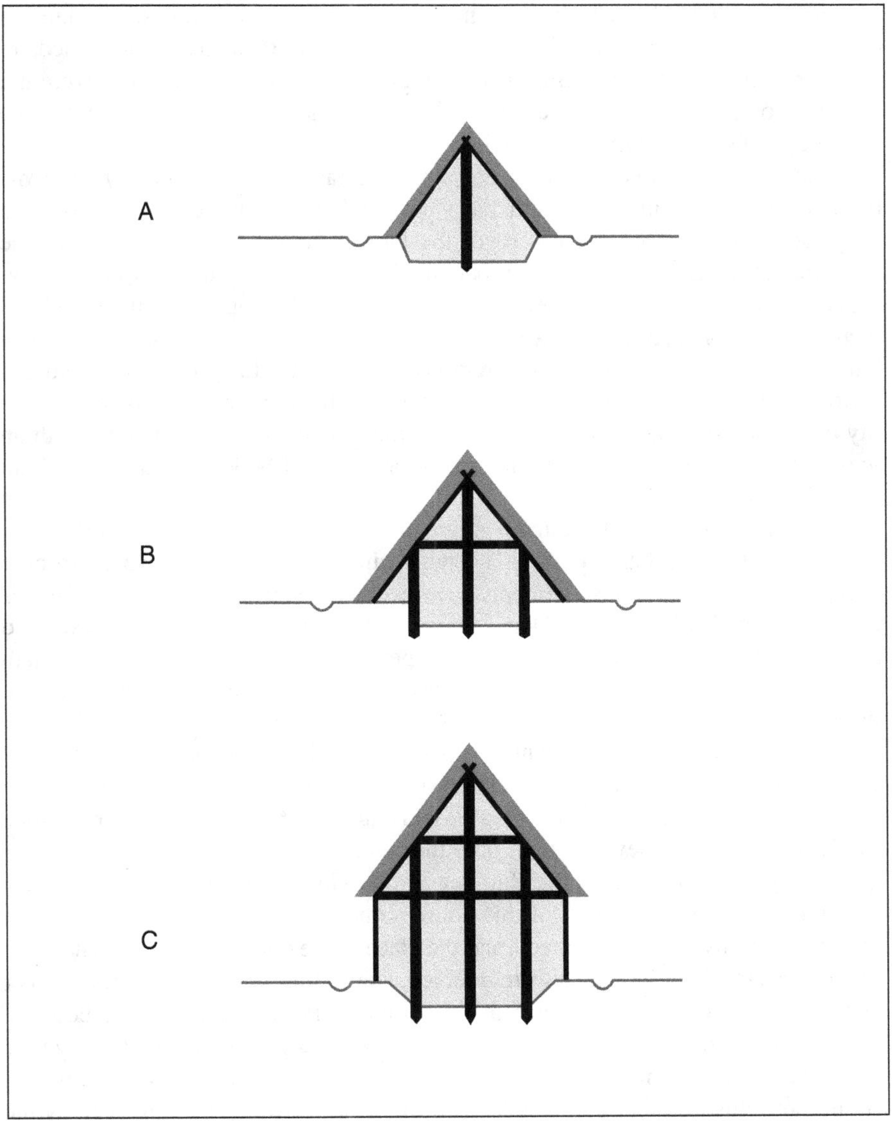

Fig. 9.14 Three different ideas on how sunken floor huts would have looked

rainwater, but such shallow ditches must have been erased by later erosion, because they are not seen during excavations (Fig. 9.14).

The small size of the structures raises the question whether they were the main dwellings of the people. The general opinion is that they were not. For instance, there is no trace of a hearth, although the original floor is preserved. Rarely, the larger ones possessed an oven. In others, sets of loom weights point to the presence of warp-weighted looms. Such finds have led to the interpretation of the huts as places where some craft was carried out. Others see in them stables and sheds to store crops or other things. Still others suggest that the sunken floor huts were not structures on their own, but were the cellar part of larger buildings. But traces of the other parts of such buildings are lacking.

Where people actually lived is not always clear. As mentioned above, some Roman Period buildings may still have provided living quarters. This has been suggested, for instance, in the case of the large *villa rustica* of Voerendaal (the Netherlands) where a group of rural people were active from the fourth to the seventh century (Fig. 9.15). But not every rural settlement belonging to this period had such obvious connections with a former *villa rustica*. Many settlements have been founded on terrains that were not previously part of a villa yard, though Roman Period remains are seldom far off. But this has probably more to do with the density of Roman occupation and the Roman choice of suitable terrains than with an actual linkage, or has to do with the ruins as a source of building materials, a kind of artificial quarry.

It is quite possible that the main farmhouses have been missed in many excavations, as they are hard to see. The huts with sunken floors leave a stronger trace. Nevertheless, some settlements have become better known. One of these is Goudelancourt-lès-Pierrepont (Dept. Aisne), a site inhabited during the sixth and seventh centuries. The site has revealed, dispersed over a terrain of 1.5 ha, a main building, outhouses, a well, sunken floor huts and ovens. The main house was a two-aisled building, measuring 8.8 × 6.0 m. A row of three posts, one in the middle of each short side and one in the centre, supported the ridge-pole of the roof. Several upright posts were set in the long walls to support the horizontal poles that bore the eaves of the roof. The walls were of wattle and daub, resting on a foundation of stone rubble. Inside, a chimney was built onto the southern long wall. The entrance was placed diagonally opposite to the chimney and provided with a porch (Fig. 9.16).

Some ten metres north-east of this house a one-aisled building of 6.5 × 4 m is situated. It seems to lack a chimney, and the place of the door is uncertain. The posts on the corners of the long side were doubled, suggesting the existence of an upper storey. This building is interpreted as an outhouse. Pottery sherds place both the main house and the outhouse in the period between AD 590 and 620/640: they may well have been contemporaneous. One other, more enigmatic, plan and a difficult to interpret cluster of postholes belong to the same period. They are seen as the remains of additional outhouses. Contemporaneity is certainly not true for every sunken floor hut in the cluster of these structures. Based on the rubbish that filled them after abandonment, they cover a wider period and some huts in the cluster were definitely not in use during the occupation of the main house. Remains of outdoor ovens and hearths are lying dispersed over the terrain. The only well was situated near the main house. The entire complex may well represent a farm of the end of the sixth century (Fig. 9.17).

The report on Goudelancourt also provides an estimate of the number of inhabitants. Only six to eight persons would have lived in the main house, members of one single family. But the author of the report, A. Nice, is of the opinion that such

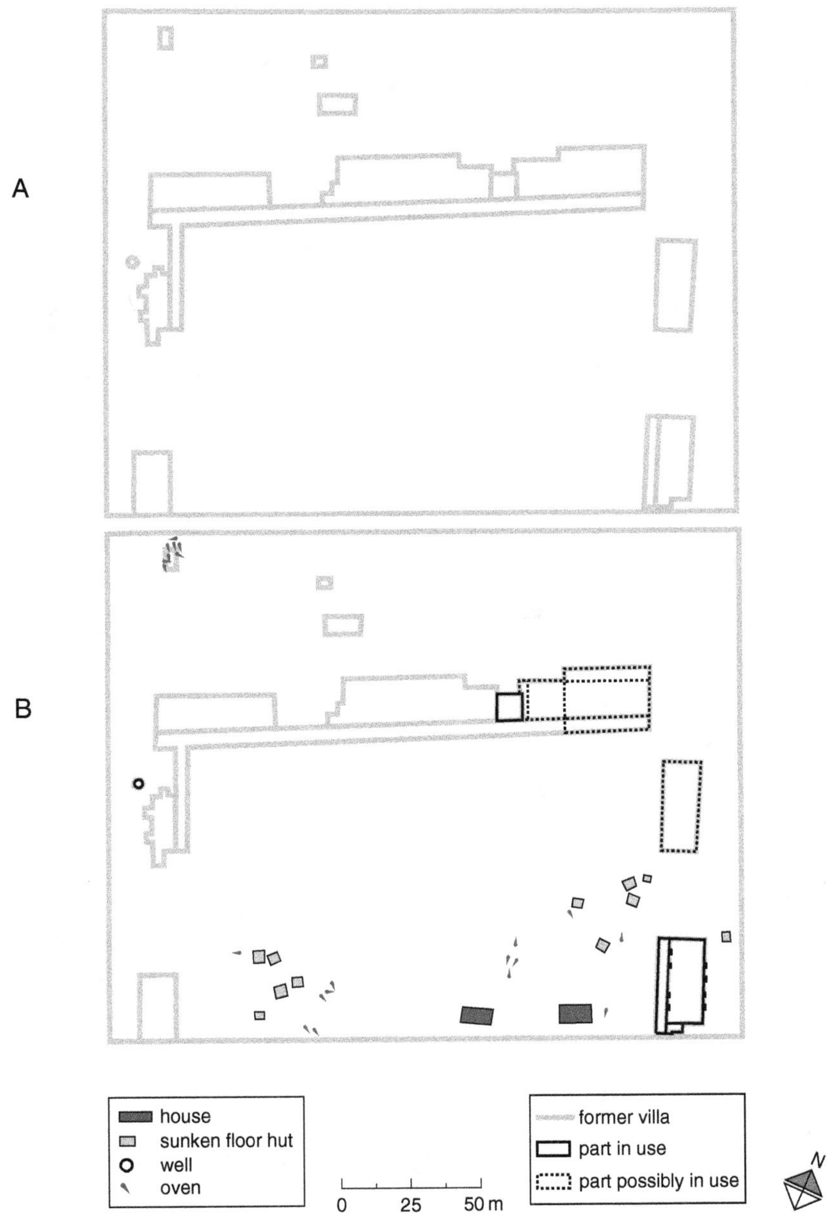

Fig. 9.15 The last use of the buildings and yard of the Roman *villa rustica* at Voerendaal, the Netherlands. **A**: outline of the buildings during the heyday of the villa, **B**: the parts in use during the 4th–7th century AD

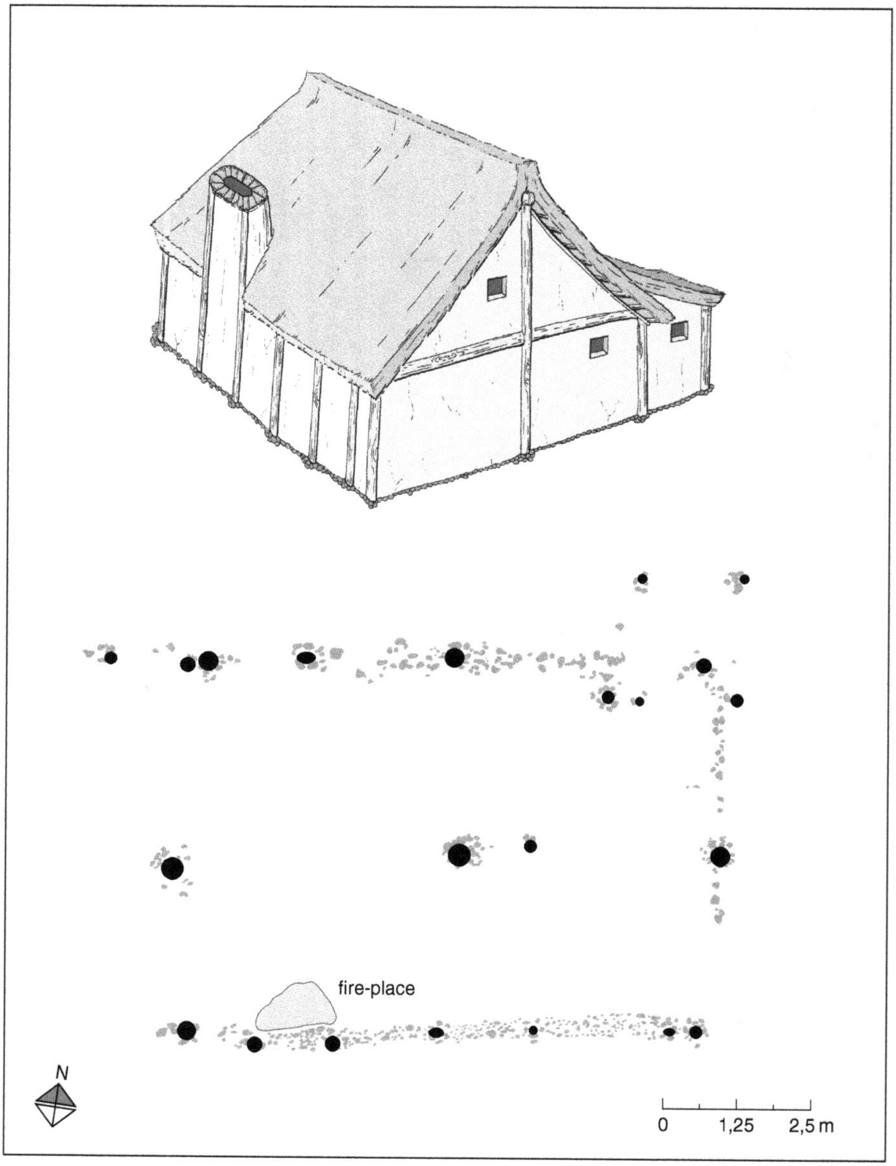

Fig. 9.16 Ground-plan of the main building excavated at Goudelancourt-lès-Pierrepont (Dept. Aisne), France and how it may have looked seen from the south-east

a farming unit could not have been run by so few people. With all the tasks to fulfill, from looking after fields and tending livestock to the production of textiles, he reckons some 15–20 inhabitants were required. Where did the others live? The remaining buildings lacked hearths, a sign that they were not destined to be lived in. It is possible that the excavated area did not cover all traces of the establishment

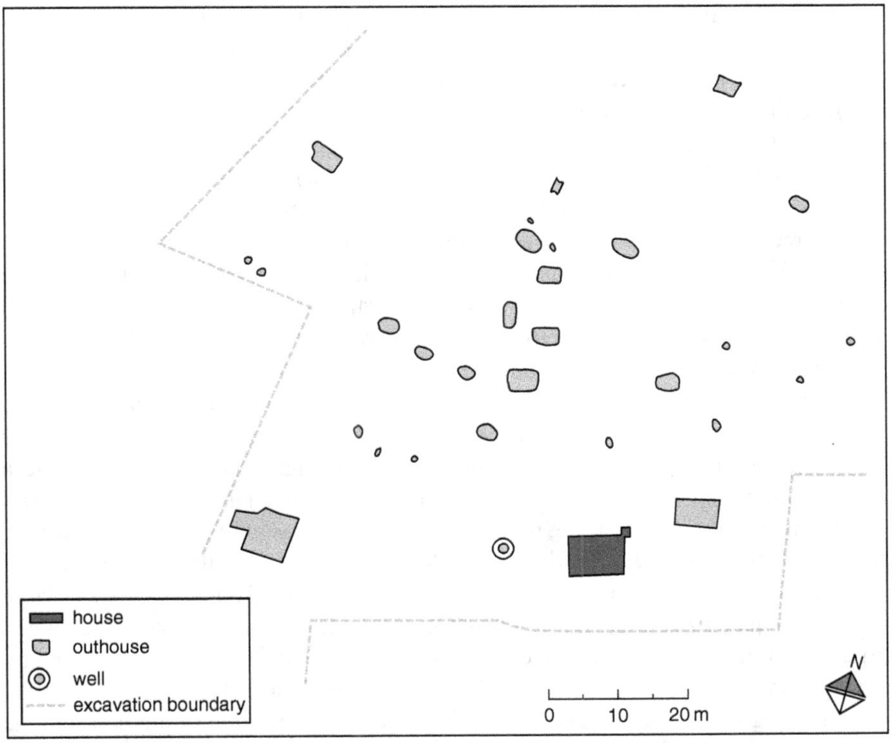

Fig. 9.17 Plan of the Goudelancourt yard with the main house, two outhouses, sunken floor huts and a well. The smallest features are outdoor ovens and kilns

and that one house was missed. Increasing the number of people living in the main house is held to be a less probable option. In a written text, albeit of the ninth century, households consisting of fewer than ten persons are the rule for such relatively small farmhouses.

A cemetery nearby must have received the members of the farm after death. It has been excavated in its entirety. On the basis of the number of inhumations and the period during which it was in use, the size of the population living at one and the same time has been estimated at 126 individuals. This implies that other farming units must have been present nearby. Traces of such units have indeed been discovered. The settlement is too dispersed to use the word 'hamlet', but people may well have worked together to get larger tasks done. The inhumations also show that life expectancy at birth was 35 years, but quite a lot of people reached an age of 60–70 years, and some reached ages of over 80.

Twenty kilometres away a similar complex has been excavated, Juvincourt-et-Damary (Dept. Aisne). In its sixth-seventh century phase it comprises a two-aisled building, measuring 11.8 × 4.8 m, with a central row of four roof-supporting posts, three other buildings and a dozen sunken floor huts. A second phase, dating from the second half of the seventh to the ninth century presents a small cluster, composed of several buildings, sunken floor huts, and underground silos. Every building was set in its own yard which was fenced off by a palisade. In this period Juvincourt-et-Damary looks like a very small hamlet. Similar clusters have been detected at a distance of 100 and 300 m.

In the Carolingian texts farms, such as described above, are indicated by the name *mansus* (plural *mansi*). They were the backbone not only of rural society, but of society as a whole (see Section 9.6). *Mansi* were parts of estates. Their activities were directed by the owners of a central, larger farm, called a domanial farm or seignorial *mansus*. Its main house was not only larger but also better built, and the yard contained more well-built outhouses. Its earliest versions are not yet very well known, but the house was sometimes at least partly built of stone in addition to the common timber and wattle and daub.

An example from the middle of the seventh to the end of the eighth century is provided by a complex excavated at Serris-les-Ruelles (Dept. Seine-et-Marne). The site revealed a complete settlement consisting of a seignorial farm built next to a large pond, a network of roads, a scatter of much smaller farms, and a graveyard. The main house of the big farm, measuring 30 × 9 m, was built in stone (at least the lower parts of its walls), and had glass windows and a tiled roof. Its western long side was embellished with a portico. Inside the house was divided into two wings or large rooms of equal size. This building was clearly the house of an aristocratic owner. Apart from the glass windows and tiles, the high status of the family is stressed by luxury items found in the waste associated with the house: fragments of marble, snippets of gold thread, bones of peacock and a high proportion of pig bones in the remains of their meals.

Next to this building, a second building built in stone was found. It was square, measuring 19 × 19 m, and seems to have had a residential function as well. But this structure was abandoned rather soon. In addition to the two structures, this part of the Serris terrain had a well. The complex was enclosed by a fence on three sides. The fourth, eastern side was delimited by a ditch which formed part of the draining system of the rather low-lying grounds. The ditch ended in the pond.

Across the ditch, buildings with a more agricultural function were excavated. Their construction was mainly in timber and wattle and daub, though some had a foundation of small stones placed in a shallow ditch. During the one-and-a-half centuries during which this seignorial farm functioned, these structures were altered several times and added to (Fig. 9.18).

At a short distance three groups of much humbler, rural dwellings have been discovered (Fig. 9.19). They are made of timber with wattle-and-daub walls and are supposed to have had thatched roofs. Their basic plan is two-aisled rectangular. Their floor surface is some 80–100 m^2. This type of structure is interpreted as the dwellings of the farmers, who were dependent on the big house. Smaller, square structures might have been granaries. The open areas between the buildings revealed sunken floor huts, underground silos, ovens and wells. Remains of ditches divided the Serris area into several parcels, but limits of individual yards could not be discerned, which may be due to the disappearance of traces of light fences or hedges. The number of contemporaneous establishments is not exactly known, but there

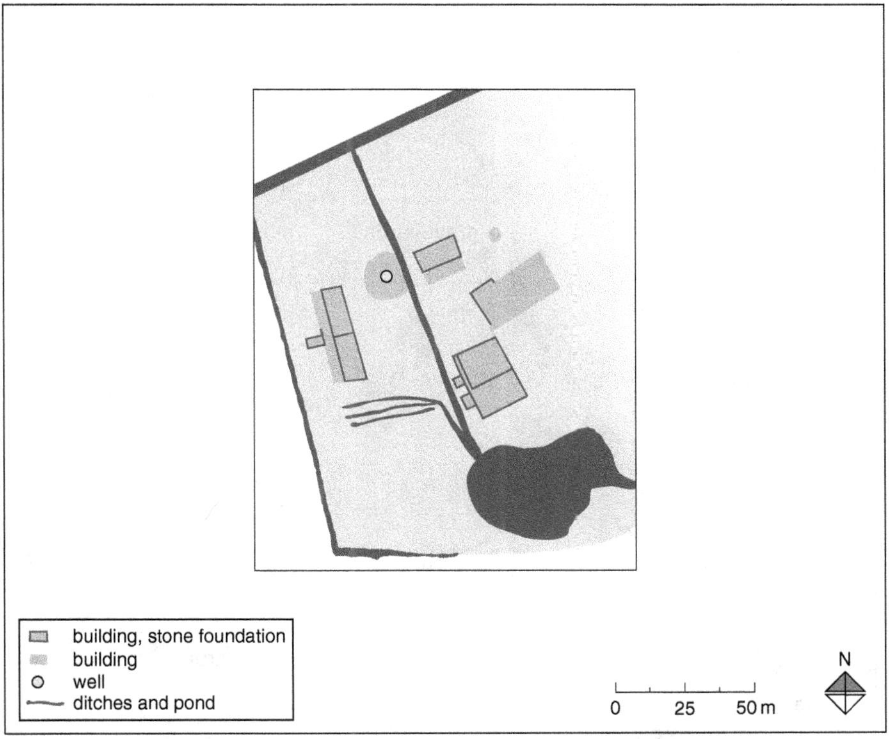

Fig. 9.18 Plan of the seignorial complex at Serris-les-Ruelles (Dept. Seine-et-Marne) France

were several of these *mansi* at the same time. A cemetery with a chapel completed the settlement. The main house was abandoned at the end of the eighth or beginning of the ninth century without a visible trace of destruction, but the rural settlement continued its existence.

A manorial house, which was less rich than the one excavated at Serris, and perhaps more representative of this kind of establishment, was found at Engis-Thier d'Olne near Liège (Belgium). Its first version consisted of a rectangular building of 8 × 4 m, divided into two rooms, just as in Serris. The foundation trench was lined with stone, but the upper structure was of wood and wattle and daub. The house was set in a yard, surrounded by a sturdy palisade, which enclosed several silos and the cemetery of the family. The cemetery even had an impressive mausoleum, the only other building detected so far within the yard. The number of graves, together with the study of the human skeletons, suggests that two or three generations of a group of 15–20 persons were buried there. This number looks high for a main building of only 8 × 4 m, but the house may have had an upper storey.

The relatively small house was abandoned in the second half of the eighth century and was replaced by a house, built with the same materials, but now measuring 13 × 12 m, and divided into three rooms. The central room had a hearth.

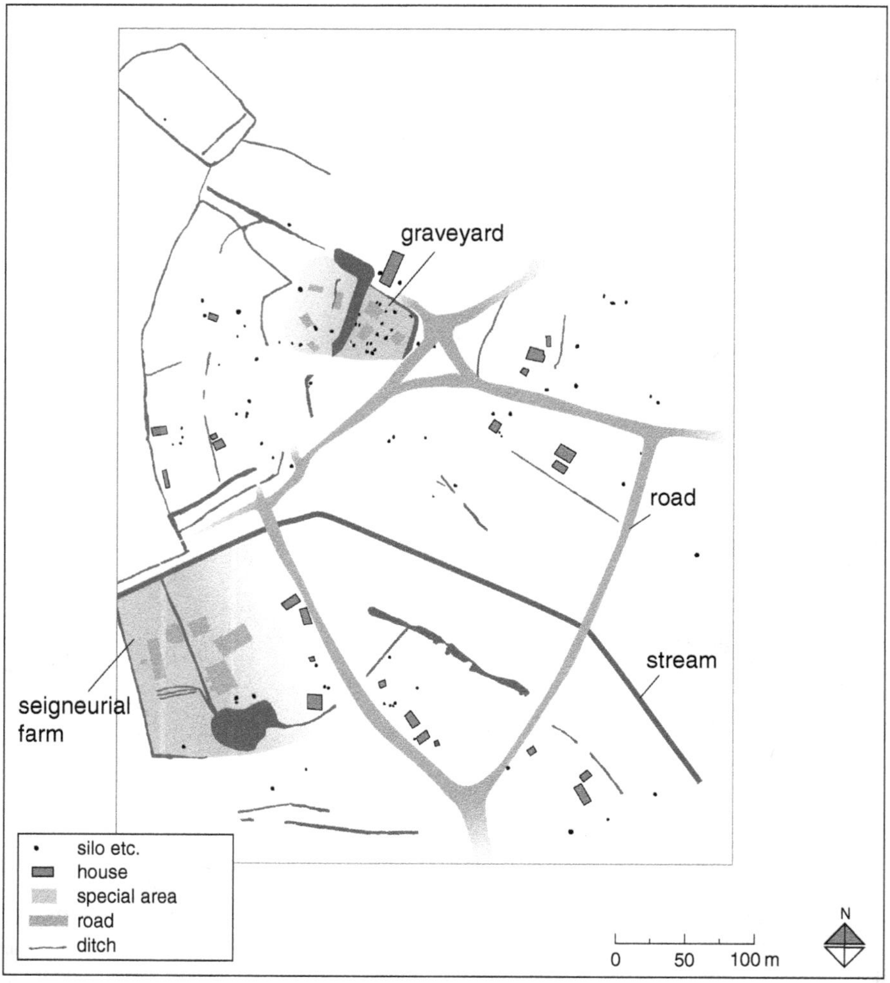

Fig. 9.19 Plan of Serris-les-Ruelles

A four-poster granary and a sunken floor hut were present nearby. The mausoleum had been replaced by a chapel. In the ninth century the main house was abandoned again for a larger house, measuring 27 × 18 m, and built of stone. It boasted at least seven rooms on the ground floor and a portico. Moreover, traces of a wooden staircase lead to the conclusion that the building had an upper storey. The building history of Thier d'Olne shows the rise of an aristocratic family. Around AD 1000 the place was abandoned, without any trace of destruction. The family moved to a proper castle elsewhere on its territory.

Aristocratic farms, such as Serris and Thiers d'Olne, are described in the *Brevium Exempla*. The inspectors found in Asnapium (now Annappes near Lille) a royal house, of stone and well-built, with three rooms. The house was completely surrounded by verandas and had, moreover, two porticos. Underneath was a cellar. The house counted eleven apartments for women. The inspectors noted 17 other houses built of wood within the courtyard with a similar number of rooms, all well constructed. In addition there was one stable, one kitchen, one mill, one granary and three barns. The yard was surrounded by a hedge and a stone entrance building with a balcony from where distributions could be given out. There was also an inner yard, enclosed by a hedge, well planned and planted with various kinds of trees.

Asnapium was certainly not one of the largest of the royal estates. Most probably it represents a modal kind of establishment. Its description evokes the Roman *villa rustica*, with a main house with its portico, outhouses, an orchard surrounded by a hedge, and the entire complex enclosed by an outer hedge. It looks as if the concept of the Roman *villa rustica* re-emerged after a century or two of abandonment. The eleven apartments for women were for female servants, charged with spinning, weaving and all other kinds of female manual work. Where these apartments fitted exactly in the building does not become clear, perhaps in an annexe, which was not deemed necessary to be defined.

9.6 The Farm in Its Setting

From the preceding section it is clear that farms, at least in the second part of the period, were organised following a hierarchical order. The basis was the farm, called *mansus* in the written texts. It was run by a single family or several families together. If the plot of land belonging to this farm, was very small, one to two hectares, it could not survive on its own. The owners had to attach themselves to larger properties. A common farm measured c. 15 ha. Nevertheless, such farms seem not have been able to operate independently either. In one way or other many of them became clustered into large estates measuring 60–180 ha. One of the farms in such a cluster grew to be dominant or was already so from the beginning. The dominant house became the manorial house. The large farm had a special name, the demesne. The use of the land became organised into what is called the domanial or manorial system. Although quite a few owners still farmed the land themselves, with the help of some farm hands and neighbours with small plots, it is the large manorial property which is commonly described in publications on Merovingian and Carolingian farming. This is not surprising, because the larger the estates, the more traces in the form of written administrative texts they left behind.

The owners of a large manorial property did not work on the land themselves. In the largest estates they did not even live there, but had stewards managing it. Characteristic of the system is its bipartite structure. On the one hand there was the demesne, consisting of the seignorial house with attached gardens and outhouses (also called *curtis dominica*) and its land (called *terra indominicata*). The *terra indominicata* comprised arable land (known as *cultura*), meadows, woods and waste land, one or more watermills, and sometimes a vineyard or a brewery. On the other hand, there was the common farm (*mansus*), which included a house, garden, arable land, meadows, sometimes also a vineyard, and the right to use wood and waste land. Demesne and *mansi* together formed a manor, also called villa. A villa at this time is not one single large farm as it was in Roman times, but a complex of several farms, one large and several smaller ones.

A common manor consisted of one demesne and 4–20 *mansi*. Most texts, however, are dealing with the very large estates of the king, other important men and especially the huge possessions of monasteries. Such owners held large estates in several parts of the kingdom, which ensured the supply of a large variety of products (Fig. 9.20).

The *terra indominicata* was worked by labourers, belonging to the demesne, and by people living in the *mansi*. The latter had, at least in the Carolingian period, the status of tenants. They held their farm in exchange for working on the land belonging to the manorial farm. These corvées included also other kinds of work. Moreover, most tenants were not free to go where they wanted, or to leave when they wanted. They were bound to the land. Nevertheless, their status was not always and everywhere the same. Some were more free than others. The judicial status of the inhabitants of *mansi* varied from free, to semi-free to true serfdom.

Some texts provide exact descriptions of what the tenants had to deliver. An example enumerates the duties of a man named Arctardus, his wife and six children, who lived c. AD 800 in a *mansus* on an estate in Villeneuve-St. Georges, belonging to the abbey of St. Germain-des-Prés near Paris. He had to plough some of the demesne lands, to maintain the enclosure of some of its meadows, and to cut wood. In addition, he had to deliver 2 muids of wine, 50 shingles, 3 chickens and 15 eggs. Every third year he had to deliver a ewe lamb with its dam. Moreover, he had to pay

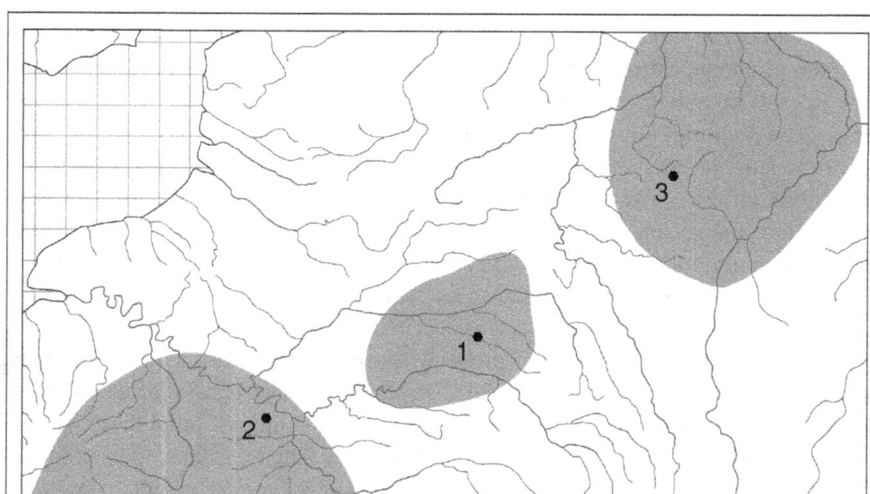

Fig. 9.20 Three important monasteries and the area in which the majority of their rural possessions were situated. 1: St. Rémi-de-Reims, 2: St. Germain-des-Prés, 3: Prüm

money as protection tax, and as tax on his livestock. A muid (*modius*) of liquids was in the surroundings of Paris equal to 274 L.

Another man, belonging to the same estate and whose wife had the status of serf, had to deliver 3 muids of wine, 1 setier of mustard, 50 osiers, 3 chickens and 15 eggs. A setier is in this case one twelfth of a muid. He had to do such corvées as he was ordered to do, and his wife had to weave cloth from the wool provided by the demesne, and to bake as many loaves of bread as she was asked to make.

At this time money was made from the selling of products, which were realised in small quantities or did not keep well. The earliest goods to be sold for money, instead of being bartered, were butter and eggs. Grain was used longest for payment in kind.

Although in the text above some ideas are expressed on how the domanial system came into existence, its exact history is not quite clear. Some elements have roots in the Roman Period with its large estates. Another element comes from the socio-political system of the Germanic (Frankish) world, in which leaders and followers swore fealty to each other. The followers (vassals) received fiefs in recognition of their aid and service. Fiefs were often land, as could be expected in a society where money played a very minor role. Lords were not able to pay their followers with coins and had to reward them with products or by granting them a piece of land. In the end there was a complete feudal pyramid of lord, vassal, sub-vassal etc.

Almost everybody lived directly off the land. If someone owned more than one estate, he travelled according to a fixed plan from one to the other. Important, rich people moved around with their retinue, consuming stocks wherever they went. For the king this was also a method to visit personally different parts of the realm. For those owners, who could not or would not move, such as the inmates of monasteries, another arrangement was needed. Their tenants had the duty to convey products to them. Remote estates were, however, a risky possession, especially in times of unrest. Towards the end of the period abbeys tended to dispose of their more distant estates. Some were for years noted down as 'non-paying'. In the long run the domanial system with vast estates came to an end altogether. Part of this was triggered by the growing use of money. The lords preferred to be paid in cash and not in products. Manorial houses converted into administrative centres, where payments were received, both in cash and in kind.

At first, farms were lying scattered over the landscape. If there was any question of clustering, it was never more than a clustering of some *mansi*, which most of the time hardly deserves the name of hamlet. Clustering of farms became more common with time, but the true rural village is a development belonging to the next period and falls outside the scope of this book. Nevertheless, the word village derives from the word villa as it was used in Carolingian times and some larger clusters of *mansi* may be called proto-villages.

The pattern of occupation was grafted on the Late Roman pattern, i.e. near main roads and waterways. This suggests a continuation of land use, at first perhaps on a modest scale, but later expanding over areas at larger distances from the original focal points. A study of place names in use at the time, tells that those expansions were realised by clearing waste land and woods. The woods were certainly no primordial woods, as has been suggested sometimes, but regrowths on land abandoned during the Late Roman period, when part of the *villae rusticae* was abandoned (see Sections 8.6 and 10.1).

Most of the population lived in the rural setting depicted so far, but some people lived in towns. The towns were very much reduced affairs compared to the Roman Period towns, although many of them were continuations of these. Towns had shrunk from the end of the third century onwards, some very important ones, such as Trier, excepted. If they had new walls of stone during the Late Roman Period, the walls enclosed a smaller surface than before. Artisanal activities were conducted outside the walls. Inside a certain ruralisation took place. During the Merovingian and Carolingian rule walls were neglected and fell into disrepair. Roman public buildings were converted into private dwellings, or used by the Church. In Trier for instance, a set of large, Late Roman public granaries (in Latin *horreum*) was converted into a monastery for nuns and became known as St. Maria in Horreo, or, later, after its second abbess, St. Irmina in Horreo. Only at the end of the eighth century did the ruralisation of the towns come to a halt, but no new walls were built.

The number of people making up rural society is difficult to assess. As mentioned in Section 9.5, a Merovingian settlement such as Goudelancourt was inhabited by some 15–20 people. Other calculations set the number of inhabitants of such agricultural units at c. 20. As long as a count of truly contemporaneous units is lacking, it is impossible to give an estimate of the population density. Nevertheless, it is generally held that the population was increasing from Merovingian times onwards. Cemeteries near Cologne in Germany point towards a population growth of 60% between the sixth and seventh centuries. The texts of the Carolingian period allow some estimates for the eighth and ninth centuries. Calculations for northern France arrive at a population density of 35 inhabitants per square kilometre.

The population increase cannot have been a smooth, steady increase, at least not always. The period knew problems with diseases and famines. Late Roman writers from outside the region mention an outbreak of the plague in Western Europe around AD 443–445, but this seems to be the only one of sufficient impact to be reported. Much later, Carolingian manuscripts provide information on measures taken by Charlemagne to cope with famines. There must have been crises in at least AD 792–793 and in 806. The measures taken were setting maximum prices for bread and grain, introducing new units of measurement, establishing help for impoverished people and requiring extra prayers to be said in churches and monasteries. Notwithstanding such crises, the population seems not to have been set back much. On the contrary, the number of inhabited sites is on the increase during the last centuries before AD 1000. The true medieval scourges with severe impact on population densities came much later.

The life of the rural population was in principle self-sustaining. Many items were made on the farm itself, where people were also engaged in the more common crafts, but some commodities had to be obtained from elsewhere. They could be had at small local markets, where pottery, finished ploughshares, sickles, textiles, cattle and horses were offered for sale. In Carolingian times such markets seem to have been held weekly. Interregional trade was based in the small urban centres, which were sometimes even founded for this purpose. Most of the trading centres were situated along waterways. Salt was one of the commonest goods to be transported over larger distances. It came either from the Atlantic coast, for instance from the salt pans north and south of the estuary of the river Loire, or from salty springs in the surroundings of Metz (Dept. Moselle). Salt was most important for the preservation of food. Grain, wine, and sometimes even honey, were also important bulk goods, traded through interregional commerce. Other types of merchandise were lumber, building stone and quern-stones from the quarries near Mayen (Germany). Most of the transport of bulk goods went over water. In between, and for delivery to customers not living near waterways, goods were transported on wagons. The main roads were kept in good order, but transport on landroutes was much more costly than transport over water.

As mentioned at the beginning of this chapter, the loess region west of the Rhine was the heartland of a much larger kingdom, later empire, and international trade was important as well. The elite wanted more than the local products. Furs, silks, other luxury textiles as well as spices and incense were very much in demand. The region was connected by trading routes over the Alps with the Mediterranean and its important centres of trade. There was much traffic and much travelling going on, at least for those who were free. River, land and sea routes linked the region with the western and eastern Mediterranean, and with the eastern and northern parts of Europe. The region was definitely part of a larger world.

Chapter 10
The Birth of the Cultural Landscape

10.1 The Vanishing of the Forest as the Main Vegetation Type

In Chapter 6 a landscape has been depicted in which forest was the dominant factor. Around 2650 BC this picture was still valid. Primary forest covered the core of the plateaus and the few waterlogged areas were not yet much affected. But in areas occupied by farming populations the forest was of a secondary nature. Such areas were mainly to be found at the edge of plateaus, near rivers, streams and their still smaller tributaries, and on the drier parts of valley bottoms. Another characteristic of these sites was the marked presence of forest-edge vegetation, which points to the existence of open space as well. But such clearances must have been relatively small, as they hardly show up in pollen analyses. The subsequent developments have been followed as closely as possible by several investigations, one that was carried out in the German Rhineland being an outstanding example. Its results are confirmed by investigations elsewhere.

Until 1100 BC the landscape remained dominated by forest. However, some changes in its composition are observed and one of these is that elm lost its place as one of the main components. A decline in elm pollen percentages is a phenomenon that is observed over larger parts of Central and Western Europe. The decline can be relatively abrupt, seems to occur more or less simultaneously around 3800 BC and is known as 'the elm decline'. In the loess region this decline is preceded by local, temporary declines, and because of this the 'event', if it is a single event, is less marked here. In the past this phenomenon has been attributed to a climatic change, i.e. the transition of the Atlantic Period to the Subboreal Period, as mentioned in Section 6.2, but this interpretation has few adherents left today, one of the arguments being that 'elm' is a complex of species with different susceptibilities to climatic factors. Others attribute the decline to an overexploitation of elm as a source of leaf fodder for livestock. Because of the scarcity of pastures, domestic animals were fed with leaves and young twigs and the reason why elm suffered and other good fodder trees obviously not, would be that elm does not develop flowers on young branches, whilst trees like ash and lime do. A frequent lopping of trees, which prevents flowering, would affect elm first. Though such practices may have played a role, the main source of the elm decline is nowadays attributed to a first wave of the elm disease, which affects all species and can wipe out complete stands.

C.C. Bakels, *The Western European Loess Belt*, DOI 10.1007/978-1-4020-9840-6_10,

Whatever the reason, the result was that from that time onwards elm held only a minor position in forest vegetation.

The second change is that the surface covered by alder carr has grown, reflecting increasing wetness of valley bottoms. The reason is not a climatic one, but is attributed to an increase in run-off due to a lowered capacity of the vegetation to hold rainwater. Repeated clearance and transformation of the surrounding forest would have lowered evaporation.

Around 1100 BC human impact on the upland became such that the recurrent, if only temporary, regenerations of the forests came to a halt. The lime forests were opened up. More rain was left to run off and reach the valleys.

Another effect of the opening-up was that a new tree species got a chance: beech. Originally, beech (*Fagus sylvestris*) was not present at all in the loess region. Normally, the tree is restricted to some specific areas in southern France, Italy and Greece, and during Pleistocene interglacials it grew only there. Only in the Holocene, the period after the most recent Ice Age, the beech is seen to spread to other regions. If its migration pattern is followed, the routes coincide somehow with the spread of farming. It follows man as an oversized weed.

Beech can grow on all kinds of soils, poor sandy soils and waterlogged soils excepted. Its seedlings do well in the shade. Given a chance, young trees grow fast and outgrow trees like oak and ash, thereby eliminating those from rejuvenating patches in the forest. In these aspects beech resembles lime. It seems that beech took over the place of lime, when lime forests suffered too much from human actions. Why lime did not recover and beech took over is not yet well understood, but may have had to do with a creeping soil deterioration or with the climatic change mentioned below, which beech may have withstood better than lime. Some authors are even of the opinion that in the long run beech would have replaced other common forest trees even in the absence of human intervention. Anyhow, the result was that around 800 BC beech had replaced lime. According to the remains of herbs, characteristic of open forest and clearances, the beech forests were floristically very similar to the former lime forests.

At approximately the same time another kind of change sets in. In the last part of the Sub-Boreal period the climate had already undergone a series of ups and downs, but now it showed a definite change towards cooler and wetter. Most of the change concerns summer temperatures, which fell to levels of 1° to 2°C below the average of the previous time. Data on temperature and rainfall have been mainly obtained in regions outside the loess region considered here, but as the changes were clearly of supra-regional importance, they must also have been felt in this region. The new climatic period is known as the Sub-Atlantic period. Its climate was not uniform throughout, but was characterised by variations. Sometimes it became slightly warmer and drier, to return within centuries to the situation that existed before (see also Section 7.5).

It has been suggested that the final success of the beech was furthered by this climatic change, but this idea no longer has many supporters. Other effects on the natural vegetation are hard to detect, because just after 800 BC people began to make a more intensive use of valley floors. They cut down stretches of carr, thus creating, as main aim or as secondary result, extensive grasslands. These grasslands were maintained by mowing. Around this time the first scythes appear (see Section 7.3). Slopes were deforested and clearances in the upland expanded until almost no vast stretches of forest were left. The water regime thereby altered considerably. Run-off increased to feed small streams, even in valleys where there had been no running water before. Plant remains show that such streams were often visited by animals that trampled the surrounding vegetation.

In the first century BC the land was a mosaic of fields, land lying waste, and remnants of secondary woodland in the uplands. The waste land and woodland were used as pasture for livestock. The low-lying areas had to a large extent been converted into pasture and rough meadow.

The question is how this reconstruction tallies with the descriptions left by the invading Romans. Julius Caesar mentions several forests. The largest forest of all Gaul was, according to him, the *Arduenna silva*. In his book *De Bello Gallico*, volume 6 part 29, the *Arduenna silva* is said to reach from the river Rhine and the frontiers of the clan called Treveri, to those of the Nervii, and to extend over more than 500 miles (Roman miles, c. 700 km). Translated into modern geography this would mean from the Rhine to almost the middle course of the river Scheldt. The name of the forest is said to derive from the Celtic *Ar Duen*, which means The Black. The text evokes an immense dark forest. The largest part is situated outside the loess region, in the Eifel and the area to its west, which is still called Ardennes. But the more western part, roughly west of the rive Meuse, transverses the loess region. It is quite possible that this part was not as dense (black) as the rest. It may even have been discontinuous, but a belt of larger forests. The cavalry of Julius Caesar could pass through by rapid marches, which would suggest sufficient open space, and not impenetrable thickets and gloomy woods. Nevertheless, the forest was not imaginary. It still existed, in a more or less fragmented state, in Merovingian and Carolingian times, when it was called Coal Forest (*Sylva Carbonaria*), because of the many charcoal kilns operating there. Remnants of this forest have been preserved until today. The main fragment is the Soignies or Sonia forest south-east of Brussels. It is mainly beech forest and covers 1654 ha.

The Roman occupation and subsequent Romanisation of farming had its impact on the landscape described so far. During the first stage of occupation change is not yet visible. Only in areas where the local population resisted Roman occupation and upheavals led the Roman army to take severe measures, is a temporary return of upland forest and alder carr seen in pollen diagrams, most probably due to a diminishing of human population numbers and/or human activities.

With the rise of the *villa rustica* (see Section 8.2) the look of the land changed, however. The landscape became organised in a less patchy and more controlled way. The uplands show a slight extension of forest, at least in rural areas, possibly due to the furthering of larger stands of trees instead of small ones. In areas where large construction works had been carried out and large centres of activity arose, the picture was different. In cases where a sufficient number of wooden construction elements could be studied, it is clear that in the course of time thinner and younger trees suitable for construction were found. Thick old trees were obviously used up. This has, for example, been observed in Roman Rouen (Dept. Seine-Maritime). The first buildings contained boards made with the age-old custom of splitting trunks radially. Later on boards of the same width had to be sawn out of trunks half as thick (Fig. 10.1). At the same time also gnarled wood had to be used. Such wood comes from trees in open stands and is usually avoided. The fact that such wood was used, implies that better wood with a straight grain was becoming scarce.

Valley bottoms were converted into true meadows destined for cutting hay. This is at least indicated by the plant remains found in sediments deposited during the Roman Period. Livestock was obviously being kept away. It was also kept away from areas with springs. In several cases it could be proven that this was so because the water was destined to feed the water ducts which supplied the *villa rustica* with clean water. Obviously, these areas had to remain as clean as possible.

Notwithstanding the human control over main parts of the landscape at that time, still another tree succeeded in establishing itself as part of the natural vegetation: the hornbeam (*Carpinus betulus*). Contrary to the beech, the immigration of the hornbeam is a natural phenomenon in the forest development of Interglacials and therefore also of the Holocene. The tree always arrives as the last of the deciduous trees. It requires good loamy soils. Its seedlings do not thrive in heavy shade, but an adult tree can stand shade readily enough, though not the heavy shade of beech.

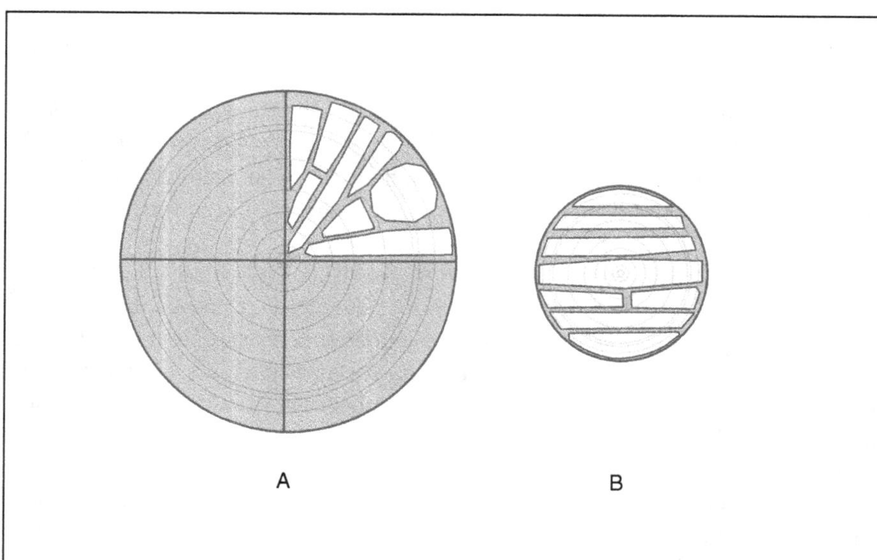

Fig. 10.1 Wooden construction material from early-medieval Rouen (Dept. Seine-Maritime), France, arranged according to their original position in a tree trunk. **A**: early buildings, **B**: later buildings

In all other types of forest vegetation hornbeam stands are able to rejuvenate in a natural way.

The collapse of the Roman Empire resulted in a regeneration of forests, both on the drier soils and on the valley bottoms. A vegetation of hazel and oak, followed by beech and hornbeam, regained even the immediate surroundings of the farms. Wet meadows converted into vegetations with tall forbs and stands of willow, proof that the meadows were not mowed anymore. Later on the valleys were colonised by alder. The return of the forest even led to the disappearance of running water in small valleys. Where central authority did not vanish altogether, the reaction of the vegetation was not as strong. And traces of human presence and farming activities nowhere vanished completely.

When the Merovingian dynasty seized control, the landscape was again opened up. Forests became lighter. Deforestation continued during Carolingian rule. Small streams reappear in small valleys. The landscape again became a patchwork of forest, fields, land lying waste, and meadows in valleys; only it was neither as patchy as in pre-Roman times, nor a copy of the Roman Period. Most reconstructions allow for something in between. An increasing use of forest by man led to regulations as to its use. It was laid down who had access, and how and for what. Pollarding of trees seems to have been practised more and more, as is deduced from wood remains. It is a way to harvest a tree for wood, at least branches. This period was the starting point of the historical landscape of the loess region, with its vast proportion of agricultural land.

10.2 Erosion

The increasing agricultural use of the landscape caused increasing erosion of the loess. A detailed study of a small valley in the German Rhineland, the valley of the Elsbach, revealed that in certain places sediment deposited on the valley bottom and obviously washed down from the adjacent upland, dated back to somewhere between 2900 BC and 2500 BC. But such sediments are very local. Continuous deposition only took place from 800 BC onwards, to continue until c. 50 BC. Sediment from between 50 BC and AD 100 is lacking, either because it was eroded by subsequent events, or because it had never been deposited. After AD 100, and presumably connected with the *villa rustica* system, an important erosion took place, resulting in a series of coarser and finer layers on the valley floor. During the last phase of Roman occupation sedimentation came to a halt. Arable land was obviously recolonised by vegetation whereby erosion was prevented. Deposition recurred in Late Merovingian times, after AD 600, when a 1 m thick layer was laid down. Carolingian activity added some more metres. The small valley became clogged with sediment.

The history of the Elsbach very well illustrates the general erosion history. Serious erosion starts everywhere around 800 BC. Although this date coincides with the change in climate described above, it is probably primarily triggered by a different use of the landscape. The temporary cessation after 50 BC is not present everywhere,

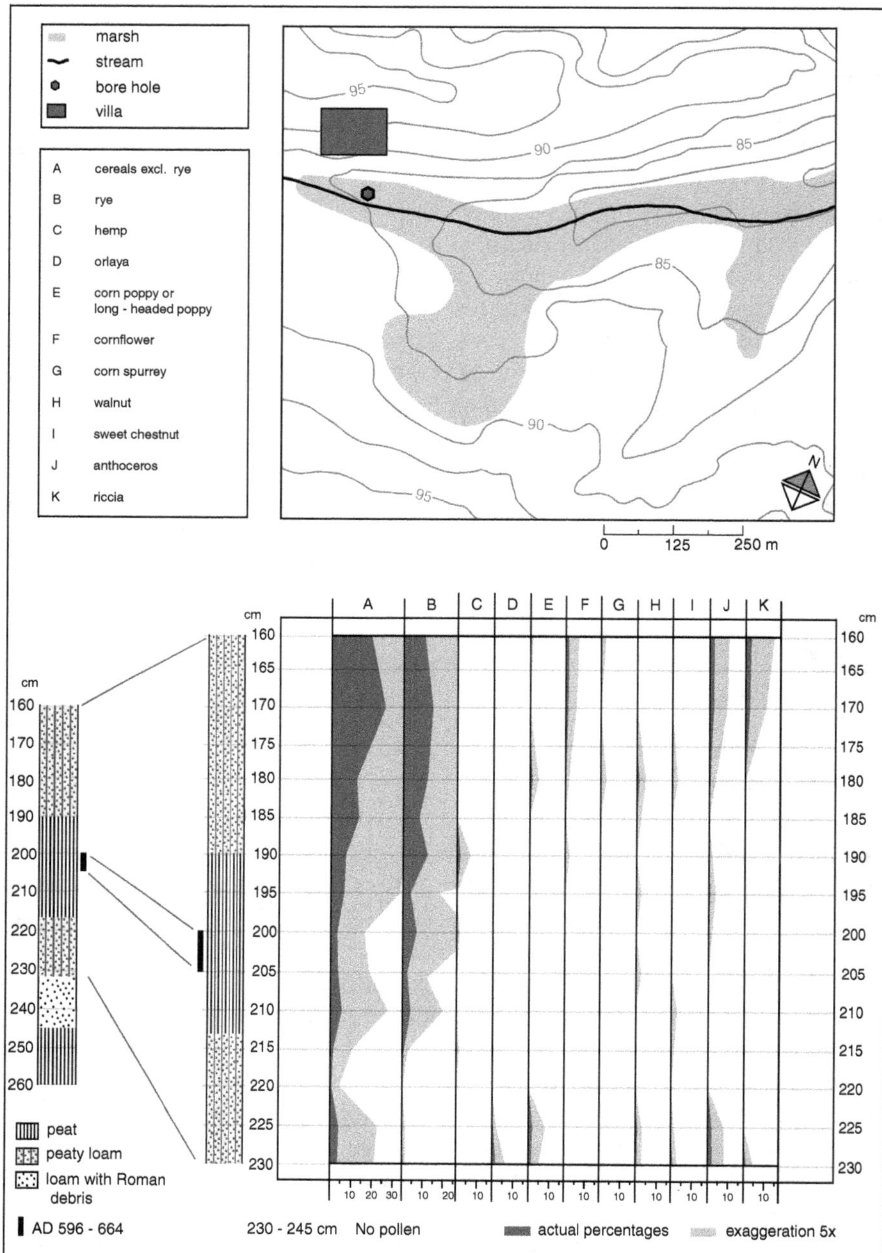

Fig. 10.2 Plan of the marsh at Voerendaal, the Netherlands, and part of the pollen diagram. The legend of the curves in the pollen diagram is placed left of the plan

as far as is known. But the impact of the *villa rustica* is. For instance, a small marsh in the immediate surroundings of the *villa rustica* at Voerendaal (the Netherlands) became covered by a 30 cm thick layer of loam, and this loam could be dated to the Roman Period. After the heyday of this villa, the sedimentation of loam stopped, to be followed by peat growth. Pollen analysis showed that farming activities with crop growing did not stop during the temporary peat formation (Fig. 10.2). Obviously farmers were present, but their way of farming, or the scale of their farming, did not lead to erosion.

From AD 700 onwards sedimentation of loam recommenced, never to stop again.

Chapter 11
Summing Up Six Millennia of Agriculture

For slightly over six thousand years people lived in what would today be called the countryside. This period started around 5300 BC and lasted until about AD 1000. People lived on the land and also directly off the land. Of course, people before 5300 BC did the same, but the great difference is that after this date food was produced, not obtained by harvesting nature. Why people started to produce their food remains obscure. It was not based on a gradual internal process. The new kind of food economy came from the outside and the main influx of ideas came from the east, from beyond the river Rhine, though there are some indications that the south and/or south-west played a minor role. Whatever the nearer sources, the ultimate source of the food-producing way of life was the Near East.

The rural village, which characterised the countryside in historical times, has never been a reality during the long stretch of time considered in this book. Only at the very end of the period did a kind of proto-village appear. Even aggregations of farmhouses which could be called hamlets, are almost absent. Small groups of houses occur during the first millennium of the period, and possibly the second as well, to disappear and reappear again quite late in early medieval times. In between, clusters of houses seem in most instances to represent something special, not just agricultural (exceptions are to be found in the last centuries BC in the north-eastern part of the region). They functioned as sacral, economical and/or political nodes. Their number is highest during the Roman occupation, but even then the common farm was on its own, and farms lay scattered over the landscape (Fig. 11.1).

The reason why members of individual households chose to live in small clusters or preferred to keep their distance should be looked for in the social side of their lives, a subject not dealt with at length in this book. Analysis of environmental constraints would not offer an explanation.

The first farmers chose to stay together in mini-aggregations of several households, possibly because they were pioneers in foreign parts. These hamlets were, in their turn, part of larger clusters, which occupied well-defined parts of the landscape. The obvious reason is that they could not, at least not socially, survive on their own. Some of these clusters grew into important units, whilst others remained quite small. The distance between the clusters varies, but amounts generally to some tens of kilometres. However, real outlyers do occur. An example is the cluster situated around

C.C. Bakels, *The Western European Loess Belt*, DOI 10.1007/978-1-4020-9840-6_11,

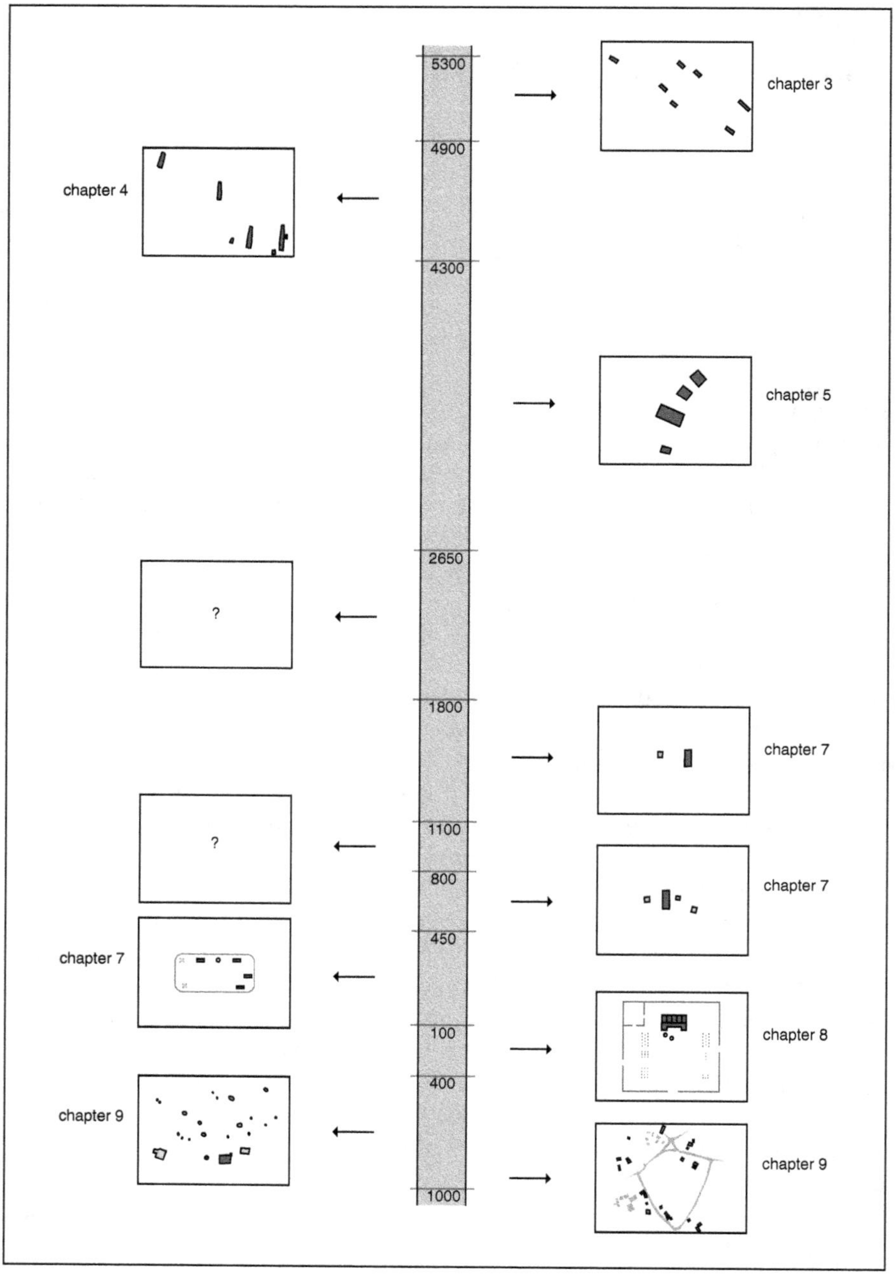

Fig. 11.1 Clustering and non-clustering of farms in the landscape through time

the upper course of the river Dendre in the Belgian Hainaut: it lies 100 km from the neighbour with which it had, according to the archaeological records, for instance the provenance of stone implements, its most frequent contacts. The 100 km gap cannot be attributed to an absence of research in the intermediate space.

As farming the land became more common, the empty spaces between most of the early pockets of rural settlement were filled in. The physical clustering at the level of the farm disappeared. Nevertheless, not all of the empty space became evenly covered with dwellings. It is at this level that environmental constraints become visible. The expansion avoids the interior of large plateaus. During the Roman occupation people ventured farthest 'inland'. The reason for staying in or in the vicinity of valleys may have been twofold. The most obvious reason is that water is indispensable to life. Though the art of digging wells, and even deep ones, was known from the beginning, wells are scarce in the archaeological records. A proper well is hard to dig and maintain, and in the centre of the large loess plateaus the water table lies very deep. Perhaps digging wells was as much avoided as possible. The construction of basins to catch rainwater has never been envisaged, as no structures of this kind have been detected.

The second reason may have been that during most of the time farming was a broad-spectrum affair. People grew several kinds of crops and kept several kinds of animals. An environment with some diversity is then preferable to a uniform landscape. River valleys, in combination with slopes and adjacent edges of plateaus offer such a varied landscape.

The gradual shift from clusters of rural dwellings to a more even distribution of occupation did not result in a situation in which farms ceased to function as part of a local network. A single farm, and a single household, cannot survive on its own. Modest numbers of individual farms continued to share interests. The expression of such units is often found in sharing a special place with a special function. This place could take the form of a large (ceremonial?) enclosure or a collective place of burial. In the centuries preceding the Roman occupation, when rural society started to display different levels of wealth and a certain kind of social stratification, individual farms became oriented towards a central place, called *oppidum*. During Roman rule the farm was part of a vast economic and political system. The situation following the collapse of this rule is not very clear, but in the end the individual farm became part of a system with several, or many, small farms dependent on a main farm. It is also the time during which clustering of farms reappeared.

During none of the millennia studied did farming communities live cut off from the outside world. The first farmers were part of a network with links to their original country of origin in Central Europe, as is proven by the provenance of some of their raw matials. Later on links with other regions remained important. Central Europe continued to play the main role, but around 1500 BC the influence of England can be observed in the north-western part of the region. Influence from the south is present too, as is seen first and foremost in the steady northwards creeping influence of the Mediterranean world in the last part of the first millennium BC. All the time the outside world brought new things, new techniques, new habits, new beliefs and presumably some new people as well. The new elements came on top of the established ones and became absorbed into daily life. Not every cultural change brought about a change in farming customs. The shift in burial customs between 1100 BC and 800 BC, for instance, brought no detectable changes in what people did on the farm.

The largest impact came from the Roman conquest. If the introduction of food production may be called the first revolution of its kind in the region, the introduction of Roman habits can be called the second. The number of crops to choose from rose from the beginning of agriculture onwards, but at no time did such a surge of new crops become available as during the Roman Period (Fig. 11.2). Moreover, a certain class of new plants required a new kind of care. The trees and vines introduced by the Romans were perennial, woody species. Prior to this, crop plants were annual herbs. With the new crops, new agricultural methods were introduced, such as grafting. And a new kind of cultured plots came into existence: the orchard and the vineyard. The new plants comprised, too, a range of kitchen herbs and vegetables. Though growing plants on garden-like plots must have been millennia-old, it is quite possible that the clear distinction between the vegetable garden and the field came into existence just then, though actual proof is lacking.

The list of animals kept on farms grew longer as well, though not as strongly as seen in the crops. The enrichment consisted of new breeds of the customary cattle, pigs, sheep/goats and horses, whilst mules, donkeys, geese, ducks, and pigeons were newcomers in the livestock (Fig. 11.3). With the introduction of new breeds in the decades before the arrival of the Romans and the following centuries the steady decline of the shoulder heights of especially cattle and pigs came to an end. This introduction may have been triggered by a new attitude to status. Whereas in the

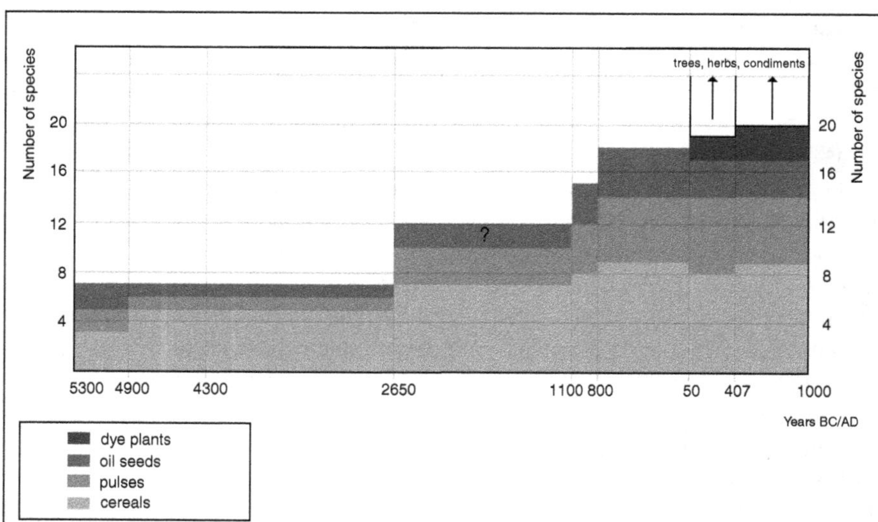

Fig. 11.2 Increase in the number of crop plants through time. The large number of plants added after AD 50 is represented by arrows indicating soaring numbers

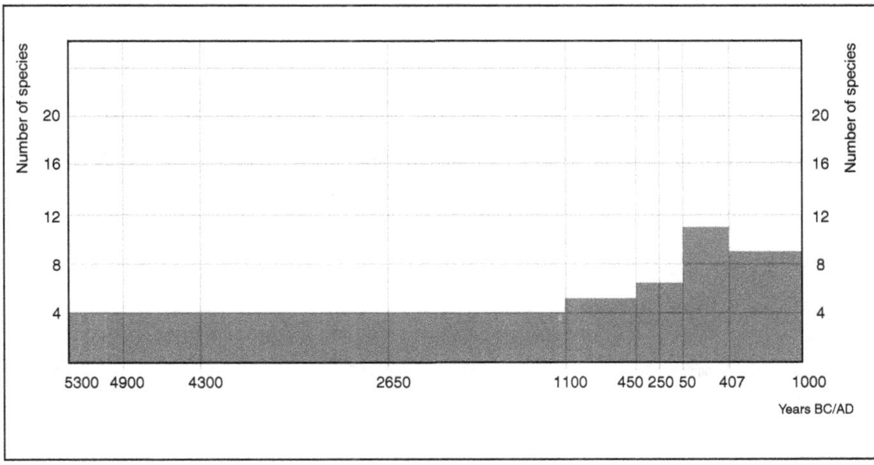

Fig. 11.3 Increase in livestock species

preceding periods the size of the herd was presumably more important than the size and appearance of individual animals, sentiments may have changed. As far as is known, sheep did not follow this trend, whilst the dataset on goats and horses is too small to allow conclusions (Fig. 11.4).

As mentioned above, the actual Roman conquest was preceded by a regular influx of Mediterranean goods in the southern part of the region. This influx included both plants and animals and this weakens the definition of 'revolution'. However, the influx had not yet reached the northern part of the region. There, the new sets of plants and animals were introduced and adopted within the span of one century.

The large stretch of time between the first and second revolution went, however, not without developments with an impact on agriculture. These were more of a technical nature. A principal change was presumably brought about by the introduction of the wheel. Transport of bulk goods and heavy loads must have become much easier. However, its impact on rural life, at least in the beginning, is difficult to detect. Harvests may have been carted to the farmyards, but this action is difficult to prove. Manure may have been carted to the fields, but even the manuring itself is hard to prove. The introduction of animal traction may have been even more important. If oxen are available to draw an ard, ploughing becomes definitely easier. The solid wooden wheel, ard and traction arrived sometime between 4300 BC and 2650 BC. They remained important for farmwork during the whole period considered in this book. The spoked wheel and the horse were for a long time part of another part of social life. Only with the arrival of the Romans did the spoked wheel and then not the horse but the mule, seem to start having a role in common transport. The horse became important only after the period considered in this book.

Another 'event' was the replacement by metal of flint and other kinds of stone. The advent of iron had probably a larger impact than the advent of bronze. Only the bronze axe developed into a common tool. Possibly sickles made of bronze

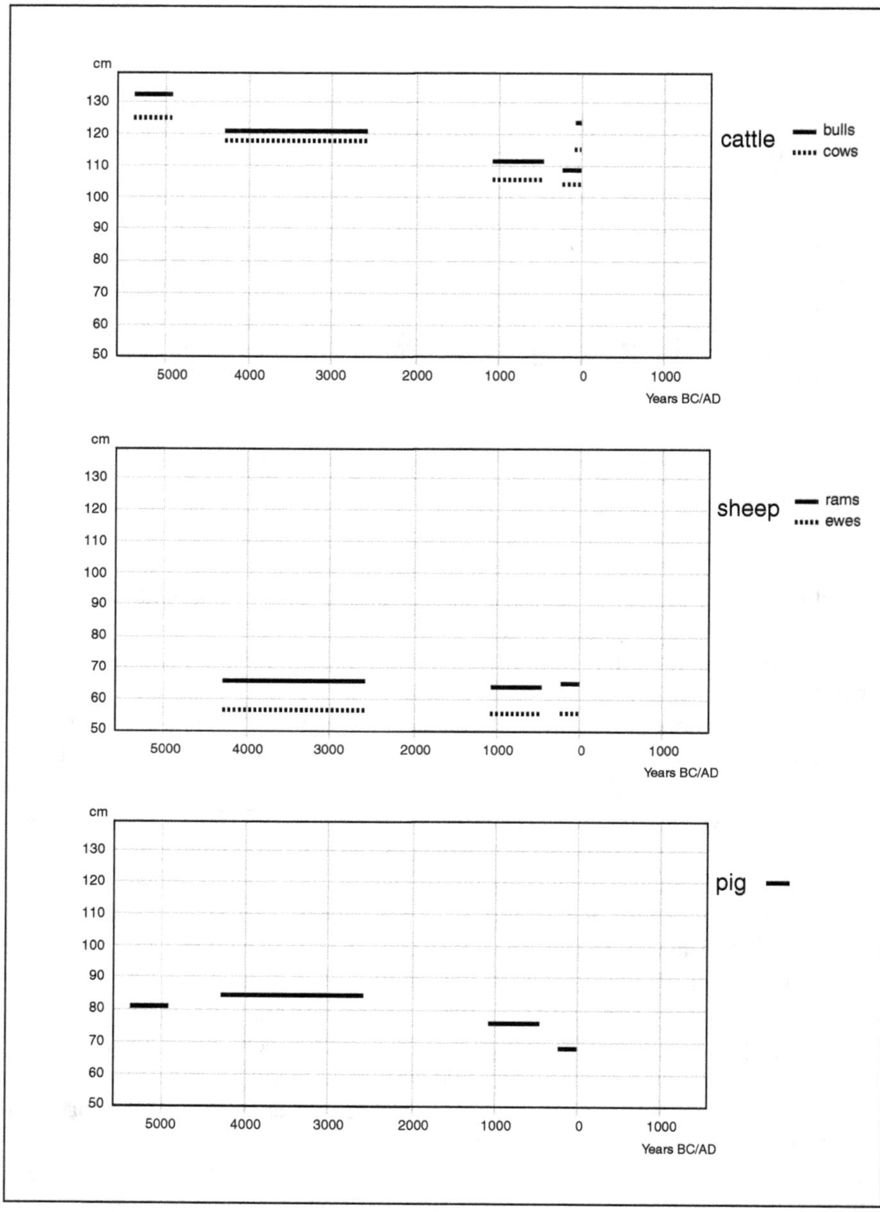

Fig. 11.4 Shoulder height through time. The graphs end around AD 0, because after that date the introduction of new breeds blurs the picture. A first indication is presented in the graph for cattle, showing the introduction of large breeds already before the Roman conquest

became common implements too. Iron, however, was used for a wider range of tools, including tips of ards and scythes. The scythe was used for mowing, not for harvesting crops. Such an implement facilitated hay-making. From 450 BC onwards hay is noticed in the archaeological records. This kind of feed made it easier to get the most valuable part of the livestock, such as draught-oxen and luxury horses through the winter, though other kinds of feed must have helped as well. The introduction of crop plants like oat and vetches may possibly be associated with the same development.

The period after the Roman occupation knew technological innovations as well. New are the true plough and the watermill. The plough gradually replaced the ard. Heavy soils could be worked with this tool. It was still drawn by oxen. Milling with the aid of running water replaced the manual work. During the Roman Period some large mills may have been operated by animal power, and some large farms may have used a watermill, though this is not yet proven for the loess region; prior to the Middle Ages milling was commonly done by humans. Other introductions were the horseshoe and the stirrup, but these were mainly important for other, non-agricultural, aspects of daily life.

To a modern passer-by the farmhouses may have looked more or less the same except for size. For six millennia the building consisted of a wooden frame with walls of wattle and daub, and a roof cover of shingles or thatch (Figs. 11.5, 11.6 and 11.7). Most were rectangular, presumably with gable ends and a gable roof. Others had an arrangement of roof-bearing posts or rounded short sides, suggesting a hipped roof. Combinations occurred as well. Only around 1500 BC were circular houses built and then only in the north-western part facing England, where such round structures were common. A variation on the rectangular house-plan is to be found in the period 4900 BC–4300 BC: the trapezoidal house. The frontal short side of this house is much wider than the rear side. As such buildings had gabled roofs, presumably sloping backwards, their facades must have been most impressive. They may even have been decorated. Fragments of plaster of such a house found in Central Europe revealed traces of an abstract red decoration on a white background. Such houses serve as a reminder that what an archaeologist can reconstruct on the basis of a ground-plan may result in a too simple, rustic image.

The arrangement of the roof-bearing posts indicates that almost all houses built in wood and wattle and daub had an internal division that was orientated lengthwise. Early houses had four aisles, but in later periods this number was reduced to three, two or even one. Transversal divisions have rarely been found. If there were any compartments present, these must have been realised by shallow-founded or above-ground arrangements. Hearths are also seldom found. They must have been present, but as the actual floor of the buildings is almost never preserved because of later erosion, traces of hearths would have vanished.

As indicated in the preceding chapters, houses of some farming communities may have been different, because constructed following circular ground-plans or with stone foundations; but these are not yet very well known. An important change in architecture came, however, with the Romans, who inspired the creation of a new kind of farmhouse: the *villa rustica*. Any sizable villa had parts built in stone

Fig. 11.5 Houses with a gable roof: life-size reconstructions of a Linearbandkeramik house (**A**) and an Early Medieval house (**B**)

Fig. 11.6 Houses with a hipped roof: life-size reconstructions of a Bronze Age house (**A**) and an Early Medieval house (**B**)

Fig. 11.7 Reconstruction of a Bronze Age settlement with a 'longhouse' and granary (**A**) and the Early Medieval hamlet of Juvencourt-et-Damary (**B**)

and roofs were obviously tiled. The interior was divided into several compartments of which many deserve the name room, and the main building was even provided with a sizable bathroom. The tradition of building in wood and wattle and daub was however preserved in smaller villas, humbler farms and outhouses. Also upper storeys may have been constructed in this way.

The Early Medieval manorial house took the Roman villa as its model, at least partly; but technical niceties such as bathrooms and water ducts were not installed.

As excavations provide ground-plans and normally not much else, the upper structure of the farmhouses remains unknown. Some plans suggest the existence of a second storey, but more detailed information on these upper storeys is sadly lacking. What ground-plans do provide, at least, is an estimation of the size of the groundfloor. The length of the houses shows more variation than the width, and size is mostly determined by length. The *villa rustica* is too different and therefore not considered in the following.

During the first phase, 5300 BC–4900 BC, ground-plans did not in general cover more than 100 m^2 (Fig. 11.8). Between 4900 BC–4300 BC houses tend to be bigger. The next phase shows some exceptionally large buildings, but as most of these houses belong to the, presumably, exceptional site at Mairy, future research may alter the picture. Nevertheless, the phase 2650 BC–1100 BC shows still larger floor-plans. After 1100 BC a change sets in. Houses became much smaller. Their groundfloor seldom covered more than 100 m^2. The reason behind these trends is sought by some authors in an increasing or decreasing number of inhabitants. The largest (i.e. longest) houses have even been described as multi-family houses, that is houses with several households under one roof. In those cases where the width of the structures

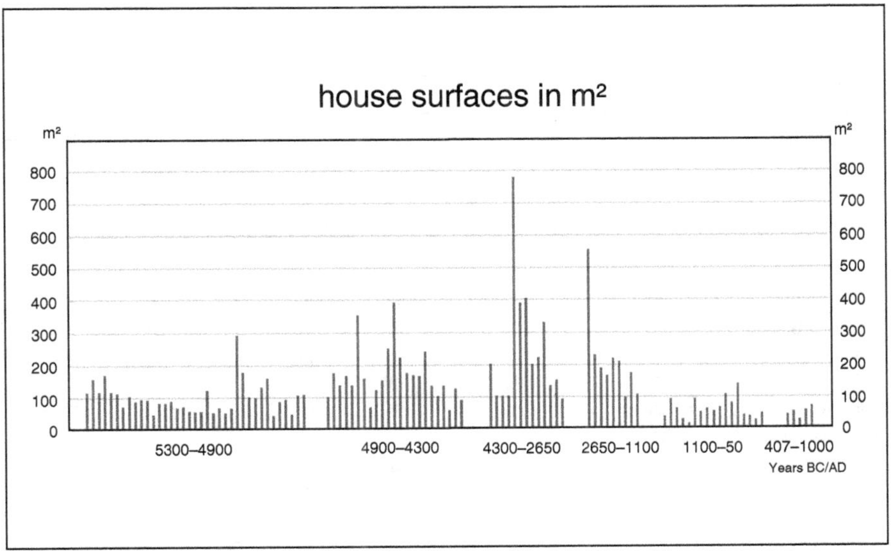

Fig. 11.8 The development of house surfaces through time

is the same from the front to the back, and where the plan suggests some internal division, this could indeed have been the case. The best examples are offered by the Michelsberg culture. But the long 4900 BC–4300 BC houses are trapezoidal in plan and just an enlarged version of the smaller houses at the same sites. It is difficult to see in the small houses a single household dwelling and in the large ones a multi-family home. Differences in status and wealth offer a better explanation for the differences in size, whilst truly exceptional sizes may have served communal purposes, but such structures are rare in the loess region.

Part of the houses may not have served for living in. This is at least proposed for the longhouses dated between 1800 BC and 1100 BC. Half of such dwellings were intended to accommodate people, and the other half to house animals (see Section 7.5). It is an explanation suggested by the arrangement of the roof-bearing posts.

With the disappearance of the large farmhouse the outhouse appears. Sometime between 1800 BC and 1100 BC the farm as a cluster of several structures arose. It is tempting to conclude that the single farmhouse had split up into separate units. From that period onwards a yard counts a main house and one or more outhouses, if only just a granary.

The yard displays a definite development through time (Table 11.1). It started as an area with a farmhouse, one or two underground silos and some pits, which ended up filled with waste. After 1800 BC other buildings appeared, first and foremost the four-poster granary. Outhouses larger than four- or six-poster granaries appear as well. The silo remained, though this kind of storage pit has not yet been detected in the southernmost part of the region in the period between 800 BC and 450 BC. Around 450 BC granaries and silos seem to have been moved out of the individual

Table 11.1 Buildings and other structures present in yards through time

Structures in yards	Main house	Other house	Outhouse	Fourp. granary	Silo	Well	Oven, kiln, etc.	Palisade
5300–4900	■	□	□	□	■	□	■/□	■/□
4900–4300	■	□	rare in n-w	□	■	□	?	?
4300–2650	supp.	□	?	?	■	?	?	?
2650–1800	?	?	?	?	?	?	?	?
1800–1100	■	□	■	■	■	□	rare	?
1100–800	■	□	?	■	■	□	rare	?
800–450	■	□	■	■	■	□	■/□	■
450–50 general	■	□	■	■	■	□	■	□
450–50 class 1–3 in s	■	■	■	■	■	□	■	■
450–50 class 4 in s	■	□	□	□	■/□	□	□	□
50–400	■	□	■	■	□	■	■	■
400–1000 general	■	■/□	■	■/□	■	shared	■	■
400–1000 seigneurial	■	■ ?	■	?	■/□	■	?	■

Black squares mean present, open squares mean absent; s is the southern part of the loess region, n-w means the north-western part of the region; supp. is supposed, fourp. is fourposter yard to a communal area, at least in the southern part of the region, but this was a temporary phenomenon.

From 450 BC onwards the yards in northern France reveal the existence of social diversity. This is expressed in their size and layout. The smallest establishments had only a main house and sometimes a silo. The large one counted one main farmhouse, some other houses and a range of outhouses and other structures.

What has not been mentioned so far as part of the yards' inventories is the outdoor oven. Traces of ovens are occasionally met with and from the earliest period onwards. But as their remains consist in most cases of crumbled pieces of fired loam, whether in their original position or not, ovens are probably underrepresented in the records. They were used for processing and preparing cereals and other kinds of food plants, for malting, for firing pottery, for bronze or iron melting, etc. What has not yet been mentioned either is the well. Wells were not part of the farmyard before the Roman Period. They were at all times rare and served, if present, a larger community.

The Roman *villa rustica* had a well. Their yards, even when of moderate size, comprised several outhouses destined for livestock, storing crops and practising crafts, but two features are absent: the underground silo and the four-poster granary. The storage of cereals appears to have been organised on a larger scale than before.

The silo reappears after the Roman Period and is therefore the most consistent element on the yard in the entire period between 5300 BC and AD 1000. The small granary reappears as well. The well remains, though in the case of small establishments the well may have been shared by several farms.

The name 'yard' implies a boundary. Traces of ditches and/or palisades suggesting a fence of some sort around the individual yard have been vaguely recognised in some settlements of the earliest farmers. Whether they were also present in the settlements of their successors is not clear. Light palisades are not deeply founded and apt to disappear from the archaeological records. The palisades that have been found are sturdier affairs surrounding complete settlements. The first confirmed boundaries defining yards date from after 800 BC. They appear as ditches. Why the yards are thus fenced off from that time onwards is not quite clear, but one of the explanations may be a change in the handling of livestock. It is the same time at which more traces of animal feed turn up as part of the waste dumped in yards. The *villa rustica* and its yard were surrounded by a clear-cut enclosure. The Early Medieval farms show light fences or hedges around their premises.

So far the loess region has been treated as an unequivocal unit in this chapter. But it is a large region and it would be surprising if nowhere, and never, would a division into subregions have appeared. As a matter of fact some regional differences did already turn up in the descriptions offered above.

During the period 5300 BC–4900 BC there are regional differences in the growing of barley, the use of the adze, the presence of houses with walls made entirely of wood and the choice of terrain to settle on. Barley (naked barley) has only been detected so far in the rather isolated clusters of the Kleine Gete and the Hainaut in Belgium, and in the Aisne valley in France. Adze blades made of rock are extremely scarce in the Paris Basin, which includes the Aisne valley. Houses with solid wooden walls are a speciality of the German Rhineland and adjacent Southern Limburg in the Netherlands. Concerning the place of settlement: in northern France people lived and farmed on higher places on the bottom of valleys. Elsewhere people lived on the edges of plateaus or on the upper part of valley slopes. A fourth regional difference is reported concerning livestock. Almost everywhere in northern France did sheep/goats come second after cattle, but in the lower Alsace pigs came second. Due to the lack of preserved animal bones elsewhere, it is difficult to assess the role of pigs in the other parts of the loess region. If the differences are taken together, it turns out that they show a regional overlap. It is difficult to point out subregions that differ in almost every aspect from other subregions.

During the period 4900 BC–4300 BC it is again barley which presents a difference. Farmers, belonging to the Villeneuve-St. Germain and Cerny cultures of the south-western part of northern France started to grow hulled barley, whilst all others grew only naked barley. The German Rhineland stands out, because in the Rössen cultural tradition all houses were built in wood, whilst the houses in the remainder of the region had no wooden walls at all. Moreover, the most marked trapezoidal houses are found in this same culture. And it looks as if some farms had outhouses. The difference in choice of a place to settle was the same as in the previous period.

The next period, 4300 BC–2650 BC, reveals that the growing of hulled barley cannot be used to designate culturally differing subregions, because this cereal has now reached the eastern parts of the Paris Basin. It is presumably a matter of slow introduction. The farmers in the remainder of the loess region still cultivated naked barley. More information on possible differences cannot be offered, because of lack of a sufficient amount of data.

The same is true for the first part of the following period, 2650 BC–50 BC. Until 1800 BC information is too hazy to allow views on the matter. Between 1800 BC and 1100 BC the existence of subregions is, however, beyond doubt. An eastern subregion, comprising the Rhineland, Southern Limburg, French Lorraine and the Moselle area, is characterised by houses with an internal division into a living area and a part with cattle boxes. In the west, near the Atlantic coast, there is a tradition of round buildings. What was in between is still largely unknown.

According to several cultural aspects the eastern subregion did in the end fall into two parts, which resulted in two separate eastern subregions during the period 1100 BC–800 BC. Whether, and how, the populations of the subregions differed in their farming practices is still unknown, but the presence or absence of cattle boxes suggests at least a difference in the handling of livestock. Another difference, now between the south and the north, is the growing of the pulses lentil and bitter vetch. But this difference is easily explained by a difference in climate.

Between 800 BC and 50 BC the large variation in the architecture of the farmhouses makes it difficult to detect possible subregions. It is perhaps too early for this. Still, there are already two phenomena that stand out. One is the clustering of granaries and silos around 450 BC, which is reported only from northern France. The other is the occurrence of classes in farms, also in northern France.

During the Roman occupation the region behaved as one unit. Some difference is observed in the cultivation of wheat, spelt wheat being the preferred crop in the northern part and in the surroundings of Amiens, whilst the remainder of France grew bread wheat. Also, in the west, near the Channel, the numbers of sheep/goats in the livestock exceed those of pigs, whilst elsewhere it was the other way round. In both cases they come after cattle. This difference persists in the following, Early Medieval period, AD 407–1000. In other aspects this period shows no clear-cut regional differences.

All in all, as far as our present knowledge goes, most diversity seems to have occurred between 2650 BC and 800 BC, but this is also the period least well known. Future research may bring a clearer view. In the long run, and seen on a European scale, the loess region behaves as one agricultural-cultural unit.

During the entire period the region had connections with the world beyond its boundaries. The part of Europe east of the river Rhine and north of the Alps was the most important as far as contacts, materials, ideas, and new people are concerned. The south comes second. Contact with southern France and the Mediterranean world has always been present, but gained importance over the millennia. The west, i.e. England, seems to have played a minor role. Of a role of the region north of the loess belt hardly anything is known.

To conclude: I have tried to present a general view. For all that, I could not always avoid the 'mere collection of disconnected card-index entries' so deplored by B.H. Slicher van Bath in his *The Agrarian History of Western Europe AD 500–1850*, Chapter 1. Too much is still unknown. The worst of all is the period between 2650 BC and 1800 BC or even 3000 BC and 1100 BC. Millennia without proper data saddle us with a big gap in a story that was intended to be continuous. It is to be hoped that this gap will be closed by future research.

Source of Figures and Tables

Figures

All figures, except for the photographs, have been drawn by Joanne Porck. Some are original but most are redrawn after, adapted from or inspired by existing figures. The source of such existing figures is given below. The provenance of the photographs is given as well.

Fig. 2.1	Photo Leiden University, C.C. Bakels.
Fig. 2.2	Photo BIAX consultancy, Zaandam, the Netherlands.
Fig. 2.3	Photo Leiden University, J. Pauptit.
Fig. 2.4	Photo Leiden University, J. Pauptit/C.C. Bakels.
Fig. 2.6	Photo A: unknown; photo B: Archol, Leiden, the Netherlands; photo C: Leiden University, C.C. Bakels.
Fig. 2.7	Photo A: Leiden University, I. Van der Jagt; photo B: H. Peeters, Amersfoort.
Fig. 2.8	Photo Leiden University, A. Verbaas.
Fig. 2.9	Photo Restaura, Haelen, the Netherlands.
Fig. 2.10	Photo Restaura, Haelen, the Netherlands.
Fig. 2.11	Photo Leiden University, J. Pauptit.
Fig. 2.12	Fig. 2.12 Photo Leiden University, J. Pauptit.
Fig. 3.2	Adze blades from Modderman PJR (1970) Linearbandkeramik aus Elsloo und Stein. Analecta Praehistorica Leidensia 3. Hafting after Weiner J, Pawlik A (1995) Neues zu einer alten Frage, Experimentelle Archäologie Bilanz 1994. Archäologische Mitteilungen aus Nordwestdeutschland Beiheft 8:132.
Fig. 3.3	Weiner J (1994) Bemerkenswerte bandkeramische Fundstücke aus Kückhoven. Archäologie im Rheinland 1993:32–34.
Fig. 3.5	Photo Leiden University, J. Pauptit/C.C. Bakels.
Fig. 3.6	Blade and photo A.L. van Gijn, Leiden University.
Fig. 3.8	After Arbogast R-M, Jeunesse C (1996) Réflexion sur la signification des groupes régionaux du Rubané: l'exemple du Rhin supérieur et du Bassin Parisien. Archäologisches Korrespondenzblatt 26:395–404.
Fig. 3.9	Photo Leiden University, G.J. Verwers.

Fig. 3.10 Ground-plans from Modderman PJR (1970) Linearbandkeramik aus Elsloo und Stein. Analecta Praehistorica Leidensia 3; Ilett M, Constantin C, Coudart A, Demoule JP (1982) the Late Bandkeramik of the Aisne Valley. Analecta Praehistorica Leidensia 15:45–61; Dubouloz J, Farruggia J-P, Ilett M, Robert B (1996) Bâtiments néolithiques non-rubanés à Berry-au-Bac 'Le Vieux-Tordoir', Aisne. Internéo 1–1996.

Fig. 3.11 After Lüning J (and Boelicke U) (1982) Research into the Bandkeramik settlement of the Aldenhovener Platte. Analecta Praehistorica Leidensia 15:1–29.

Fig. 3.12 From Wijk IM van (2002) Een archeologische begeleiding in de bandkeramische nederzetting van Elsloo. ARCHOL Rapport 22.

Fig. 3.13 After Weiner J (1998) Drei Brunnenkästen, aber nur zwei Brunnen: eine neue Hypothese zur Baugeschichte des Brunnen von Erkelenz-Kückhoven. In: Brunnen der Jungsteinzeit, Materialien zur Bodendenkmalpflege 11:95–111.

Fig. 3.14 After Velde P van der (2007) The early Bandkeramik settlement of Geleen-Janskamperveld. Analecta Praehistorica Leidensia 39, p. 385; Hachem L (1994) Structuration spatiale d'un village du Rubané Récent, Cuiry-lès-Chaudardes (Aisne). In: Espaces physiques sociaux dans l'analyse interne des sites du Néolithique à l'Age du Fer. 119e congrès CTHS, Amiens, 245–261.

Fig. 4.1 Photo Leiden University, J. Pauptit/C.C. Bakels.

Fig. 4.2 From Constantin C, Sidera I, Demarez L (1991) Deux sites du Groupe de Blicquy, Anthropologie et Préhistoire 102:29–54.

Fig. 4.3 Ground-plans from Bakels CC (1992) Das Neolithikum. In: Spurensicherung-Archäologische Denkmalpflege in der Euregio Maas-Rhein, Mainz am Rhein, 58–79; Bostyn F (2003) Néolithique ancien en Haute-Normandie: Le village Villeneuve-Saint-Germain de Poses 'Sur la Mare' et les sites de la Boucle du Vaudreuil. Travaux 4 de la Société Préhistorique Française; Dubouloz J, Farruggia J-P, Ilett M, Robert B (1996) Bâtiments néolithiques non-rubanés à Berry-au-Bac 'Le Vieux-Tordoir', Aisne. Internéo 1–1996.

Fig. 4.4 After Lüning J (1982) Siedlung und Siedlungslandschaft in bandkeramische und Rössener Zeit. Offa 39:9–33.

Fig. 4.5 After Bostyn F (2003) Néolithique ancien en Haute-Normandie: Le village Villeneuve-Saint-Germain de Poses 'Sur la Mare' et les sites de la Boucle du Vaudreuil. Travaux 4 de la Société Préhistorique Française.

Fig. 5.1 Inspired by Médard F (2003) Vestiges textiles et activités de filage sur le site néolithique d'Arbon-Bleiche 3 (TG, Suisse). Bulletin de la Société Préhistorique Française 100:375–391.

Fig. 5.2 Blade B from Schreurs J (2005) Het Midden-Neolithicum in Zuid-Nederland. In: Deeben J, Drenth E, Oorsouw M-F van, Verhart L (eds.) De Steentijd van Nederland, 301–332.

Fig. 5.3	After a replica in Hansen H-O (1969) Experimental ploughing with a Døstrup ard replica. Tools and Tillage 1:67–92.
Fig. 5.4	Photo Jos Fielmich.
Fig. 5.6	After Arbogast R-M (1994) Premiers élevages néolithiques du nord-est de la France. Études et Recherches Archéologiques de l'Université de Liège, Fig. 67.
Fig. 5.7	Inspired by Sherratt A (2006) La traction animale et la transformation de l'Europe néolithique. In: Pétrequin P, Arbogast R-M, Pétrequin A-M, van Willigen S van Bailly M (eds.) Premiers chariots, premiers araires. CRA monographies 29:329–358.
Fig. 5.10	Ground-plans after Dubouloz J, Martinez R, Mordant D (1989) 3789 avant J.-C. en Bassin parisien, Nemours, Fig. 49; Marolle C (1989) Le village Michelsberg des Hautes Chanvières à Mairy (Ardennes), Gallia Préhistoire 31:93–158, Fig. 6.
Fig. 5.11	After Dubouloz J, Martinez R, Mordant D (1989) op cit. 3789 avant J.-C. en Bassin parisien. Nemours: Fig. 49.
Fig. 5.12	After Marolle C (1998) Le site Michelsberg des 'Hautes Chanvières' avec bâtiments et enceinte à Mairy, Ardennes, France. In: Schlichterle H, Biel J, Strobel M (eds.) Die Michelsberger Kultur und ihre Randgebiete. Stuttgart, 21–28.
Fig. 5.14	Slightly modified after Dubouloz J, Mordant D, Prestreau M (1991) Les enceintes néolithiques du Bassin parisien. In: 'Identité du Chasséen, Actes du Colloque International de Nemours 1989. Mémoires du Musée de Préhistoire de l'Ile-de-France 4:211–119, Fig. 5.
Fig. 6.1	Graph by Prof. Dr. W.H. Zagwijn.
Fig. 6.4	After Damblon F, Buydens C, Hauzeur A (2001–2002) Analyse anthracologique des occupations néolithiques du site d' Altwies-'Op dem Boesch' (Grand-Duché de Luxembourg). Bulletin de la Société Préhistorique Luxembourgoise 23–24:181–207.
Fig. 7.2	Blade B from Bostyn F, Blancquaert G , Lanchon Y (1992) Un enclose triple du Bronze ancient à Fréthun (Pas-de-Calais). Bulletin de la Société Préhistorique Française 89:393–412.
Fig. 7.3	Hoe blade from Malrain F, Matterne V, Méniel P (2002) Les paysans gaulois (III siècle – 52 av. J.-C.), Errance, Paris p. 48.
Fig. 7.4	Data from Hingh AE de (2000) Food production and food procurement in the Bronze Age and Early Iron Age (2000–500 BC). Archaeological Studies Leiden University 7.
Fig. 7.5	Blade A from Fontijn DR (2001–2002) Sacrificial landscapes. Analecta Praehistorica Leidensia 33/34, p. 180.
Fig. 7.6	Blades A and C from Malrain F, Matterne V, Méniel P (2002) op cit. p. 76.
Fig. 7.7	Based on tombstones found in Buzenol and Arlon in Belgium, reunited as suggested by C. Massart.
Fig. 7.9	From Gransar F (2000) Le stockage alimentaire sur les établissements ruraux de l'Âge du Fer en France septentrionale: complémentarité des

structures et tendances évolutives. In: Marion S, Blanquaert G (eds.), Les installations agricoles de l'Âge du Fer en France septentrionale. École Normale Supérieure, Paris, 277–297.

Fig. 7.10 From Gransar F (2000) op cit.

Fig. 7.11 After Gransar F (2000) op cit.

Fig. 7.12 After Joachim H-E (1985) Zu eisenzeitlichen Reibsteinen aus Basaltlava, den sog. Napoleonshüten. Archäologisches Korrespondenzblatt 15:359–369.

Fig. 7.14 With consent of the Landesmuseum Stuttgart, where the original is to be found.

Fig. 7.15 After Buchez N, Gransar F, Matterne V, Pernaud J-M, Yvinec J-H (2002) L'habitat de La Tène ancienne sur la Z.A.C. Centre-ville de Bussy-Saint-Georges (Seine-et-Marne) – 2e partie. Revue Archéologique du Centre de la France 41:35–55.

Fig. 7.16 Left part from Malrain F, Matterne V and Méniel P (2002) op cit. p. 178.

Fig. 7.17 From Fichtl S (1994) Les Gaulois du Nord de la Gaule (150–20 av. J.-C.), Errance, Paris, p. 78.

Fig. 7.18 Ground-plans A and B from Hingh AE de (2000) op cit.; plan C from Koenig M-P (2002) Le gisement protohistorique de Rosières-aux-Salines (Meurthe-et-Moselle). Actes des congrès nationaux des sociétés historiques et scientifiques 127:91–147.

Fig. 7.19 a After Hingh AE de (2000) op cit; b After Koenig M-P (2002) op cit.

Fig. 7.21 From Desfossés Y, Martial E, Vallin L (1992) Le site d'habitat du Bronze moyen du 'Château d'eau'à Roeux (Pas-de-Calais). Bulletin de la Société Préhistorique Française 89:343–391.

Fig. 7.22 From Hingh AE de (2000) op cit.

Fig. 7.23 After Gouge P and Seguier J-M (1994) L'habitat rural de l'Âge du Fer en Bassée et la confluence Seine-Yonne (Seine-et-Marne): un état des recherches. In: Les installations agricoles de l'Âge du Fer en Ile-de-France. École Normale Supérieure, Paris, 45–69.

Fig. 7.24 After Kranendonk P (1992) Einige metallzeitliche Baubefunde im Umsiedlungsstandort Altdorf. Archäologie im Rheinland 1992:38–41.

Fig. 7.25 From Auxiette G, Brun P, Gransar F, Hénon B, Naze Y, Pommepuy C, Robert B (1994) Bucy-Le-Long 'Le Grand Marais'. Les Fouilles protohistoriques dans la Vallée de l'Aisne 22:163–194.

Fig. 7.26 Obtained from F. Gransar.

Fig. 7.27 From Malrain F, Matterne V and Méniel P (2002) op cit. p. 153.

Fig. 7.28 After Malrain F, Matterne V, Méniel P (2002) op cit. p. 143.

Fig. 7.29 After Malrain F, Matterne V, Méniel P (2002) op cit. p. 193.

Fig. 7.30 After Blancquaert G, Desfossés Y (1992) L'occupation du Bronze final du 'Chemin des Vaches' à Fresnes-lès-Montauban (Pas-de-Calais). Bulletin de la Société Préhistorique Française 89:429–438.

Fig. 7.33 After Brun P (1994) From the Hallstatt to La Tène Period in the perspective of the Mediterranean World Economy. In: Kristiansen K,

Jensen J (eds.) Europe in the First Millennium BC, Collis, Sheffield, 57–65.

Fig. 7.34 Photos private collection and Rijksmuseum van Oudheden, Leiden, the Netherlands.

Fig. 7.35 Photo R. Agache – SRA Picardie.

Fig. 8.2 Inspired by material sent by J. Jacomet.

Fig. 8.5 Photo H. Thörnig, drawing J. Dewald, object in Rheinisches Landesmuseum Trier.

Fig. 8.6 Provided by V. Matterne.

Fig. 8.7 Photo W.J. Kuijper.

Fig. 8.8 Composed of photos Rheinisches Landesmuseum Bonn, copyright LVR Bonn.

Fig. 8.10 From Koehler A (2003) Vergers antiques dans les campagnes périurbaines: le cas de Reims. Revue Archéologique de Picardie 2003 (1–2):37–45.

Fig. 8.11 Photo RMN/copyright Jean Schormans; mosaic in Musée d'Archéologie Nationale, Saint-Germain-en-Laye, France.

Fig. 8.12 Partly after Gilles KJ (1991) Eine weitere römische Weinkelter aus Brauneberg. Funde und Ausgrabungen im Bezirk Trier 23:20–32.

Fig. 8.13 After Toupet C, Lemaître P (2003) Vignobles et exploitations viticoles antiques dans le Nord de la Gaule, l'exemple de Bruyères-sur-Oise. Revue Archéologique de Picardie 2003 (1–2):209–226.

Fig. 8.14 After Gilles KJ (1991) op cit.

Fig. 8.15 Photo and copyright Römisches Museum Augsburg.

Fig. 8.16 Photo and copyright Rheinisches Landesmuseum Trier.

Fig. 8.17 Photo Institut Archéologique de Luxembourg, Arlon, Belgium, copyright Institut Archéologique du Luxembourg Arlon.

Fig. 8.18 After King A (1999) Diet in the Roman world: a regional inter-site comparison of the mammal bones. Journal of Roman Archaeology 12:168–202.

Fig. 8.19 Partly after Ferdière A, Malrain F, Matterne V, Méniel P, Nissen Jaubert A (2006). Histoire de l'Agriculture en Gaule 500 av. J.-C.–1000 apr. J.-C., Errance, Paris, p. 126.

Fig. 8.20 After Kooistra LI (1996). Borderland farming, possibilities and limitations of farming in the Roman Period and Early Middle Ages between the Rhine and the Meuse, Van Gorcum, Assen, pp. 131–133.

Fig. 8.24 Plan provided by the Cercle archéologique Hesbaye-Condroz (C.A.H. C.-Amay).

Fig. 8.25 After Vanvinckeroye W (1975) Tongeren Romeinse stad. Provinciaal Gallo-Romeins Museum, Tongeren.

Fig. 8.26 After Bayard D, Massy J-L (1984) Le développement d'Amiens romain du Ier s. av. J.C. au IVe s. ap. J.C.. Revue Archéologique de Picardie (3–4):89–112.

Fig. 8.27 Photo P. Quaedflieg and P. Rutten of a model in The Museum of Natural History Maastricht.

Fig. 9.2 Photo Leiden University J. Pauptit/C.C. Bakels.

Fig. 9.5 Photo Leiden University, C.C. Bakels.

Fig. 9.7 Photo Détail de la tapissserie de Bayeux – 11th century, avec autorisation spéciale de la Ville de Bayeux.

Fig. 9.8 Photo University Library Utrecht, the Netherlands.

Fig. 9.9 From J.-H. Yvinec at www.museedestempsbarbares.fr

Fig. 9.10 From Thouvenot S, Yvinec J-H (2006). Les habitats laténiens, gallo-romains et médiévaux de Varesnes 'Le bois du Lombril' (Oise). Revue Archéologique de Picardie 2006 (1–2):71–126.

Fig. 9.11 Photo University Library Utrecht, the Netherlands.

Fig. 9.12 Photo RMN/ copyright Jean-Gilles Berizzi; statue in Musée du Louvre, Paris.

Fig. 9.13 From Nice A (1994). L'habitat mérovingien de Goudelancourt-lès-Pierrepont (Aisne). Revue Archéologique de Picardie 1994 (1–2): 21–63.

Fig. 9.15 Adapted from Kooistra LI (1996) op cit., p. 134.

Fig. 9.16 Ground-plan from and reconstruction adapted from Nice A (1994) op cit.

Fig. 9.17 After Nice A (1994) op cit.

Fig. 9.18 After a plan drawn and provided by F. Gentili (INRAP).

Fig. 9.19 After a plan drawn and provided by F. Gentili (INRAP).

Fig. 10.1 From Bernard V (1998). L'homme, le bois et la fôret dans la France du Nord entre le Mésolithique et le Haut Moyen-Age. BAR International Series 733.

Fig. 11.5 Photo A: D. Olthof, locality Alphen a/d Rijn, the Netherlands; photo B: C.C. Bakels, locality Marle, France.

Fig. 11.6 Photo A: H. Splinter, locality Alphen a/d Rijn, the Netherlands; photo B: C.C. Bakels, locality Marle, France.

Fig. 11.7 Photo E. Mols, locality Wilhelminaoord, the Netherlands; photo B: C.C. Bakels, locality Marle, France

Tables

Some tables are based on original work done by the author. The remainder is partly or entirely based on work published by others; sources are mentioned below.

Table 3.1 The data from the German Rhineland incorporated in this table come from Knörzer K-H (1971), Urgeschichtliche Unkräuter im Rheinland, ein Beitrag zur Entstehungsgeschichte der Segetalgesellschaften. Vegetatio 23:89–111.

Table 3.2 From Bakels CC (1991) Tracing crop processing in the Bandkeramik Culture. In: Renfrew J (ed.) New Light on Early Farming, Edinburgh University Press, Edinburgh, 281–288.

Table 4.1 Partly based on Knörzer K-H (1971) op cit.

Table 7.1 Partly based on Hingh AE de (2000) Food production and food procurement in the Bronze Age and Early Iron Age (2000–400 BC). Archaeological Studies Leiden University 7, Leiden; Matterne V (2001) Agriculture et alimentation végétale durant l'âge du Fer et l'époque gallo-romaine en France septentrionale, Monique Mergoil, Montagnac; Roymans N (1985) Carbonized grain from two Iron Age storage pits at Neerharen-Rekem. Archaeologia Belgica 1:97–105.

Table 7.2 as in Table 7.1.

Table 7.3 as in Table 7.1.

Table 7.4 as in Table 7.1.

Table 8.1 Table 8.1 Based on Bakels C, Jacomet S (2003) Access to luxury foods in Central Europe during the Roman period: the archaeobotanical evidence. World Archaeology 34 (3):542–557.

Table 8.2 Data from Zeist W van (1970) Prehistoric and Early Historic food plants in the Netherlands. Palaeohistoria 14:41–173; Pals JP, Hakbijl T (1992) Weed and insect infestation of a grain cargo in a ship at the Roman fort of Laurium in Woerden (Province of Zuid-Holland). Review of Palaeobotany and Palynology 73:287–300; Knörzer K-H (1981). Römerzeitliche Pflanzenfunde aus Xanten. Archaeo-Physika 11:3–176; Vanderhoeven A, Vynckier G, Ervynck A, Cooremans B (1992) Het oudheidkundig bodemonderzoek aan de Kielenstraat te Tongeren (prov. Limburg). Archeologie in Vlaanderen 2:89–146; Matterne V (2001) op cit.; König M (1999) Ein umfangreicher spätantiker Getreidefund aus Trier. Funde und Ausgrabungen im Bezirk Trier 31:87–94.

Table 8.4 From Gilles K-J (1991) Eine weitere römische Weinkelter aus Brauneberg, Funde und Ausgrabungen im Bezirk Trier 23 (= Kurtrierisches Jahrbuch 31):20–32.

Table 9.1 Based on the text-edition by Gareis K (1895) Die Landgüterordnung Kaiser Karls des Großen (Capitulare de villis vel curtis imperii), J. Guttentag, Berlin. English names mostly from Polunin O (1969) Flowers of Europe, Oxford University Press, London.

Table 9.2 From Devroey J-P (1989) Entre Loire et Rhin: les fluctuations du terroir de l'épeautre au Moyen Age. In: Devroey J-P, Van Mol J-J (eds.) L'homme et son terroir, l'épeautre (Triticum spelta) histoire et ethnologie, Dire, Treignes, 89–105.

Table 9.3 From Bakels CC (2005) Crops produced in the southern Netherlands and northern France during the early medieval period: a comparison. Vegetation History and Archaeobotany 14:394–399.

Table 9.4 From Devroey J-P (1989) op cit.

Table 9.6 From Slicher van Bath BH (1963) The agrarian history of western Europe A.D. 500–1850, Arnold, London, p. 67.

Glossary

1 Words and Meanings

Aisle = free space in the longitudinal axis of a house, either between the wall and a row of roof-bearing posts, or between two rows of roof-bearing posts.

Alloy = a substance obtained by dissolving two or more metals or non-metals in the original metal. This result is obtained by mixing and melting. Alloying a metal changes its properties.

Anaerobic = in the absence of oxygen or air.

Annual plant = plant living one year or less.

Articulated = in the correct anatomical order.

Corvée = a day's unpaid labour.

Culm = the hollow stem of grasses, cereals included.

Daub = a sticky material made of a mix of wet clay or heavy loam acting as binder, sand or some other coarse material for bulk, and a fibrous material like chopped straw and chaff for reinforcement. In combination with wattle, daub is a building material for making walls.

Dioecious = having the male and female reproductive organs in separate flowers on separate plants.

Diploid = having a pair of each type of chromosome, i.e. two sets of chromosomes.

Ellenberg indicator values = H. Ellenberg (Göttingen, Germany) has expressed the ecological behaviour of plants by figures representing nine degrees of behaviour with regard to the main environmental factors. Used in this book are his figures in relation to soil acidity (R) and the ammonia or nitrate supply (N). R: 1 only in very acid soils, 3 mostly in acid soils, 5 mostly in weakly acid soils, 7 mostly in neutral soils, 9 only in neutral or alkaline soils. N: 1 only in soils very poor in mineral nitrogen, 3 mostly in poor soils, 5 mostly in intermediate soils, 7 mostly in soils rich in mineral oxygen, 8 nitrogen indicator, 9 only in soils very rich in mineral nitrogen (indicating pollution, manure deposits or similar). Reference: Ellenberg H (1979), Zeigerwerte der Gefäßpflanzen Mitteleuropas. *Scripta Geobotanica* 9.

Fealty = to swear fealty is to pledge allegiance to another person.

Feral = a feral organism is one which has escaped from culture and has returned, partly or entirely, to its wild state.

Fief = a good, for instance land or the rights to rents, given in exchange for loyality.

Gable roof = also known as gabled roof: a roof which slopes down on two sides from the ridge, leaving a triangle on the facade.

Glume = one of a pair of dry membranous bracts at the base of the inflorescence, esp. the spikelet of grasses.

Hexaploid = having six sets of chromosomes (see also diploid). The genetic term is often used in connection with wheat, because the classification of wheats according to the grade of ploidy is easier than the assignation to species (see also tetraploid).

Hipped roof = a roof which slopes down on all four sides instead of ending against the triangle of the short facade (see also gable roof).

Hulled cereals = the kernels of hulled cereals are enclosed by tough husks which are not loosened by simple threshing (see also naked cereals).

Muid = a measure of capacity. Its size varies with the product and the region.

Naked cereals = the kernels of naked cereals are covered by thin and loosely adhering husks; threshing releases the naked kernels, and such cereals are also known as free-threshing cereals (see also hulled cereals).

Perennial plant = plant living for more than two years.

Phytosociology = also called Plant sociology, concerns the study of plant communities (units of vegetation).

Plane = horizontal plane of an excavation.

Pollarding = pollarding is a woodland management method. The stem of trees is cut off c. 2 m above the ground level to encourage the growth of lateral branches. Pollarding produces poles which are not damaged by browsing animals.

Rachis = the main axis or stem of the spike of cereals.

Scion = shoot or twig of a plant used to form a graft.

Shifting cultivation = an agricultural system in which plots of land are cultivated for a restricted number of years until the soil loses fertility and are then abandoned, whilst crop production is shifted to newly prepared plots elsewhere. After abandonment the land is left to be reclaimed by natural vegetation.

Slash-and-burn = the practice of slash-and-burn consists of felling and burning of forests to create fields for agriculture or pasture for livestock. It is often part of shifting cultivation and is often used as a synonym for shifting cultivation.

Spikelet = the ultimate flower cluster unit in the grasses family. This includes the cereals. A spikelet comprises one or more florets enclosed by two glumes.

Spindle whorl = a small disc, usually of pottery or stone, with a hole in the centre for the insertion of the end of a spindle. It acts as a fly-wheel, maintaining the momentum of the spindle during spinning.

Taxon = the name for an organism or a group of organisms.

Tetraploid = having four sets of chromosomes (see also diploid and hexaploid).

Thermophyllous = having leaves only in the warmer part of the year; deciduous.

Travois = an elongated triangular frame used to drag loads, consisting of a platform mounted on two long poles. One end of the poles touches the ground whilst the other is tilted upwards to rest on the shoulders of the human or animal dragger, directly or via a shoulder harness.

Triangular diagram = also called ternary graph, plots three variables which sum up to a constant, in this book 100%. In the below figure, A presents a single case (1) with 60% cattle, sheep/goats 30% and pigs 10%, resulting in one dot; B shows five of such cases (dots) which are united by a grey area. The graphs in the book present only such grey areas; the underlying dots are omitted.

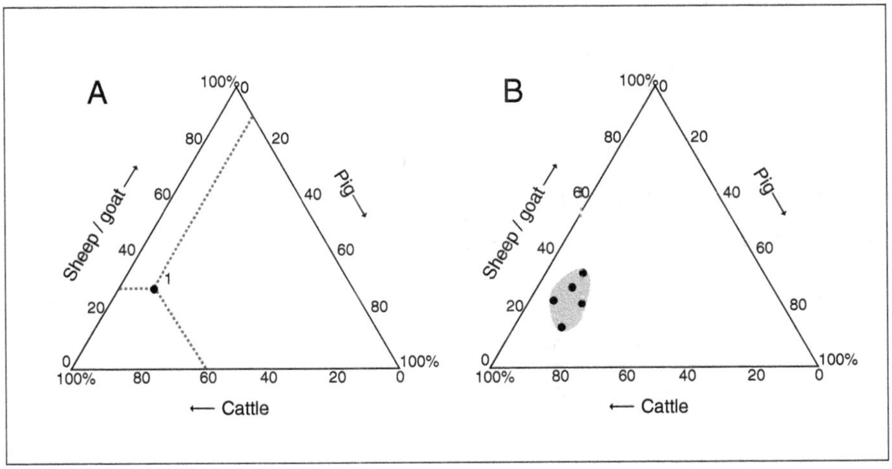

Wattle = a frame made by weaving thin sticks and twigs over thick sticks, used for making fences and walls.

2 Scientific Names of Crop Plants Mentioned in the Text

Almond = *Prunus dulcis* (Mill.) D.A. Webb

Aniseed = *Pimpinella anisum* L.

Apple = *Malus pumila* Mill. (syn. *Malus domestica* Borkh.)

Barley = *Hordeum vulgare* L.

Beet = *Beta vulgaris* L.

Bitter vetch = *Vicia ervilia* (L.) Willd.

Black pepper = *Piper nigrum* L.

Bottle gourd = *Lagenaria siceraria* (Molina) Standl.

Bread wheat = *Triticum aestivum* L.

Broad bean = *Vicia faba* L. var. *maior*

Broomcorn millet = *Panicum miliaceum* L.

Celery = *Apium graveolens* L.

Cherry = *Prunus avium* L. or *Prunus cerasus* L.

Cherryplum = *Prunus cerasifera* Ehrh.

Chickling pea = *Lathyrus sativus* L.

Chickpea = *Cicer arietinum* L.

Common vetch = *Vicia sativa* L.

Coriander = *Coriandrum sativum* L.

Date = *Phoenix dactylifera* L.

Dyer's rocket = *Reseda luteola* L.

Einkorn wheat = *Triticum monococcum* L.

Emmer wheat = *Triticum dicoccum* Schübl.

Fennel = *Foeniculum vulgare* Mill.

Fig = *Ficus carica* L.

Flax = *Linum usitatissimum* L.

Foxtail millet = *Setaria italica* (L.) P. Beauvois

Garlic = *Allium sativum* L.

Gold of pleasure = *Camelina sativa* (L.) Crantz

Grape = *Vitis vinifera* L.

Grass pea = *Lathyrus sativus* L.

Ground cherry = *Prunus fruticosa* Pall.

Hemp = *Cannabis sativa* L.

Horse bean = *Vicia faba* L. var. minor

Hulled barley = *Hordeum vulgare* L.

Laurel = *Laurus nobilis* L.

Lentil = *Lens culinaris* Medik.

Linseed = *Linum usitatissimum* L.

Macaroni wheat = *Triticum durum* Desf.

Madder = *Rubia tinctorum* L.

Medlar = *Mespilus germanica* L.

Melon = *Cucumis melo* L.

Millet = either *Panicum miliaceum* or *Setaria italica*

Mulberry = *Morus nigra* L.

Naked barley = *Hordeum vulgare* L. var. *nudum*

Oat = *Avena sativa* L.

Olive = *Olea europaea* L.

Opium poppy = *Papaver somniferum* L. and its variety *setigerum*

Pea = *Pisum sativum* L.

Peach = *Prunus persica* (L.) Batsch

Pear = *Pyrus communis* L.

Pepper = *Piper nigrum* L.

Pine = *Pinus pinea* L.

Pistachio = *Pistacia vera* L.

Plum = *Prunus domestica* L.

Pomegranate = *Punica granatum* L.

Quince = *Cydonia oblonga* Mill.

Rice = *Oryza sativa* L.

Rivet wheat = *Triticum turgidum* L.

Rye = *Secale cereale* L.

Rye-brome = *Bromus secalinus* L.

Safflower = *Carthamus tinctorius* L.

Service tree = *Sorbus domestica* L.

Sour cherry = *Prunus cerasus* L.

Spelt wheat = *Triticum spelta* L.

Sweet cherry = *Prunus avium* L.

Sweet chestnut = *Castanea sativa* Mill.

Walnut = *Juglans regia* L.

Weld = *Reseda luteola* L.

Woad = *Isatis tinctoria* L.

Bibliography

Chapter 1

Slicher van Bath BH (1963) The Agrarian History of Western Europe A.D. 500 – 1850. Edward Arnold, London

Chapter 3

Arbogast R-M, Jeunesse C (1996) Réflexion sur la signification des groupes régionaux du Rubané: l'exemple du Rhin supérieur et du Bassin Parisien. Archäologisches Korrespondenzblatt 26:395–404

Bakels CC (1978) Four Linearbandkeramik settlements and their environment: a paleoecological study of Sittard, Stein, Elsloo and Hienheim. Analecta Praehistorica Leidensia 11:1–248

Bakels CC (1982) The settlement system of the Dutch Linearbandkeramik. Analecta Praehistorica Leidensia 15:31–43

Bakels CC, Rousselle R (1985) Restes botaniques et agriculture du Néolithique Ancien en Belgique et aux Pays-Bas. Helinium 25:37–57

Bogaard A (2004) Neolithic farming in Central Europe, an archaeobotanical study of crop husbandry practices. Routledge, London and New York

Coudart A (1998) Architecture et société néolithique. Editions de la Maison des Sciences de l'Homme, Paris

Hachem L (1994) Structuration spatiale d'un village du Rubané Récent, Cuiry-lès-Chaudardes (Aisne), analyse d'une catégorie de rejets domestiques: la faune. In: Espaces physiques sociaux dans l'analyse interne des sites du Néolithique à l'Age du Fer, 119e congrès CTHS, Amiens, pp. 245–261

Ilett M, Constantin C, Coudart A, Demoule JP (1982) The Late Bandkeramik of the Aisne valley: environment and spatial organisation. Analecta Praehistorica Leidensia 15:45–61

Knörzer K-H, Gerlach R, Meurers-Balke J, Kalis AJ, Tegtmeier U, Becker WD, Jürgens A (1999) Pflanzenspuren, Archäobotanik im Rheinland: Agrarlandschaft und Nutzpflanzen im Wandel der Zeiten. Rudolf Habelt, Bonn

Lüning J (1988) Frühe Bauern in Mitteleuropa im 6. und 5. Jahrtausend v. Chr. Jahrbuch des Römisch-Germanischen Zentralmuseums Mainz 35:27–93

Lüning J (2000) Steinzeitliche Bauern in Deutschland, die Landwirtschaft im Neolithikum. Universitätsforschungen zur prähistorischen Archäologie 58. Rudolf Habelt, Bonn

Méniel P (1984) Contribution à l'histoire de l'élevage en Picardie, du Néolithique à la fin de l'Age du Fer. Revue Archéologique de Picardie, numéro spécial

Zohary D, Hopf M (2000) Domestication of Plants in the Old World, third edition. University Press, Oxford

Chapter 4

Bakels C (2007) Nature or Culture ? Cereal crops raised by neolithic farmers on Dutch loess soils. In: Colledge S, Conolly J (eds.) The origins and spread of domestic plants in southwest Asia and Europe. Publications of the Institute of Archaeology, University College London, pp. 343–347

Beyer B, Gechter-Jones J, Gerlach G, Weiner J (eds.) (1998) Brunnen der Jungsteinzeit. Materialien zur Bodendenkmalpflege im Deutschland 11. Rudolf Habelt, Bonn

Constantin C, Mordant D, Simonin D (eds.) (1997) La culture de Cerny, nouvelle économie, nouvelle société au Néolithique. Mémoires du Musée de Préhistoire d'Ile-de-France 6, Nemours

Coudart A (1998) Architecture et société néolithique. Editions de la Maison des Sciences de l'Homme, Paris

Ilett M, Constantin C, Farruggia J-P (1995) Bâtiments voisins du Rubané et du Groupe de Villeneuve-Saint-Germain sur le site de Bucy-le-Long "La Fosse Tounise" (Aisne). Revue archéologique de Picardie numéro spécial 9:17–39

Ilett M, Plateaux M (1995) Le site néolithique de Berry-au-Bac "Le Chemin de la Pêcherie" (Aisne). Monographie du Centre de Recherches Archéologiques 15. CNRS Editions, Paris

Lüning J (1982) Siedlung und Siedlungslandschaft in bandkeramischer und Rössener Zeit. Offa 39:9–33

Méniel P (1984) Contribution à l'histoire de l'élevage en Picardie, du Néolithique à la fin de l'Age du Fer. Revue Archéologique de Picardie, numéro spécial

Chapter 5

Arbogast R-M (1994) Premiers élevages néolithiques du nord-est de la France. Etudes et Recherches Archéologiques de l'Université de Liège 67

Bakels CC (1997) The beginnings of manuring in western Europe. Antiquity 71:442–445

Bakker J-A, Kruk J, Lanting AE, Milisauskas S (1999) The earliest evidence of wheeled vehicles in Europe and the Near East. Antiquity 73:778–790

Biel J, Schlichterle H, Strobel M, Zee A (eds.) (1998) Die Michelsberger Kultur und ihre Randgebiete: Probleme der Entstehung, Chronologie und des Siedlungswesens. Materialhefte zur Archäologie in Baden-Württemberg 43, Theiss, Stuttgart

Dubouloz J, Mordant D, Prestreau M (1991) Les enceintes 'néolithiques' du Bassin Parisien. In: Identité du Chasséen, Actes du Colloque International de Nemours 1989, Mémoires du Musée de Préhistoire d'Ile-de-France 4:211–229

Hansen H-O (1969) Experimental ploughing with a Døstrup ard replica. Tools and Tillage 1(2): 67–92

Méniel P (1984) Contribution à l'histoire de l'élevage en Picardie, du Néolithique à la fin de l'Age du Fer. Revue Archéologique de Picardie, numéro spécial

Pétrequin P, Arbogast R-M, Pétrequin A-M, Van Willigen S, Bailly M (2006) Premiers chariots, premiers araires, la diffusion de la traction animale en Europe pendant les 4e et 3e millénaires avant notre ère. CRA monographies 29, CNRS éditions, Paris

Rasmussen P (1989) Leaf-foddering of Livestock in the Neolithic: archaeobotanical evidence from Weier, Switzerland. Journal of Danish Archaeology 8:51–71

Sherratt A (1981) Plough and pastoralism: aspects of the secondary products revolution. In: Hodder I, Isaac G, Hammond N (eds.) Pattern of the Past. Cambridge University Press, Cambridge, pp. 261–305

Chapter 6

Bakels CC (1992) The botanical shadow of two early Neolithic settlements in Belgium: carbonized seeds and disturbances in a pollen record. Review of Palaeobotany and Palynology 73:1–19

Bernard V (1998) l'Homme, le Bois et la Forêt dans la France du Nord entre le Mésolithique et le Haut Moyen-Age. BAR International Series S733

Knörzer K-H, Gerlach R, Meurers-Balke J, Kalis AJ, Tegtmeier U, Becker WD, Jürgens A (1999) Pflanzenspuren, Archäobotanik im Rheinland: Agrarlandschaft und Nutzpflanzen im Wandel der Zeiten. Rudolf Habelt, Bonn

Schalich J (1973) Langweiler 2, Boden- und Landschaftsgeschichte. In: Farruggia J-P, Kuper R, Lüning J, Stehli P (eds.) Der bandkeramische Siedlungsplatz Langweiler 2. Rheinische Ausgrabungen 13:5–16

Vera FWM (2000) Grazing Ecology and Forest History. CABI Publishing, Waltingford

Zagwijn WH (1994) Reconstruction of climate change during the Holocene in western and central Europe based on pollen records of indicator species. Vegetation History and Archaeobotany 3:65–88

Chapter 7

Brun P (1994) From the Hallstatt to La Tène Period in the Perspective of the Mediterranean World Economy. In: Kristiansen K, Jensen J (eds.) Europe in the First Millennium BC. Sheffield Archaeological Monographs 9:57–65

Buchsenschutz O, Méniel P (eds.) (1994) Les installations agricoles de l'âge du Fer en Île-de-France. Études d'histoire et d'archéologie 4. Presses de l'École Normale Supérieure, Paris

Buchsenschutz O, Mordant C (2005) Architectures protohistoriques en Europe occidentale du Néolithique final à l'âge du Fer. Éditions du Comité des Travaux Historiques et Scientifiques, Paris

Desfossés Y (avec la collaboration de M. Philippe) (2002) Angleterre et France à l'âge du Bronze, les contacts transmanche. Archéologia Juillet-Août 2002:46–57

Hingh AE de (2000) Food production and food procurement in the Bronze Age and Early Iron Age (2000–500 BC). Archaeological Studies Leiden University 7

Joachim HE (1985) Zu eisenzeitlichen Reibsteinen aus Basaltlava, den sog. Napoleonshüten. Archäologisches Korrespondenzblatt 15:359–369

Malrain F, Matterne V, Méniel P (2002) Les Paysans Gaulois. Errance, Paris

Marion S, Blancquaert G (eds.) (2000) Les installations agricoles de l'âge du Fer en France septentrionale. Études d'histoire et d'archéologie 6. Presses de l'École Normale Supérieure, Paris

Matterne V (2001) Agriculture et alimentation végétale durant l'âge du Fer et l'époque gallo-romaine en France septentrionale. Éditions Monique Mergoil, Montagnac

Méniel P (1984) Contribution à l'histoire de l'élevage en Picardie, du Néolithique à la fin de l'Age du Fer. Revue Archéologique de Picardie, numéro spécial

Olivier L (2004) Nouvelles recherches sur l'or blanc des Celtes (Exploitation du sel à l'âge du Fer). Archéologia Juillet-Août 2004:35–44

Simons A (1989) Bronze- und eisenzeitliche Besiedlung in den Rheinische Lössbörden. BAR International Series 467

Chapter 8

Bakels C, Dijkman W (2000) Maastricht in the first millennium AD, the archaeobotanical evidence. Archaeologica Mosana 2, Maastricht.

Bayard D, Collart J-L (eds.) (1996) De la ferme indigène à la villa romaine, la romanisation des campagnes de la Gaule. Revue archéologique de Picardie numéro spécial 11

Bromwich J (2003) The Roman remains of northern and eastern France, a guidebook. Routledge, London and New York

Ervynck A, Vanderhoeven A (1997) Tongeren (Belgium): changing patterns of meat consumption in a Roman civitas capital. Anthropozoologica 25–26:457–464

Fischer Th (1999) Die Römer in Deutschland. Konrad Theiss, Stuttgart

Gaitzsch W (1990) Der römische Gutshof im "Gewährbau" bei Niederzier, Modell einer Landsiedlung in der Germania inferior. Archäologie in Nordrhein-Westfalen, P. von Zabern, Mainz, pp. 235–240

Gilles K-J (1991) Eine weitere römische Weinkelter aus Brauneberg. Funde und Ausgrabungen im Bezirk Trier 23:20–32

King A (1999) Diet in the Roman world: a regional inter-site comparison of the mammal bones. Journal of Roman Archaeology 12:168–202

Knörzer K-H, and Meurers-Balke J (1990) Die Wirtschafts- und Nutzungsflächen eines römischen Gutshofes, eine Rekonstruktion aufgrund des botanischen Befundes. Archäologie in Nordrhein-Westfalen, P. von Zabern, Mainz, pp. 242–246

Kooistra LI (1996) Borderland Farming, possibilities and limitations of farming in the Roman Period and the Early Middle Ages between the Rhine and the Meuse. Van Gorcum, Assen

Kremer B (1999) Wasserversorgung aus dem Tunnel, der römische qanat von Mehring. Funde und Ausgrabungen im Bezirk Trier 31:37–50

Lepetz S, Matterne V (eds.) (2003) Cultivateurs, éleveurs et artisans dans les campagnes de la Gaule romaine; matières premières et produits transformés. Revue Archéologique de Picardie 2003 1/2

Lewitt T (2003) Vanishing villas': what happened to élite rural habitation in the West in the 5th–6th c? Journal of Roman Archaeology 16:260–274

Matterne V (2001) Agriculture et alimentation végétale durant l'âge du Fer et l'époque gallo-romaine en France septentrionale. Mergoil, Montagnac, p. 310

Ossel P van (1992) Etablissements ruraux de l'Antiquité tardive dans le nord de la Gaule. Gallia 51e supplément

Ossel P van, Ouzoulias P (2000) Rural settlement economy in Northern Gaul in the Late Empire: an overview and assessment. Journal of Roman Archaeology 13:133–160

White KD (1970) Roman farming. Thames and Hudson, London

Chapter 9

Bakels CC (2005) Crops produced in the southern Netherlands and northern France during the early medieval period: a comparison. Vegetation History and Archaeobotany 14:394–399

Devroey J-P (1989) Entre Loire et Rhin: les fluctuations du terroir de l'épeautre au Moyen Age. In: Devroey J-P, Mol J-J van (eds.) L'épeautre (Triticum spelta) histoire et ethnologie. Editions Dire, Treignes, pp. 89–105

Devroey J-P (2006) Puissants et misérables, système social et monde paysan dans l'Europe des Francs (VIe-IXe siècles). Mémoires de l'Académie royale de Belgique, Classe de Lettres 40

Eckhart KA (ed.) (1969) Lex Salica, Monumenta Germaniae Historica, Legum sectio I, Legum nationum Germanicarum Tomus 4

Foucray B (1996) Les Ruelles de Serris – habitats aristocratique et paysan du Haut Moyen-Age (fin VII/Xe siècle). Ruralia I, Památki archeologické Supplementum 5:203–210

Gareis K (1895) Die Landgüterordnung Kaiser Karls des Grossen (Capitulare de villis vel curtis imperii) Textausgabe. Guttentag, Berlin

Gentili A, Valais A (2007) Composantes aristocratiques et organisation de l'espace au sein de grands habitats ruraux du Haut Moyen Âge. In: Depreux F, Bougard F, Le Jan R (eds.) Les Élites et leurs espaces, Mobilité, rayonnement, domination (du VIe au XIe siècle). Brepols, Turnhout, pp. 99–134

Henning J (2004) Germanisch-romanische Agrarkontinuität und –diskontinuität im nordalpinen Kontinentaleuropa – Teile eines Systemwandels? Beobachtungen aus archäologischer Sicht. In: Akkulturation- RGA-E Band 41:396–435

McCormick M (2001) Origins of the European Economy: communications and commerce AD 300–900, Cambridge University Press, Cambridge

Nice A (1994) L'habitat mérovingien de Goudelancourt-lès-Pierrepont (Aisne). Revue Archéologique de Picardie 1994 (1/2):21–63

Pertz GH (ed) (1836, new edition 1965) Brevium Exempla ad describendas res ecclesiasticas et fiscales. Monumenta Germaniae Historica, Legum sectio II, Capitularia regum francorum Tomus 1, Hiersemann/Kraus, Stuttgart/New York

Peytremann E (2003) Archéologie de l'habitat rural dans le nord de la France du IVe au XIIe siècle. Mémoires publiés par l'Association française d'Archéologie mérovingienne 13, Saint-Germain-en-Laye

Pounds NJG (1967) Northwest Europe in the Ninth Century: its geography in light of the polyptiques. Annals of the Association of American Geographers 57(3):439–461

Slicher van Bath BH (1963) The Agrarian History of Western Europe A.D. 500 – 1850. Edward Arnold, London

Verhulst A (2002) The Carolingian Economy. University Press, Cambridge

Witvrouw J, Gava G, Dardenne L, Gava S (2003) Le Thier d'Olne à Engis, centre domanial du Haut Moyen Age. http://cahc.free.fr/fouilles/thierdolne.php

Yvinec J-H (1999) Étude archéozoologique du site de Dury "Le Moulin" (Somme). Revue Archéologique de Picardie 1999 (1/2):247–256

Zohary D, Hopf M (2000) Domestication of Plants in the Old World, third edition. University Press, Oxford

Chapter 10

Becker W-D (2005) Das Elsbachtal: die Landschaftsgeschichte vom Endneolithikum bis ins Hochmittelalter. Von Zabern, Mainz

Bernard V (1998) l'Homme, le Bois et la Forêt dans la France du Nord entre le Mésolithique et le Haut Moyen-Age. BAR International Series S733

Index

Note: Entries that appear in tables are denoted by a "t" and those in figures by an "f" along with the locators, e.g., 169t, 161f.

Animals

Cultivation methods and Tools

Cultural setting

Plants

Products

Structures

GPSR Compliance
The European Union's (EU) General Product Safety Regulation (GPSR) is a set of rules that requires consumer products to be safe and our obligations to ensure this.

If you have any concerns about our products, you can contact us on

ProductSafety@springernature.com

In case Publisher is established outside the EU, the EU authorized representative is:

Springer Nature Customer Service Center GmbH
Europaplatz 3
69115 Heidelberg, Germany

www.ingramcontent.com/pod-product-compliance
Lightning Source LLC
Chambersburg PA
CBHW070914300825
31900CB00005B/168
9781402098390